Lecture Notes Electrical Engineering

Volume 15

Juan Andrade Cetto · Jean-Louis Ferrier ·
José Miguel Costa dias Pereira ·
Joaquim Filipe (Eds.)

Informatics in Control Automation and Robotics

Selected Papers from the International Conference on Informatics in Control Automation and Robotics 2006

Springer

Juan Andrade Cetto
Ramon y Cajal Postdoctoral Fellow
Institut de Robotica i Informatica
Industrial, CSIC-UPC
Llorens Artigas, 4-6
08028 Barcelona
Spain

Professeur Jean-Louis Ferrier
LISA - ISTIA
62, avenue Notre Dame du Lac
49000 Angers
France

José Miguel Costa Dias Pereira
Instituto Politécnico de Setúbal
Largo Defensores da República, 1
2910-470 Setúbal
Portugal

Joaquim Filipe
INSTICC
Av. D. Manuel I
27A 2°Esq.
2910-595 Setúbal
Portugal

ISBN: 978-3-642-09803-1

e-ISBN: 978-3-540-79142-3

Cover design: eStudio Calamar S.L.

Printed on acid-free paper

9 8 7 6 5 4 3 2 1

springer.com

Preface

The present book includes a set of selected papers from the third "International Conference on Informatics in Control Automation and Robotics" (ICINCO 2006), held in Setúbal, Portugal, from 1 to 5 August 2006, sponsored by the Institute for Systems and Technologies of Information, Control and Communication (INSTICC).

The conference was organized in three simultaneous tracks: "Intelligent Control Systems and Optimization", "Robotics and Automation" and "Systems Modeling, Signal Processing and Control".

The book is based on the same structure.

Although ICINCO 2006 received 309 paper submissions, from more than 50 different countries in all continents, only 31 where accepted as full papers. From those, only 23 were selected for inclusion in this book, based on the classifications provided by the Program Committee. The selected papers also reflect the interdisciplinary nature of the conference. The diversity of topics is an important feature of this conference, enabling an overall perception of several important scientific and technological trends. These high quality standards will be maintained and reinforced at ICINCO 2007, to be held in Angers, France, and in future editions of this conference.

Furthermore, ICINCO 2006 included 7 plenary keynote lectures and 1 tutorial, given by internationally recognized researchers. Their presentations represented an important contribution to increasing the overall quality of the conference, and are partially included in the first section of the book. We would like to express our appreciation to all the invited keynote speakers who took the time to contribute with a paper to this book, namely, in alphabetical order: Oleg Gusikhin (Ford Research & Adv. Engineering), Norihiro Hagita (ATR Intelligent Robotics and Communication Labs), Gerard T. McKee (University of Reading) and William J. O'Connor, University College Dublin.

On behalf of the conference organizing committee, we would like to thank all participants. First of all to the authors, whose quality work is the essence of the conference and to the members of the program committee, who helped us with their expertise and time.

As we all know, producing a conference requires the effort of many individuals. We wish to thank all the people from INSTICC, whose work and commitment were invaluable.

February 2008

Juan A. Cetto
Jean-Louis Ferrier
José Dias Pereira
Joaquim Filipe

Conference Committee

Conference Chair

Joaquim Filipe, Polytechnic Institute of Setúbal / INSTICC, Portugal

Honorary Chair

Hojjat Adeli, The Ohio State University, U.S.A.

Program Co-chairs

Juan Andrade Cetto, Universitat Autònoma de Barcelona, Spain
Jean-Louis Ferrier, University of Angers, France
José Dias Pereira, Polytechnic Institute of Setúbal, Portugal

Organising Committee

Paulo Brito, INSTICC, Portugal
Marina Carvalho, INSTICC, Portugal
Helder Coelhas, INSTICC, Portugal
Bruno Encarnação, INSTICC, Portugal
Vítor Pedrosa, INSTICC, Portugal
Mónica Saramago, INSTICC, Portugal

Programme Committee

Eugenio Aguirre, Spain
Frank Allgower, Germany
Fouad Al-Sunni, Saudi Arabia
Yacine Amirat, France
Luis Antunes, Portugal
Peter Arato, Hungary
Helder Araújo, Portugal
Gustavo Arroyo-Figueroa, Mexico
Marco Antonio Arteaga, Mexico
Nikos Aspragathos, Greece
Miguel Ayala Botto, Portugal
Robert Babuska, The Netherlands
Mark Balas, U.S.A.
Bijnan Bandyopadhyay, India
Ruth Bars, Hungary
Karsten Berns, Germany
Patrick Boucher, France
Guido Bugmann, U.K.
Edmund Burke, U.K.
Kevin Burn, U.K.
Clifford Burrows, U.K.
Luis M. Camarinha-Matos, Portugal
Marco Campi, Italy
Jorge Martins de Carvalho, Portugal
Alicia Casals, Spain
Christos Cassandras, U.S.A.
Raja Chatila, France
Tongwen Chen, Canada
Albert M. K. Cheng, U.S.A.
Sung-Bae Cho, Korea
Ryszard S. Choras, Poland
Carlos Coello Coello, Mexico
António Dourado Correia, Portugal
Yechiel Crispin, U.S.A.
Keshav Dahal, U.K.
Danilo De Rossi, Italy
Angel P. del Pobil, Spain
Guilherme DeSouza, U.S.A.
Rüdiger Dillmann, Germany
Denis Dochain, Belgium

Robert Richardson, U.K.
John Ringwood, Ireland
Juha Röning, Finland
Agostinho Rosa, Portugal
Hubert Roth, Germany
António Ruano, Portugal
Erol Sahin, Turkey
Antonio Sala, Spain
Abdel-Badeeh M. Salem, Egypt
Ricardo Sanz, Spain
Medha Sarkar, U.S.A.
Nilanjan Sarkar, U.S.A.
Jurek Sasiadek, Canada
Carlos Sagüés, Spain
Daniel Sbarbaro, Chile
Klaus Schilling, Germany
Chi-Ren Shyu, U.S.A.
Bruno Siciliano, Italy
João Silva Sequeira, Portugal
Mark Spong, U.S.A.
Tarasiewicz Stanislaw, Canada
Aleksandar Stankovic, U.S.A.
Gerrit van Straten, The Netherlands
Raúl Suárez, Spain
Ryszard Tadeusiewicz, Poland
Tianhao Tang, China
Daniel Thalmann, Switzerland
Gui Yun Tian, U.K.
Ivan Tyukin, Japan
Cees van Leeuwen, Japan
Annamaria R. Varkonyi-Koczy, Hungary
Bernardo Wagner, Germany
Axel Walthelm, Germany
Jun Wang, China
Lipo Wang, Singapore
Alfredo Weitzenfeld, Mexico
Dirk Wollherr, Germany
Sangchul Won, Korea
Kainam Thomas Wong, Canada
Jeremy Wyatt, U.K.
Alex Yakovlev, U.K.
Hujun Yin, U.K.
Anibal Zanini, Argentina
Yanqing Zhang, U.S.A.
Dayong Zhou, U.S.A.
Albert Zomaya, Australia
Detlef Zuehlke, Germany

Auxiliary Reviewers

Alejandra Barrera, Mexico
Levent Bayindir, Turkey
Domingo Biel, Spain
Stephan Brummund, Germany
F. Wilhelm Bruns, Germany
Roman Buil, U.S.A.
Yang Cao, China
Raquel Cesar, Portugal
Ying Chen, U.S.A.
Paulo Coelho, Portugal
Gert van Dijck, Belgium
Liya Ding, U.S.A.
Didier Dumur, France
Adriano Fagiolini, Italy
Daniele Fontanelli, Italy
Jeff Fortuna, U.S.A.
Istvan Harmati, Hungary
Sunghoi Huh, Italy
Feng Jin, China
Abhinaya Joshi, U.S.A.
Balint Kiss, Hungary
Yan Li, China
Gonzalo Lopez-Nicolas, Spain
Patrick De Mazière, Belgium
Rafael Muñoz-Salinas, Spain
Ana Cristina Murillo, Spain
Ming Ni, U.S.A.
Soumen Sen, Italy
Razvan Solea, Portugal
Onur Soysal, Turkey
Wei Tan, China
Giovanni Tonietti, Italy
Ali Emre Turgut, Turkey
Jörg Velten, Germany
Anne von Vietinghoff, Germany
Youqing Wang, China
Yunhua Wang, U.S.A.
Bo Xiong, U.S.A.
Bailly Yan, France
Feng Zhao, U.S.A.

Invited Speakers

Mihaela Ulieru, The University of New Brunswick, Canada
Oleg Gusikhin, Ford Research & Adv. Engineering, U.S.A.
Norihiro Hagita, ATR Intelligent Robotics and Communication Laboratories, Japan
Hojjat Adeli, The Ohio State University, U.S.A.
Mark d'Inverno, University of Westminster, U.K.
William J. O'Connor, University College Dublin, Ireland
Gerard T. McKee, The University of Reading, U.K.

Contents

Invited Papers

Part I: Intelligent Control Systems and Optimization

Part II: Robotics and Automation

Part III: Signal Processing, Systems Modeling and Control

Contributors

Shadi Abras
Laboratoire Leibniz-Institut IMAG, CNRS, UMR5552, 46, Avenue Félix Viallet, France, e-mail: Shadi.Abras@imag.fr

Antonio Adán
Escuela Superior de Informatica, Universidad de Castilla La Mancha, 13071 Ciudad Real, Spain, e-mail: Antonio.Adan@uclm.es

M. Antón-Rodríguez
Departamento de Teoría de la Señal, Comunicaciones e Ingeniería Telemática, ETSIT Universidad de Valladolid, Campus Miguel Delibes s/n, 47011 Valladolid, España, e-mail: mirant@tel.uva.es

Carlos Ariño
Systems Engineering and Control Department, Univ. Politécnica de Valencia, Cno. Vera s/n, 46022 Valencia, Spain

José Raul Azinheira
IDMEC – IST, Instituto Superior Técnico, Av. Rovisco Pais, 1047-001 Lisbon, Portugal, e-mail: jraz@dem.ist.utl.pt

Luca Baglivo
CISAS, Centre of Studies and Activities for Space, Via Venezia 1, 35131 Padova, Italy, e-mail: luca.baglivo@unipd.it

L. F. Baptista
Escola Nutica Infante D. Henrique, Department of Marine Engineering, Av. Eng. Bonneville Franco, 2770-058 Paço de Arcos, Portugal, e-mail: luisbaptista@enautica.pt

N. F. S. Bóia
Technical University of Lisbon, Instituto Superior Técnico, Department of Mechanical Engineering, GCAR/IDMEC, Avenida Rovisco Pais, 1049-001 Lisboa Codex, Portugal

D. Boto-Giralda
Departamento de Teoría de la Señal, Comunicaciones e Ingeniería Telemática, ETSIT Universidad de Valladolid, Campus Miguel Delibes s/n, 47011 Valladolid, España, e-mail: danbot@tel.uva.es

Fernando Gomez Bravo
Departamento de Ingeniería Electrónica, Sistemas Informáticos y Automática, Univ. de Huelva, Campus de la Rábida, Crta. Huelva-Palos de la Frontera s/n, 21819 Huelva, Spain, e-mail: fernando.gomez@diesia.uhu.es

Olivier Bruneau

Laboratoire Vision et Robotique - Ecole Nationale Supérieure d'Ingénieurs de Bourges, 10 Boulevard Lahitolle 18020 Bourges, France, e-mail: olivier.bruneau@ensi-bourges.fr

Philippe Charbonnaud

Laboratoire Génie de Production, EA 1905 69042 Heidelberg, Germany; Ecole Nationale d'Ingénieurs de Tarbes, 47, avenue d'Azereix, BP 1629, 65016 Tarbes Cedex, France, e-mail: Philippe.Charbonnaud@enit.fr

Pascale Chiron

Laboratoire Génie de Production, EA 1905 69042 Heidelberg, Germany; Ecole Nationale d'Ingénieurs de Tarbes, 47, avenue d'Azereix, BP 1629, 65016 Tarbes Cedex, France, e-mail: Pascale.Chiron@enit.fr

Yechiel J. Crispin

Department of Aerospace Engineering, Embry-Riddle University, Daytona Beach, FL 32114, USA, e-mail: crispinj@erau.edu

Mariolino De Cecco

Department of Mechanical and Structural Engineering, University of Trento, Via Mesiano 77, Trento, Italy, e-mail: mariolino.dececco@unipd.it

Eric Duviella

Laboratoire Génie de Production, EA 1905 69042 Heidelberg, Germany; Ecole Nationale d'Ingénieurs de Tarbes, 47, avenue d'Azereix, BP 1629, 65016 Tarbes Cedex, France, e-mail: Eric.Duviella@enit.fr

F. J. Díaz-Pernas

Departamento de Teoría de la Señal, Comunicaciones e Ingeniería Telemática, ETSIT Universidad de Valladolid, Campus Miguel Delibes s/n, 47011 Valladolid, España, e-mail: pacper@tel.uva.es

J. F. Díez-Higuera

Departamento de Teoría de la Señal, Comunicaciones e Ingeniería Telemática, ETSIT Universidad de Valladolid, Campus Miguel Delibes s/n, 47011 Valladolid, España, e-mail: josdie@tel.uva.es

Alicia Esparza

Systems Engineering and Control Dept., Univ. Polit´ecnica de Valencia, Cno. Vera s/n, 46022 Valencia, Spain, e-mail: alespe@isa.upv.es

Dimitar Filev

Ford Research & Advanced Engineering 2101 Village Rd., Dearborn, Michigan, USA, e-mail: dfilev@ford.com

G. J. García

Physics, Systems Engineering and Signal Theory Department, University of Alicante, Alicante, Spain, e-mail: gjgg@ua.es

Juan A. Gómez-Pulido
Dept. Informática, Univ. Extremadura, Escuela Politécnica, Campus Universitario s/n, 10071, Cáceres, Spain, e-mail: jangomez@unex.es

Toon Goedemé
PSI – VISICS, Katholieke Universiteit Leuven, Belgium,
e-mail: tgoedeme@esat.kuleuven.be

Luc Van Gool
PSI – VISICS, Katholieke Universiteit Leuven, Belgium,
e-mail: vangool@esat.kuleuven.be

Oleg Gusikhin
Ford Research & Advanced Engineering 2101 Village Rd.,
Dearborn, Michigan, USA, e-mail: ogusikhi@ford.com

Eija Haapalainen
Intelligent Systems Group, Department of Electrical and Information Engineering
P.O. Box 4500, FIN-90014 University of Oulu, Finland,
e-mail: Eija.Haapalainen@ee.oulu.fi

Norihiro Hagita
ATR Intelligent Robotics and Communication Laboratories, Kyoto, Japan,
e-mail: hagita@atr.jp

Uwe D. Hanebeck
Intelligent Sensor-Actuator-Systems Laboratory, Institute of Computer Science and Engineering, Universität Karlsruhe (TH), Karlsruhe, Germany,
e-mail: uwe.hanebeck@ieee.org

Benedetto Intrigila
Dipartimento di Matematica Pura ed Applicata, Università di Roma "Tor Vergata", Italy, e-mail: intrigil@mat.uniroma2.it

Hiroshi Ishiguro
ATR Intelligent Robotics and Communication Laboratories, Kyoto, Japan; Osaka University, Osaka, Japan, e-mail: ishiguro@ams.eng.osaka-u.ac.jp

Mireille Jacomino
Laboratoire d'Automatique de Grenoble, CNRS, UMR5528, BP 46, France,
e-mail: Mireille.Jacomino@inpg.fr

Heli Junno
Intelligent Systems Group, Department of Electrical and Information Engineering,
P.O. Box 4500, FIN-90014 University of Oulu, Finland,
e-mail: Heli.Junno@ee.oulu.fi

Takayuki Kanda
ATR Intelligent Robotics and Communication Laboratories, Kyoto, Japan,
e-mail: kanda@atr.jp

Andreas Kräußling

Research Establishment for Applied Sciences (FGAN), Neuenahrer Straße 20, 53343 Wachtberg, e-mail: a.kraeussling@fgan.de

Shunichi Kubo

Department of Information and Computer Sciences, Toyohashi University of Technology, 69042 Heidelberg, Germany; 1-1 Hibarigaoka, Tempaku-cho, Toyohashi-shi, Aichi 441-8580, Japan, e-mail: skubo@signal.ics.tut.ac.jp

Kazuhiro Kuwabara

ATR Intelligent Robotics and Communication Laboratories, Kyoto, Japan; Ritsumeikan University, Shiga, Japan, e-mail: kuwabara@is.ritsumei.ac.jp

Larissa Labakhua

University of Algarve, Escola Superior de Tecnologia/ADEE, Faro, Portugal, e-mail: llabak@ualg.pt

Perttu Laurinen

Intelligent Systems Group, Department of Electrical and Information Engineering, P.O. Box 4500, FIN-90014, University of Oulu, Finland, e-mail: Perttu.Laurinen@ee.oulu.fi

Fátima S. Leite

Institute of Systems and Robotics, University of Coimbra, Coimbra, Portugal e-mail: fleite@mat.uc.pt

Kurosh Madani

Laboratoire Images, Signaux, et Systèmes Intelligents (LISSI EA / 3956), Université Paris-XII, IUT de S′enart, Avenue Pierre Point, 77127 Lieusaint, France, e-mail: madani@univ-paris12.fr

Daniele Magazzeni

Dipartimento di Informatica, Università di L'Aquila, Italy, e-mail: magazzeni@di.univaq.it

Enrico Marcuzzi

CISAS, Centre of Studies and Activities for Space, Via Venezia 1, 35131 Padova, Italy, e-mail: enrico.marcuzzi@unipd.it

J. M. M. Martins

Technical University of Lisbon, Instituto Superior Técnico, Department of Mechanical Engineering, GCAR/IDMEC, Avenida Rovisco Pais, 1049-001 Lisboa Codex, Portugal, e-mail: martins@dem.ist.utl.pt

Gerard McKee

Active Robotics Laboratory, School of Systems Engineering, The University of Reading, Reading, Berkshire, UK, RG6 6AY, e-mail: g.t.mckee@reading.ac.uk

Igor Melatti

Dipartimento di Matematica Pura ed Applicata, Università di Roma "Tor Vergata", Italy, e-mail: melatti,tronci@di.uniroma1.it

Pilar Merchán

Escuela de Ingenierías Industriales,Universidad de Extremadura, Avda. Elvas, s/n, 06071Badajoz, Spain, e-mail: pmerchan@unex.es

Takahiro Miyashita

ATR Intelligent Robotics and Communication Laboratories, Kyoto, Japan, e-mail: miyasita@atr.jp

Alexandra Moutinho

IDMEC – IST, Instituto Superior Técnico, Av. Rovisco Pais, 1047-001 Lisbon, Portugal, e-mail: moutinho@dem.ist.utl.pt

Masanori Natsui

Department of Information and Computer Sciences, Toyohashi University of Technology, 69042 Heidelberg, Germany; 1-1 Hibarigaoka, Tempaku-cho, Toyohashi-shi, Aichi 441-8580, Japan, e-mail: natsui@signal.ics.tut.ac.jp

ThanhVu H. Nguyen

Code 5741, Naval Research Laboratory, Washington, DC, 20375-5320, USA, e-mail: thanhvu.nguyen@nrl.navy.mil

Urbano Nunes

Institute of Systems and Robotics, University of Coimbra, Coimbra, Portugal, e-mail: urbano@isr.uc.pt

William J. O'Connor

213 Mechanical Engineering,UCD Belfield, Dublin 4, Ireland, e-mail: william.oconnor@ucd.ie

Carlos Paiz

Heinz Nixdorf Institute, University of Paderborn, Fürstenallee 11, 33102 Paderborn, Germany, e-mail: paiz@hni.upb.de

Beatriz Paniagua-Paniagua

Dept. Informática, Univ. Extremadura, Escuela Politécnica, Campus Universitario s/n, 10071, Cáceres, Spain, e-mail: bpaniagua@unex.es

L. Payá

Physics, Systems Engineering and Signal Theory Department, University of Alicante, Alicante, Spain, e-mail: laura.paya@ua.es

Giuseppe Della Penna

Dipartimento di Informatica, Universit`a di L'Aquila, Italy, e-mail: dellapenna@di.univaq.it

Sylvie Pesty

Laboratoire Leibniz-Institut IMAG, CNRS, UMR5552, 46, Avenue Féelix Viallet, France, e-mail: Sylvie.Pesty@imag.fr

Stéphane Ploix

Laboratoire d'Automatique de Grenoble, CNRS, UMR5528, BP 46, France, e-mail: Stephane.Ploix@inpg.fr

Christopher Pohl

Heinz Nixdorf Institute, University of Paderborn, Fürstenallee 11, 33102 Paderborn, Germany, e-mail: pohl@hni.upb.de

J. Pomares

Physics, Systems Engineering and Signal Theory Department, University of Alicante, Alicante, Spain, e-mail: jpomares@ua.es

Mario Porrmann

Heinz Nixdorf Institute, University of Paderborn, Fürstenallee 11, 33102 Paderborn, Germany, e-mail: porrmann@hni.upb.de

Maria Isabel Ribeiro

Institute for Systems and Robotics, Instituto Superior Técnico, Av. Rovisco Pais 1, 1049-001 Lisboa, Portugal, e-mail: mir@isr.ist.utl.pt

Kathrin Roberts

Intelligent Sensor-Actuator-Systems Laboratory, Institute of Computer Science and Engineering

Universität Karlsruhe (TH), Karlsruhe, Germany, e-mail: roberts@ira.uka.de

Rui Rodrigues

Institute of Systems and Robotics, University of Coimbra, Coimbra, Portugal, e-mail: ruicr@isec.pt

Jose V. Roig

Systems Engineering and Control Dept., Univ. Politécnica de Valencia, Cno. Vera s/n, 46022 Valencia, Spain, e-mail: jvroig@isa.upv.es

Juha Röning

Intelligent Systems Group, Department of Electrical and Information Engineering, P.O. Box 4500, FIN-90014 University of Oulu, Finland, e-mail: Juha.Roning@ee.oulu.fi

Nestor Rychtyckyj

Global Manufacturing Engineering Systems, Ford Motor Company, Dearborn, Michigan, USA, e-mail: nrychtyc@ford.com

Christophe Sabourin

Laboratoire Images, Signaux, et Systèmes Intelligents (LISSI EA / 3956), Université Paris-XII, IUT de S´enart, Avenue Pierre Point, 77127 Lieusaint, France, e-mail: sabourin@univ-paris12.fr

Antonio Sala

Systems Engineering and Control Department, Univ. Politécnica de Valencia, Cno. Vera s/n, 46022 Valencia, Spain, e-mail: asala@isa.upv.es

Santiago Salamanca
Escuela de Ingenierías Industriales,Universidad de Extremadura, Avda, Elvas, s/n, 06071Badajoz, Spain, e-mail: ssalaman@unex.es

Felix Sawo
Intelligent Sensor-Actuator-Systems Laboratory, Institute of Computer Science and Engineering, Universität Karlsruhe (TH), Karlsruhe, Germany,
e-mail: sawo@ira.uka.de

João Silva Sequeira
Institute for Systems and Robotics, Instituto Superior Técnico Av. Rovisco Pais 1, 1049-001, Lisbon, Portugal, e-mail: jseq@isr.ist.utl.pt

Masahiro Shiomi
ATR Intelligent Robotics and Communication Laboratories, Kyoto, Japan,
e-mail: m-shiomi@atr.jp

James F. Smith III
Code 5741, Naval Research Laboratory, Washington, DC, 20375-5320, USA,
e-mail: james.smith@nrl.navy.mil

J. M. G. Sáda Costa
Technical University of Lisbon, Instituto Superior Técnico, Department of Mechanical Engineering, GCAR/IDMEC, Avenida Rovisco Pais, 1049-001 Lisboa Codex, Portugal, e-mail: sadacosta@dem.ist.utl.pt

Juan M. Sánchez-Pérez
Dept. Informática, Univ. Extremadura, Escuela Politécnica, Campus Universitario s/n, 10071, Cáceres, Spain, e-mail: sanperez@unex.es

Yoshiaki Tadokoro
Department of Information and Computer Sciences, Toyohashi University of Technology, 69042 Heidelberg, Germany; 1-1 Hibarigaoka, Tempaku-cho, Toyohashi-shi, Aichi 441-8580, Japan, e-mail: tadokoro@signal.ics.tut.ac.jp

Alberto Tofani
Dipartimento di Informatica, Università di L'Aquila, Italy,
e-mail: tofani@di.univaq.it

F. Torres
Physics, Systems Engineering and Signal Theory Department, University of Alicante, Alicante, Spain, e-mail: Fernando.Torres@ua.es

Enrico Tronci
Dipartimento di Matematica Pura ed Applicata, Università di Roma "Tor Vergata", Italy, e-mail: tronci@di.uniroma1.it

Lauri Tuovinen
Intelligent Systems Group, Department of Electrical and Information Engineering, P.O. Box 4500, FIN-90014 University of Oulu, Finland,
e-mail: Lauri.Tuovinen@ee.oulu.fi

Tinne Tuytelaars
PSI – VISICS, Katholieke Universiteit Leuven, Belgium,
e-mail: tuytelaa@esat.kuleuven.be

Alberto Vale
Institute for Systems and Robotics, Instituto Superior Técnico, Av. Rovisco Pais 1, 1049-001 Lisboa, Portugal, e-mail: vale@isr.ist.utl.pt

Miguel A. Vega-Rodríguez
Dept. Informática, Univ. Extremadura, Escuela Politécnica, Campus Universitario s/n, 10071, Cáceres, Spain, e-mail: mavega@unex.es

Mirco Zaccariotto
CISAS, Centre of Studies and Activities for Space, Via Venezia 1, 35131 Padova, Italy

Invited Papers

Intelligent Vehicle Systems: Applications and New Trends

Oleg Gusikhin[1], Dimitar Filev[1] and Nestor Rychtyckyj[2]

[1]Ford Research & Advanced Engineering 2101 Village Rd., Dearborn, Michigan, USA
ogusikhi@ford.com, dfilev@ford.com
[2]Global Manufacturing Engineering Systems, Ford Motor Company, Dearborn, Michigan, USA
nrychtyc@ford.com

Abstract. Most people usually do not consider the car sitting in their driveway to be on the leading edge of new technology. However, for most people, the personal automobile has now become their initial exposure to new intelligent computational technologies such as fuzzy logic, neural networks, adaptive computing, voice recognition and others. In this chapter we will discuss the various intelligent vehicle systems that are now being deployed into motor vehicles. These intelligent system applications impact every facet of the driver experience and improve both vehicle safety and performance. We will also describe recent developments in autonomous vehicle design and demonstrate that this type of technology is not that far away from deployment. Other applications of intelligent system design apply to adapting the vehicle to the driver's preferences and helping the driver stay aware. The automobile industry is very competitive and there are many other new advances in vehicle technology that cannot be discussed yet. However, this chapter provides an introduction into those technologies that have already been announced or deployed and shows how the automobile has evolved from a basic transportation device into an advanced vehicle with a host of on-board computational technologies.

Keywords. Computational intelligence, vehicle systems.

1 Introduction

Although the automotive industry has always been a leading force behind many engineering innovations, this trend has become especially apparent in recent years. The competitive pressure creates an unprecedented need for innovation to differentiate products and reduce cost in a highly saturated automotive market to satisfy the ever increasing demand of technology savvy customers for increased safety, fuel economy, performance, convenience, entertainment, and personalization. With innovation thriving in all aspects of the automotive industry, the most visible advancements are probably in the area of vehicle controls enabled by the proliferation of on-board electronics, computing power, wireless communication capabilities, and sensor and drive-by-wire technologies.

The increasing sophistication of modern vehicles is also accompanied by the growing complexity of required control models. Therefore, it is not surprising that numerous applications of methodologies generally known as "intelligent", "soft computing", "computational intelligence", and "artificial intelligence" have become increasingly popular in the implementation of vehicle systems. In this chapter, we focus on applications of computational intelligence methodologies such as Fuzzy Logic, Neural Networks, Machine Learning, Knowledge Representation, Probabilistic and Possibilistic Reasoning as building blocks for intelligent vehicle systems. These examples are drawn from published sources with credible evidence of successful vehicle implementation, or research sponsored by automotive enterprises. This chapter does not provide an exhaustive bibliographical review, but limits the number of references that are necessary to illustrate relevant examples of applications of intelligent technologies.

In this review we describe the introduction of different methods of computational intelligence for vehicle control in chronological order. In the next section we review one of the first applications of computational intelligence for vehicle control: fuzzy-neural controls. Section 3 describes automotive applications of speech recognition, while Sect. 4 discusses the varied uses of on-board vehicle diagnostics. In Sect. 5 we describe applications of intelligent vehicle technologies which also include a discussion on the technology needed for autonomous vehicles. Section 6 discusses the emerging field of application of driver-aware technologies that monitor and mitigate adversary driver conditions, such as fatigue, impairment, stress or anger. The final section summarizes the chapter and presents our conclusions.

2 Fuzzy-Neural Systems Control

Fuzzy logic and neural networks were the first computational intelligence techniques implemented in the vehicle as viable alternatives to the classical control methods that may be infeasible, inefficient or uneconomical. The first commercial applications of fuzzy logic for speed control and continuous variable transmission date back to 1988 [37] [38].

Fuzzy logic controllers take advantage of human knowledge of the control behavior. The control process is described inside a set of "IF-THEN" rules that also includes probabilistic fuzzy variables for control values. In a fuzzy logic controller, the crisp sensor inputs are converted to the fuzzy variables that are processed against the rule base. A combined result is then converted back into a specific crisp control value.

There are a number of reviews outlining the advantages and production implementations of fuzzy logic in control of different vehicle systems, including anti-lock breaking systems (ABS), engine control, automatic transmissions, anti-skid steering, and climate control [4] [43]. In recent years, the proliferation of hybrid vehicles (e.g. vehicles that combine combustion engines and electric motors) created the potential for a new application area of fuzzy logic control for vehicle subsystems [32]. These examples demonstrate that incorporating expert rules expressed through fuzzy logic simplifies complex control models.

In addition, fuzzy logic allows the modeling of such inherently ambiguous notions as driver behavior in an efficient and effective way. Exploring this feature of fuzzy logic, Takahashi [38] presents the concept of vehicle control, where the driver plays the role of the human sensor for the control system. In this case, the driving environment and driver intentions might be predicted by analyzing the operations executed by the driver, such as pedal inputs and steering maneuvers. Furthermore, this control system makes it possible to infer driver classification (for example "defensive", "medium", "sporty" [45]) and adjust the characteristics of the engine, transmission and other vehicle subsystems to the driver preferences.

While fuzzy logic allows for the representation of the knowledge of human experts in the form of rules, neural networks allow for the capture of expertise through training. Often both techniques are combined together. Hayashi et al. [14] describes a Neuro-Fuzzy Transmission Control system developed at Isuzu Motors. This system combines both a Fuzzy Logic module and Neural Nets. Fuzzy Logic is used to estimate the automobile load and driver intentions from both the input shaft speed and accelerator position displacement. The Neural Net module determines the optimal gear-shift position from the estimated load, driver intentions, vehicle speed and accelerator pedal displacement. The Neural Net is trained using a standard gear-shift scheduling map, uphill driving data, and knowledge from an experienced driver.

The efficient control of vehicle subsystems depends on the accuracy and completeness of the feedback data from the system parameters. However, in many cases, the direct measurement of such system parameters is impractical due to complexity, noise and the dynamic nature of the system. Marko et al. [20] demonstrates that neural networks could be trained to emulate "virtual", ideal sensors that enhance diagnostic information from existing sensors on production vehicles.

The most prominent application area of neural-network based sensors is the on-line diagnostics of engine combustion failures, featured in the Aston Martin DB9 engine control system [1]. The importance of this application is enhanced by the fact that engine misfires are the leading contributors to excessive vehicle emissions and fuel consumption. In general, the identification of engine misfires can be done through the observation of crankshaft dynamics. However, the complexity of these dynamics can easily lead to misinterpretation. Neural Networks, trained by artificially inducing a combustion failure, can classify a misfire with a high level of accuracy based on indirect data, such as engine speed, load, crankshaft acceleration, and phase of the cylinder firing sequence [21] [28].

3 Speech Recognition

Speech technology is another important type of an in-vehicle AI application. The importance of an in-vehicle speech interface is related to requirements for non-destructive hands-free control of the ever increasing number of auxiliary functions offered in vehicles, such as telephones, entertainment, navigation, and climate control systems.

One of the first vehicle speech dialog systems, called Linguatronic, was introduced by Mercedes-Benz in their S-class car line in 1996 [15]. The speech recognizer used in Liguatronic is speaker-independent and based on the Hidden Markov Model (HMM) combined with the Dynamic Time Warping (DTW) word recognizer for a user definable telephone directory [6].

Most of the systems available today are based on a single utterance command and control paradigm. Such systems typically require the memorization of all commands from the manual that are often expressed in an artificial (non-natural) language. To address these limitations, automotive companies and suppliers have been actively pursuing research and development of the next generation of in-vehicle intelligent dialog systems [22] [27]. For example, Pieraccini et al. [27] presents a multimodal conversational interface prototype that was implemented on the Ford Model U Concept Vehicle shown at the 2003 North American International Auto Show in Detroit, Michigan. This system adopts a conversational speech interface coupled with a touch screen display. The speech recognition engine makes use of dynamic semantic models that keep track of current and past contextual information and dynamically modify the language model in order to increase accuracy of the speech recognizer.

4 On-Board Diagnostics and Prognostics

While intelligent systems in service diagnostics have been in use since the 1980s, vehicle on-board diagnostics and prognostics define an emerging area of computational intelligence applications. Each new vehicle currently contains a large number of processors that control the operation of various automotive subsystems, such as the engine, lights, climate control, airbags, anti-lock braking systems, traction control, transmissions, stereo systems and others. Each of these processors runs software that deals with faults and abnormal behavior in the various subsystems. This software has three main goals:

- Detection of faults
- Ability to operate when a fault has been triggered
- Ability to provide diagnostic information that can be used to locate the fault by a service technician.

Vehicle fault information is aggregated in the On Board Diagnostic (OBD) system that is a standard component of every modern vehicle. The fault detection algorithms (predominantly model based) provide input to the OBD that is used to evaluate the health of individual vehicle subsystems for on-board monitoring and to support off-line diagnostic maintenance systems. There has also been considerable work done to apply model-based systems and qualitative reasoning to support on-board diagnostics [36]. This work includes the development of the Vehicle Model-Based Diagnosis (VMBD) project in Europe. This project involves running model-based diagnosis on demonstrator vehicles to analyze problems with emissions in a diesel engine. In this case, a model was developed that represented the turbocontrol subsystem in the engine and a solution to a problem was found using a consistency-based diagnosis system. The model of the system is not a single model of the entire system, but instead contains a library of component models. Qualitative models capture the interdependencies and physical effects of the airflow and pressure that is present in the engine. The concept of model based diagnostics is further refined and developed by combining it with a dynamic Bayesian network [33] [34] [35]. The network model is applied to approximate the fault dynamics, interpret the residuals generated by multiple models and to determine fault probabilities. This approach was piloted for on-board diagnosis of the Anti-lock Braking System (ABS) and Electronic Stability

Program (ESP) of a Daimler Chrysler pilot vehicle and demonstrated an effective way to detect faults from multiple model residuals.

Fault prognostics recently became an important feature of on board diagnostic systems. The goal of this technology is to continually evaluate the diagnostics information over time in order to identify any significant potential degradation of vehicle subsystems that may cause a fault, to predict the remaining useful life of the particular component or subsystem and to alert the driver before such a fault occurs. Most of the work in this direction is inspired by the recent progress in Condition Based and Predictive Maintenance [7] [9]. Presently available on selected military vehicles [13], a prognostic capability is envisioned as becoming a substantial extension of OBD systems and vehicle telematics [5].

Model based prognostics assume models that are used to calculate the residuals between the measured and model predicted features, estimate the measure of degradation, and to evaluate the remaining useful life of the component. Model based prognostics use the advantages of first principle models and provide an accurate representation of the particular vehicle subsystems [18] [19]. Alternatively, learning based prognostic techniques are data driven and employ black box type models, e.g. neural networks, Support Vector Machines, fuzzy models, statistical models, and other approximators to identify the trend of change in the features, and can consequently predict fault scenarios [12] [13].

An open scalable Integrated Diagnostic/Prognostic System (IDPS) architecture for real time diagnostics and prognostics was proposed in [41]. Diagnostics is performed by a fuzzy inference engine and static wavelet neural network that is capable of recognizing the occurrence of a fault mode and identifying the fault. Prognostic functionality includes a virtual sensor to provide fault dimensions and a prediction module employing a dynamic wavelet neural network for fault trending and estimation of remaining useful life of bearings.

As the complexity of vehicles increases, the need for intelligent diagnostics tools, such as the ones described above becomes more critical.

5 Intelligent Vehicle Technologies

Intelligent Vehicle Technology is a concept typically associated with the development of autonomous vehicle functionality. The key attributes of intelligent vehicles include the following:

- the ability to sense the vehicle's own status as well as its environment;
- the ability to communicate with the environment;
- the ability to plan and execute the most appropriate maneuvers [42].

Intelligent vehicle technologies are a rapidly growing field pursued by the automotive industry, academia and government agencies [42] [2]. The general interest in intelligent vehicle technologies is also fuelled by a number of competitions for unmanned ground vehicles (UGV) around the world: the annual Intelligent Ground Vehicle Competition (see http://www.igvc.org) sponsored by the International Association for Unmanned Vehicle Systems held since 1993; the Defense Advanced Research Projects Agency (DARPA) Grand challenge (see http://www.grandchallenge.org/) started in 2004; and the European Grand-Robot Trail (see http://www.elrob.org/) held its first annual contest in May 2006. Today, the DARPA Grand challenge is probably the most publicized event with its grand prize of $2 million in 2005. In 2005, the teams had to complete a 132 mile race through the Nevada Mojave desert in less than 10 hours. Interestingly, a number of teams in the 2005 DARPA Grand Challenge based their design on existing production vehicles. For instance, the winning team from Stanford in collaboration with Volkswagen used a specially modified "drive-by-wire" diesel "Toureg" R5. Furthermore, the team "Gray" that completed the race in fourth place used a standard 2005 Ford Escape Hybrid integrated with other off-the-shelf instrumentation and control technologies. Team "Gray" specifically mentioned in their technical paper [40] that the team approached the Grand Challenge from the standpoint of being integrators rather than developers of such technology. These examples clearly demonstrate how close existing automotive products are in regards to the implementation of intelligent vehicle functionality.

Although the autonomous vehicle is not currently a goal of the automotive companies, the elements of this technology are quickly finding their way into passenger vehicles to provide driver assistance in critical moments. The applications of intelligent vehicle technologies to the automotive sector are often seen as the next generation of vehicle safety systems. Specifically, for applications within the automotive industry, Richard Bishop [2] defines "Intelligent Vehicle systems" as systems that sense the driving environment and provide information or vehicle control to assist the driver in optimum vehicle operation.

Today different data about the driving environment can be obtained through any combination of sources such as on-board video cameras, radars, lidars (light detecting and ranging, the laser-based analog to radar), digital maps navigated by global positioning systems, communication from other vehicles or highway systems. The on-board system analyzes this data in real-time and provides a warning to the driver or even takes over control of the vehicle. Examples of intelligent vehicle technologies existing today include lane departure warning, adaptive cruise control, parallel parking assistants, crash warning and automated crash avoidance.

In general, intelligent vehicle systems do not necessarily employ the full scale of computational intelligence techniques. However, it is clear that intelligent systems when combined with the conventional systems and control techniques can play a significant role to facilitate or even enable the implementation of many of the intelligent vehicle functionalities. For instance, analysis of images from video cameras calls for the application of traditional AI techniques such as machine vision and pattern recognition. The fusion of the disjointed data from multiple sources benefits from the application of neural networks in a similar fashion to the virtual sensor development in engine control. The implementation of real-time response to the changes in driving conditions may take advantage of fuzzy logic. For example, Tascillo et al. [39] describes the prototype of a system that identifies and classifies objects in close proximity using a neural net approach to select the best course of action to avoid an accident. Nigro and Rombaut [25] proposes a rule-based system incorporating linguistic variables to recognize driving situations. Engstrom and Victor [8] developed real-time recognition of the driving context (e.g. city, highway, suburban driving) using neural networks. Miyahara et al. [23] presents a vision-based target tracking system based on the range window algorithm and pattern matching. Schlenoff et al. [31] discusses the use of ontology to enhance the capabilities and performance of autonomous vehicles, particularly in navigation planning. These are only few examples from the vast on-going research using computational intelligence techniques to address intelligent vehicle functionality.

The integration of vehicle control systems and fusion of a different type of information provides another new dimension for building intelligent vehicle systems. For example, algorithms that combine engine and navigation (GPS) data create the opportunity for the development of predictive models and control strategies that optimize fuel efficiency and vehicle performance. In [29] [30] an intelligent control method using fuzzy logic is applied to improve traditional Hybrid Electric Vehicle (HEV) control. A rule-base with a fuzzy reasoning mechanism is used as a lower level controller to calculate the operating point of the internal combustion engine based on the current speed, engine efficiency and emission characteristics and driver required torque. A second fuzzy controller works as a predictor for the future state of the vehicle using information about the speed and elevation of the sampled route that is provided by the navigation system. The role of the second (supervisory) fuzzy controller is to anticipate changes in the vehicle state and to implement predefined heuristics based on the battery charge/discharge rate and on the estimated changes in the road and traffic conditions (e.g. downhill/uphill, city/highway). Fuzzy logic is then used in conjunction with the conventional HEV control system to provide additional flexibility and information fusion that result in substantial fuel economy and emission reduction.

6 Driver-Aware Technologies

In the past decade there has been an increased interest in technologies that monitor and mitigate driver conditions, such as fatigue, impairment, stress or anger that adversely affects the driver's vigilance and reduces their ability to safely operate the vehicle.

There are two main approaches for real-time detection of driver conditions: by monitoring the deviations in driver's performance in the vehicle operation and by monitoring the driver's bio-physical parameters [16]. The first approach involves the analysis of steering wheel movements, acceleration, braking, gear changing, lane deviation and distance between vehicles. The second approach measures and analyses bio-physical parameters of the drivers such as features of the eyes (such as eye closure rating, called PERCLOS), face, head, heart, brain electrical activity, skin conductance and respiration, body posture, head nodding, voice pitch, etc. These measurements can be conducted by using video camera, optical sensors, voice/emotion recognition, and steering wheel sensors.

There has been substantial research addressing the issues of driver drowsiness and fatigue. Many of the proposed systems rely on a number of soft computing methods, such as sensor fusion, neural networks, and fuzzy logic. For example, Ward and Brookhuis [44] describes project SAVE (System of effective Assessment of the driver state and Vehicle control in Emergency situations) and a subsequent project AWAKE (effective Assessment of driver vigilance and warning to traffic risK Estimation) undertaken in Europe in the late 1990s with the aim of real-time detection of driver impairment and the engagement of emergency handling maneuvers. In SAVE the data from the vehicle sensors is first classified using neural networks and then the final diagnostics is performed using fuzzy logic.

Ford has been extensively studying the efficacy of different methods to identify and provide remedies for drowsy drivers using VIRtual Test Track Experiment (VIRTTEX). Kozak et al. [17] describes the analysis of different methods to provide lane departure warning for drowsy drivers including steering wheel torque and vibration, rumble strip sound, and heads up display.

The emerging area of affective computing [26] opened up a new opportunity to monitor and mitigate the adversary driver behaviors based on negative emotions such as stress and anger. In fact, Prof. Picard considers that the automotive industry will be the first to apply truly interactive affective computing to products for safety reasons [3]. "Sensors can decide the driver's emotional condition. A stressed driver might need to be spoken to in a subdued voice or not interrupted at all."

However, the attention to affective technologies in the automotive industry encompasses more than just safety issues. The success of humanoid robots leaves no doubt of the importance of emotional intelligence for building machines and systems that can appeal to people. The description of the modern vehicle as a highly computerized machine that continuously interacts with the driver seems to be a reasonable candidate for the massive realization of the concept of emotional intelligence. It is reasonable to expect that a vehicle that is implanted with emotional intelligence ability can be appealing to the customer and may stimulate the creation of an emotional bond between the vehicle and the driver.

Toyota's POD (Personalization on Demand) concept vehicle [24] that was developed in collaboration with Sony is an intelligent vehicle control system that is able to estimate the driver's emotion and also exhibits its own emotional behavior corresponding to the vehicle status. The POD vehicle is inspired by the idea of affective computing and represents the first vehicle spin-off of humanoid robot technology [11]. From a systems perspective it implements a cognitive model that is similar to the cognitive emotional engine of Sony's Aibo companion robot [10] but with vehicle specific sensors and actuators. Its main components include three AI modules that are derived from the architecture of Aibo robot – Perception Module, Cognitive Behavior Module, and Control Module.

POD's Perception Module detects variations in driving conditions; monitors the steering wheel, accelerator and brakes, the pulse, the face and the perspiration level of the driver. Soft sensors screen driver's preferences, including driving style, music and other favorites. The result is a set of features that describe the current status of the driver and vehicle. A nonlinear mapping with predefined thresholds maps the feature set into 10 different emotional states.

POD's Cognitive Behavior Module estimates the new state based on the current and the previous state and pulls the set of behaviors (reactions) that correspond to this new state. This is the reaction of the POD vehicle to current emotional state of the vehicle and the driver. POD's behaviors are event driven software agents that create actions based on the information from the sensors and the other behaviors. The agents exemplify different behaviors; some of those behaviors are blended in ten different emotions, including happiness, surprise, sadness, etc. The cognitive module functions as an evolving adaptive controller that continually monitors the vehicle systems and driver's status and generates actions that maximize safety and comfort objective functions. POD's cognition module learns from the driver's habits and actions and evolves the behavior agents accordingly. The result of this is that POD's emotions continually evolve and reflect the current status of the vehicle and the driver.

POD's Control Module implements the actions associated with the selected behaviors by activating specific actuators. Actuators include color changing LED panels on the front, servomotors that change the positions of the headlamps, grille, and side mirrors that communicate the current emotional status of the vehicle. The POD actuators display warnings, chose the right music, control the A/C. The emotional state of the vehicle is expressed and communicated by controlling the shutters, antenna, vehicle height, windshield color, and ornament line.

7 Conclusions

In this chapter we have reviewed the major areas of intelligent system applications that are utilized in motor vehicles. The goal of this chapter was to focus on the technologies that are actually deployed inside the customer vehicle and interact with the driver. The modern passenger car or truck is an extremely sophisticated and complex piece of machinery that plays a critical role in the lives of many consumers. It is also much more than a mechanical transportation device and is often the center of passionate debate among consumers. There are few other industries that are as competitive as the automobile industry and this often results in very fast implementation of new technologies.

We discussed many approaches to intelligent system design that impact the driver with the intention of improving the overall driving experience. It has been shown that not all new technologies are readily embraced by drivers and the auto manufacturers have learned that "talking cars" and other intrusive technologies are not always welcome. Therefore, the automobile manufacturers must balance the benefits of introducing new technologies with the possible consumer backlash if the technology application is rejected. All of the applications described in this chapter have been deployed or tested and they show the wide range of technologies that have been adapted into the cars and trucks that we drive.

It is quite clear that the AI and intelligent systems have become a valuable asset that has many important uses in the automotive industry. The use of intelligent systems and technologies results in applications that provide many benefits to both the auto manufacturers and their customers. We believe that this trend will increase into the future as we move toward the age of intelligent vehicles and transportation systems.

References

1. Aston Martin DB9 Brochure, 2005, http://www.astonmartin.com/thecars/db9.
2. Bishop, R., 2005. *Intelligent Vehicle Technology and Trends*, Artech House, Inc. Norwood, MA.
3. Bostrom, J., 2005. Emotion-Sensing PCs Could Feel Your Stress. *PC World*, April.
4. Boverie, S., Demaya, B., Le Quellec, J.M., Titli, A., 1993. Contribution of Fuzzy Logic Control to the Improvement of Modern Car Performances. *Control Engineering Practice,* Vol. 1, Issue 2.

5. Breed, D., 2004. *Telematics System for Vehicle Diagnostics.* US Patent # 6,738,697.
6. Buhler, D., Vignier, S., Heisterkamp, P., Minker, W., 2003. Safety and Operating Issues for Mobile Human-Machine Interfaces. In *Proceedings IUI'03.* Miami, FL, January 12–15.
7. Djurdjanovic, D., Lee, J., Ni, J., 2003. Watchdog Agent—an Infotronics-Based Prognostics Approach for Product Performance Degradation Assessment and Prediction. *Advanced Engineering Informatics,* Vol. *17,* pp. 109–125.
8. Engstrom, J., Victor, T. 2005. *System and Method for Real-Time Recognition of Driving Patterns.* US Patent Application # 20050159851.
9. Filev, D., Tseng, F., 2002. Novelty Detection Based Machine Health Prognostics, In *Proceedings of the 2006 Int. Symposium on Evolving Fuzzy Systems.* Lancaster, UK, September 2006, pp. 193–199.
10. Fong, T.W., Nourbakhsh, I., Dautenhahn, K., 2003. A Survey of Socially Interactive Robots. *Robotics and Autonomous Systems*, 42(3–4).
11. Fujita, M., Kitano, H., 1998. Development of an Autonomous Quadruped Robot for Robot Entertainment. *Autonomous Robots* Vol. 5, Issue 1, pp. 7–18.
12. Greitzer, F.L., Hostick, C.J., Rhoads, R.E., Keeney, M., 2001. Determining how to Do Prognostics, and then Determining what to Do with It. In *Proceedings of AUTOTESTCON 2001.* Valley Forge, Pennsylvania, August 20–23.
13. Greitzer, F., Pawlowski, R., 2002. Embedded Prognostics Health Monitoring. In *Instrumentation Symposium, Embedded Health Monitoring Workshop*, May.
14. Hayashi, K., Shimizu, Y., Dote, Y., Takayama, A., Hirako, A., 1995. Neuro Fuzzy Transmission Control for Automobile with Variable Loads. *IEEE Transactions on Control Systems Technology*, Vol. 3, Issue 1, March.
15. Heisterkamp, P., 2001. Linguatronic: Product-Level Speech System for Mercedes-Benz Car. In *Proceedings of the First International Conference on Human Language Technology Research.* Kaufmann, San Francisco, CA.
16. Horberry, T., Hartley, L., Krueger, G.P., Mabbott, N., 2001. Fatigue Detection Technologies for Drivers: a Review of Existing Operator-Centred Systems. In *Human Interfaces in Control Rooms, Cockpits and Command Canters.* People in Control. The Second International Conference on (IEE Conf. Publ. No. 481).
17. Kozak, K., Greenberg, J., Curry, R., Artz, B., Blommer, M., Cathey, L., 2006. Evaluation of Lane Departure Warnings for Drowsy Drivers. To appear in *Proceedings of Human Factors and Ergonomics Society.* San Francisco, October.
18. Luo, J., Namburu, M., Pattipati, K., Liu Qiao Kawamoto, M., Chigusa, S., 2003a. Model-Based Prognostic Techniques [Maintenance Applications]. In *Proceedings of AUTOTESTCON 2003.* IEEE Systems Readiness Technology Conference, Anaheim, CA, September, pp. 330–340
19. Luo, J., Tu, F., Azam, M., Pattipati, K., Willett, P., Qiao, L., Kawamoto, M., 2003b. Intelligent Model-Based Diagnostics for Vehicle Health Management. In *Proceedings of the SPIE.*, Orlando, FL, Vol. 5107, pp. 13–26.
20. Marko, K.A., James, J.V., Feldkamp, T.M., Puskorius, G.V., Feldkamp, L.A., Prokhorov, D., 1996a. Training recurrent Networks for Classification: Realization of Automotive Engine Diagnostics. In *Proceedings of the World Congress on Neural Networks (WCNN'96).* San Diego, CA, pp. 845–850.
21. Marko, K., James, J., Feldkamp, T., Puskorius, G., Feldkamp, L., 1996b. Signal Processing by Neural Networks to Create Virtual Sensors and Model-Based Diagnostics. In *Proceedings of the 1996 International Conference on Artificial Neural Networks.* Vol. 1112, pp. 191–196, Springer.
22. Minker, W., Haiber, U., Heisterkamp, P., Scheible, S., 2002. Intelligent Dialogue Strategy for Accessing Infotainment Applications in Mobile Environments. In *Proceedings of the ISCA Tutorial and Research Workshop on Multi-Modal Dialogue in Mobile Environments.* Kloster Irsee, Germany.

23. Miyahara, S., Sielagoski, J., Koulinitch, A., Ibrahim, F., 2006. Target Tracking by a Single Camera based on Range Window Algorithm and Pattern Matching. In *Proceeding of 2006 SAE World Congress*. SAE, Detroit, Michigan, April 3–6.
24. Mori et al., 2004. *Vehicle Expression Control System, Vehicle Communication System, and Vehicle Which Performs Expression Operation*, US Patent #6,757,593.
25. Nigro, J.M., Rombaut, M., 2003. IDRES: A Rule-Based System for driving Situation Recognition with Uncertainty Management. *Information Fusion*, Vol. 4, pp. 309–317.
26. Picard, R., 1997. *Affective Computing*, MIT Press.
27. Pieraccini, R., Dayanidhi, K., Bloom, J., Dahan, J.-G., Phillips, M., Goodman, B.R., Venkatesh, P.K., 2004. Multimodal Conversational Systems for Automobiles. *Communications of the ACM (CACM)*, Vol. 47, Issue 1, pp. 47–49.
28. Puskorius, G., Feldkamp, L.A., 2001. Parameter-Based Kalman Filter Training: Theory and Implementation. In S. Haykin (ed.) *Kalman Filtering and Neural Networks*. John Wiley & Sons, Inc., New York.
29. Rajagopalan, A., Washington, G., 2002. Intelligent Control of Hybrid Electric Vehicles Using GPS Information. SAE Paper, 2002-01-1936.
30. Rajagopalan, A., Washington, G., 2003. Development of Fuzzy Logic and Neural Network Control and Advanced Emission Modeling for Parallel Hybrid Vehicles. National Renewable Energy Laboratory Report #NREL/SR-540-32919.
31. Schlenoff, C., Balakirsky, S., Uschold, M., Provine, R., Smith, S., 2004. Using Ontologies to Aid Navigation Planning in Autonomous Vehicles. In P. McBurney, S. Parsons (eds.). *The Knowledge Engineering Review*, Vol. 18, Issue 3, pp. 243–255, Cambridge University Press.
32. Schouten, N.J., Salman, M.A., Kheir, N.A., 2003. Fuzzy Logic Control for Parallel Hybrid Vehciles. *IEEE Transactions on Control Systems Technology*, Vol. 10, Issue 3, May, pp. 460–468.
33. Schwall M., Gerdes, J.C., 2001. Multi-Modal Diagnostics for Vehicle Fault Detection. In *Proceedings of IMECE2001*. ASME International Mechanical Engineering Congress and Exposition. New York, NY, November 11–16.
34. Schwall, M.L., Gerdes, J.C., 2002. A Probabilistic Approach to Residual Processing for Vehicle Fault Detection. In *Proceedings of the American Controls Conference*, Anchorage, AL, pp. 2552–2557.
35. Schwall, M.L., Gerdes, J.C., Baker, B., Forchert, T., 2003. A Probabilistic Vehicle Diagnostic System Using Multiple Models. In *Proceedings of the Fifteenth Conference on Innovative Applications of Artificial Intelligence*. Acapulco, August 12–14, pp. 123–128.
36. Struss P., Price C., 2003. Model-Based Systems in the Automotive Industry. *AI Magazine*, Vol. 24, pp. 17–34.
37. Takahashi, H., 1988. Automatic Speed Control Device Using Self-Tuning Fuzzy Logic. In *Proceedings IEEE Workshop on Automotive Applications of Electronics*, pp. 65–71.
38. Takahashi, H., 1995. Fuzzy Applications for Automobiles. In Kaoru Hirota, Michio Sugeno (eds) *Industrial Applications of Fuzzy Technology in the World*. World Scientific.
39. Tascillo, A., DiMeo, D., Macneille, P., Miller, R., 2002. Predicting a Vehicle or Pedestrian's Next Move with Neural Networks. In *Proceedings of the International Joint Conference of Neural Networks*. Honolulu, Hawaii, May 12–17, pp. 2310–2314.
40. Trepagnier, P.G., Kinney, P.M., Nagel, J.E., Dooner, M.T., Pearce, J.S., 2005. *Team Gray Technical Paper*, http://www.darpa.mil/grandchallenge05/ TechPapers/GreyTeam.pdf.
41. Vachtsevanos, G., Wang, P., 2001. Fault prognosis using dynamic wavelet neural networks. In *Proceedings of the AUTOTESTCON 2001*, IEEE Systems Readiness Technology Conference, Philadelphia, PA, pp. 857–870.
42. Vlacic, L., Parent, M., Harashima, F., 2001. *Intelligent Vehicle Technologies*. Butterworth-Heinemann.
43. Von Altrock, C., 1997. "Fuzzy Logic in Automotive Engineering", Circuit Cellar ink., Issue 88, November 1997.

44. Ward, N.J., Brookhuis, K., 2001. Recent European Projects on Driver Impairment. In *Proceedings of International Symposium on Human Factors in Driving Assessment, Training and Vehicle Design*. Aspen, Colorado, August 14–17.
45. Weil, H.-G., Probst, F.G., 1994. Fuzzy Expert System for Automatic Transmission Control. In R.J. Marks II (ed.) *IEEE Technology Update Series: Fuzzy Logic Technology and Applications*, IEEE, pp. 88–93.

Symbiosis of Human and Communication Robots

Norihiro Hagita[1], Hiroshi Ishiguro[1,2], Takahiro Miyashita[1], Takayuki Kanda[1]
Masahiro Shiomi[1] and Kazuhiro Kuwabara[1,3]

[1]ATR Intelligent Robotics and Communication Laboratories, Kyoto, Japan
[2]Osaka University, Osaka, Japan
[3]Ritsumeikan University, Shiga, Japan
hagita@atr.jp, ishiguro@ams.eng.osaka-u.ac.jp, miyasita@atr.jp
kanda@atr.jp, m-shiomi@atr.jp, kuwabara@is.ritsumei.ac.jp

Abstract. This chapter discusses the possibilities of symbiosis with human and communication robots from the viewpoint of communication media. Recently communication robots have come into greater use as next-generation communication media by being networked with humans, PCs and ubiquitous sensors (stationary and wearable). The "Network Robots", a new framework for integrating ubiquitous network and robot technologies, is a step towards providing infrastructure to make robots into communication media. In this chapter, the development of communication robots at ATR is introduced along with the results of two field experiments at elementary school and a science museum as notable examples of communication robots in the real world.

Keywords. Communication robot, network robot system, human robot interaction.

1 Introduction

How could robots become next-generation media of communication? We can guess a possible scenario by examining the history of media usage. The history of media strongly suggests that we human beings have an inherent motivation to disseminate our feelings and experiences to each other using any and all kinds of communication media. For example, paper and the printing technology invented by Gutenberg many centuries ago still give us many chances to communicate with others. The Internet, which was invented only several decades ago, now becomes an indispensable communication media allowing person-to-person, person-to-community, and community-to-community communication. It allows people to communicate anywhere, anytime, and with anyone using PCs, cell phones, and PDAs.

People tend to adopt more easy-to-use media intrinsically for disseminating and sharing their experiences. In this regard, communication robots may become a plausible media of communication, since we can ask them and communicate with them without, for example, typing commands as if they were human partners. However, their communication ability, such as speech recognition, is insufficient in practical use and needs to be improved. Recent developments in ubiquitous networks may increase the possibility of improving this ability and living with robots by having robots cooperate with PCs and ubiquitous sensors via networking. Therefore, by making use of networks, communication robots will become next-generation communication media.

This chapter discusses the possibilities of symbiosis with human beings and communication robots based on humanoids. Several research issues for symbiosis are discussed. Communication robots developed at ATR shows that they can improve their communication abilities by a network robot system, a new framework for integrating ubiquitous network and robot technologies, especially with ubiquitous tags and sensors.

2 Towards Symbiosis with Humans

Let us consider an everyday communication robot living together with humans. The robot should recognize and understand a succession of scenes, including persons, objects, and the environment, and perform daily activities while communicating with humans. In the case of a humanoid robot, it recognizes scenes using vision, audio and tactile sensors and generates human-like behaviors by its arms, hands, neck, eyes, etc. Having the same parts as a human body, it might be called a "physical existence media" of communication and give us an impression which is different from present communication media, such as cell phones, PCs and PDAs. For example, when the eyes of a robot look into a person and follows him as shown in Fig. 1, we may feel a strong impression as if we were with another person there. Since the communication functions in humanoid robots will be applied in part to other types of communication robots, i.e., robots in virtual space and ones embedded in the environment, our discussion will embrace all communication robots. Therefore, this chapter focuses on communication robots based on humanoids.

There are three kinds of research issues for symbiosis. The first issue is how to allow a robot to naturally communicate with human. One significant function of communication robots is to talk while gesturing and gazing as physical existence media as if they could talk to each other. Current communication media hardly do that. For example, let us imagine how to respond if a cell phone asks us "please, hug me." We may have no idea of how to hug it, since it has no hands and body.

Previous AI technology tended to focus on individual communication skills such as seeing, hearing, talking, and thinking. However, perceptual abilities are yet insufficient for real world environments. One possible way of improving these abilities is to communicate with ubiquitous sensors, PCs and Internet information via networks, since they can obtain missing/additional information easily, such as human and object identifications, missed events, individual information on the environment, etc.

The second issue is to realize network robots that allow communication with other robots, ubiquitous sensors (stationary and wearable) and PCs for improving individual communication skills and obtaining missing/additional information in the environment. In a network robot system, three different types of robots are identified, that is, a "visible type," a "virtual type" and an "unconscious type", as shown in Fig. 2 [1].

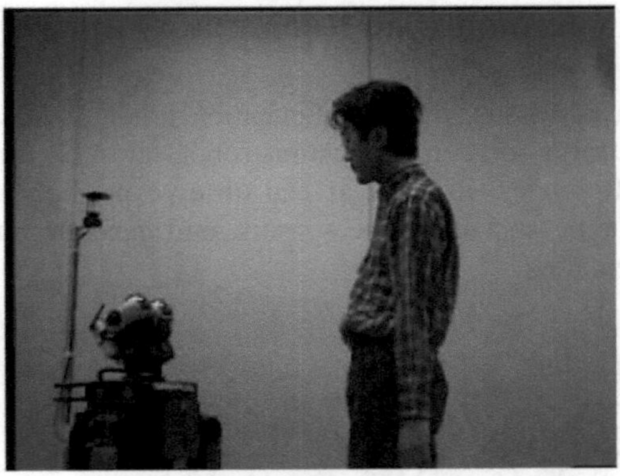

Fig. 1 Eye contact.

Fig. 2 Three types of network robots.

The "visible type" or "real existence" robot has a concrete body in the real world. Typical examples are humanoid and pet robots. In contrast, the "virtual type" robot works in a virtual (cyber) world. It has a graphical representation, and interacts with human users through a display. The "unconscious type" robot is embedded in an environment, such as roads, towns, rooms, and equipment. Examples include a "robotic room" that monitors people within it and provides support using actuators that are integrated with the room. These different types of robots are connected via a network and collaborate with each other. Together with various sensors embedded in the environment, network robots can provide services that cannot be realized with a single robot.

The third issue is to carefully analyze whether humans could accept robots or not in a society. In particular, we need to conduct field experiments in various public places such as exhibition halls, schools, downtown, and busy streets as well as at private space such as home.

3 Developing Communication Robots at ATR

As for the first research issue, the communication robot “*Robovie*” series has been developed as a platform for communication robots at ATR since 1999. Figure 3 shows a series of Robovies. Robovie II can observe perceptual information using vision, audio and tactile and ultrasonic sensors, and generate human-like behaviors using human-like actuators [2].

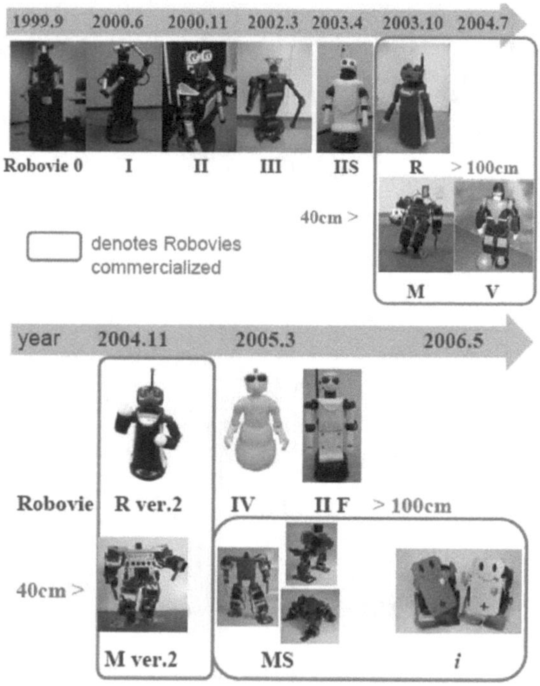

Fig. 3 “Robovie” series for communication robots.

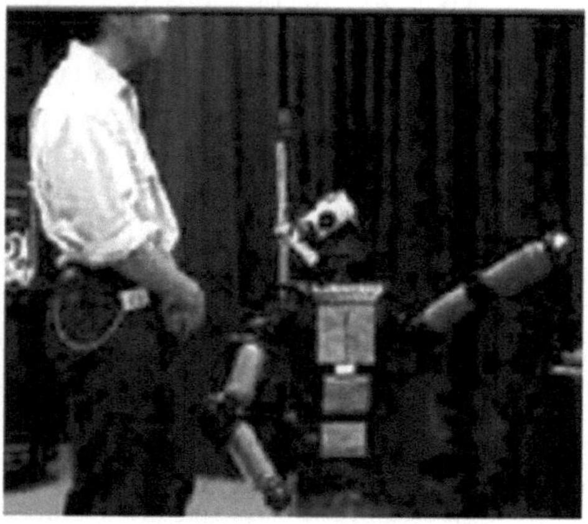

Fig. 4 An example of “joint attention” behavior.

(**a**) Balancing (**b**) Doing a headstand

Fig. 5 Robovie-M with Robovie Maker.

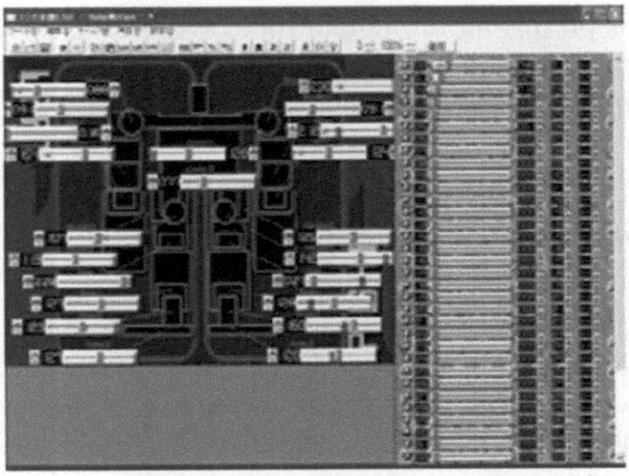

Fig. 6. Behavior editor "Robovie Maker®".

Its height and weight are 114 cm and 39 kg respectively. It includes a 3-joint head, two 4-joint arms, skin sensors, mobility, omni-directional image recognition, voice dialog capability, and ultrasonic distance sensor. It can roughly recognize human faces using an omni-directional camera, maintain eye-contact with a specific person, and recognize about three hundred Japanese or fifty English words, in addition to speech-synthesizing about three hundred sentences. More than one hundred behaviors can be registered in advance. Eye contact is automatically changed depending on the communication situation.

In general, human communication even in a meeting consists of a chain of short-time interactions, which includes the parts of introduction ("hello"), development ("shake your hand"), turn ("where do you come from?") and conclusion ("bye"). A network of situated modules [3] has been developed in order to tackle this problem. The situated modules fall into more than one hundred behaviors, such as "hello", "shake your hand", "please hug me", "play with me", "bye", etc. The state space construction enables a robot to make appropriate internal representation by itself from only sensor information. As a result, Robovie II can continue to communicate with humans autonomously by representing friendly interactive behaviors such as greeting, nodding, gazing, kissing, singing, hugging, etc., while running on batteries. For example, "joint attention", in which a robot gazes at a human and then points at an object is a significant behavior for a robot to serve as an assistant in a room or on the street. In joint attention, humans tend to glance in the direction of the object pointed at by the robots as shown in Fig. 4. This behavior may help humans pay attention to specific objects or events. Various arm movements will help humans decide the appropriate route or the path to a place of interest.

Robovie-M has been commercialized and currently on the market. Robovie-M, with a 29 cm height and 1.9 kg weight, is available in an assembly kit for use as an educational tool for students, researchers, and engineers. Figure 5 shows an assembled Robovie-M. Robovie-M has 22 servomotors and one acceleration sensor. It can also be used as a platform for physical existence media or various demonstrations targeted for researchers, engineers, etc.

We also developed a software program, called "Robovie Maker®," that can easily generate a lot of behaviors for Robovie-M using a PC and a mouse (Fig. 6). A set of text files describing robot behaviors is generated. This software will become a meaningful step towards standardization of communication robot behavior description. The standardization is promising for human-robot symbiosis, since it will enable us to exchange a variety of behaviors of many kinds of communication robots by downloading them from websites in the near future.

4 Network Robots at an Elementary School

Field experiments in the real world often point out intrinsic research issues. Experiments on symbiosis with pupils and Robovie-II at an elementary school were conducted in order to find some clues for all of the research issues. For example, we easily encounter the problem of the so-called cocktail-party effect in speech recognition. That is, since many pupils attempt to come near the robot as a newcomer and say many words at the same time, it could not recognize who is speaking and what he/she says. This is related to the first research issue. Work on blind source separation has been known as a possible solution for the cocktail-party effect. However, we can consider alternate approach of using network robots since the robot can obtain additional/missing information from ubiquitous sensors and tags. That is related to the second research issue. As for the third research issue, we are analyzing the relationship between pupils and Robovie-II in an 18-day field trial held at a Japanese elementary school, as shown in Fig. 7 [3].

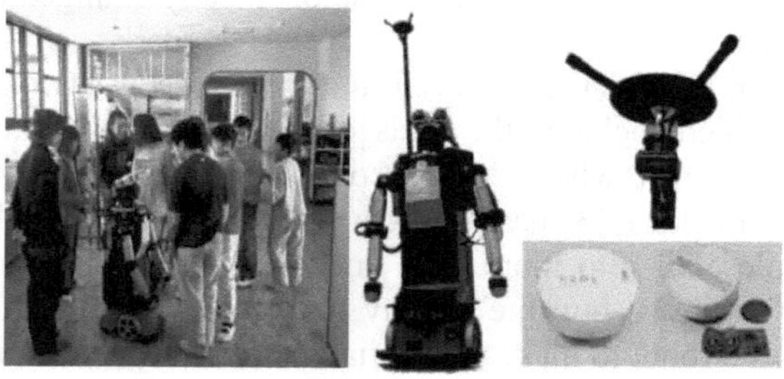

Fig. 7 Robovie-II at a school and wireless tags.

Another interesting trial was included in the experiment. The basic idea is to examine the proposition that children might learn from robots as they learn from other children. Two English-speaking Robovie-IIs communicated with first- and six-grade pupils at the perimeter of their respective classroom. Using wireless identification tags and sensors, these robots identified and interacted with children who came near them. The robots gestured and spoke English with the children, using a vocabulary of about 300 sentences for speaking and 50 words for recognition. The children were given a brief picture-word matching English test at the start of the trial, after one week and after two weeks. Interactions were counted using the tags, and video and audio were recorded.

The result shows that interaction with the robot was frequent in the first week, and then it fell off sharply by the second week. Nonetheless, some children continued to interact with the robot. Interaction time during the second week predicted improvements in English skill at the post-test, controlling for pre-test scores.

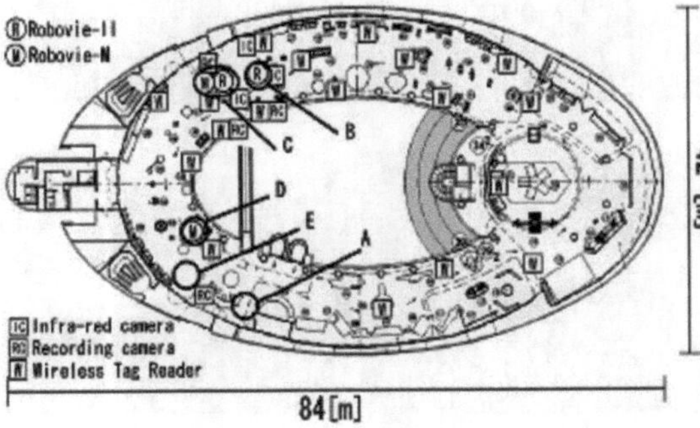

Fig. 8 The map of the Osaka Science Museum.

5 Guide Robots in a Science Museum

We installed a network robot system [4] in the Osaka Science Museum (Fig. 8) to guide visitors and to motivate them to study science. RFID readers, infrared cameras, and video cameras were used to acquire rich sensory information for monitoring and recognizing the behavior of visitors in an environment. The Robovie II autonomously interacted with people using gestures and utterances that came from humans. Robovie M was characterized by its human-like physical expression. Their task is to guide visitors to exhibits.

Figure 9 shows an example of this behavior. When bringing visitors to the exhibition place related to the telescope. Robovie said to visitors, “I am taking you to an exhibit, please follow me!” (a), and approaches the telescope booth (b,c). Robovie suggested looking through it and then explained its inventor (d).

Two stationary robots (Robovie II and Robovie-M) casually talk about the exhibits like humans with accurate timing because they are synchronized with each other using an Ethernet network. The topic itself is intelligently determined by the ubiquitous sensors. By knowing the previous visiting course of a visitor, the robots can try to interest him in an exhibit and he overlooked by starting a conversation on that exhibit. Figure 10 shows robots talking. The flow and an example of dialogue are given below:

(1) Robovie-M explains an exhibit.
(2) Robovie II asks Robovie-M a question about it. For example, “Who made it?”
(3) Robovie-M answers the question and expounds on his answer.

(Robovie-M): "That chair can float, even if a person is sitting on it."
(Robovie II): "That's incredible! How does it do that?"
(Robovie-M): "By magnetic power."
(Robovie II): "I wonder if I can sit on that…"
(Robovie-M): "I doubt it."

(**a**) (**b**) (**c**) (**d**)

Fig. 9 Robovie guiding visitors to the telescope exhibition place.

(**a**) Talking to each other (**b**) Talking to visitors

Fig. 10 Scenes of the robots who talk to each other.

By the end of the two-month period, the number of visitors reached 91,107, the number of subjects who wore RFID tags was 11,927, and the number of returned questionnaires was 2,891. The experimental results show that the new approach positively affected people's experience in their visits to the Science Museum. We compared the effects of the exhibits-guidance and free-play interactions. As a result, the free-play interaction helped to build visitors' interest in science, while Robovie mostly received higher subjective impressions, which is probably due to its novelty.

6 Conclusions

This chapter discussed symbiosis with humans and communication robots from the viewpoint of communication media. Three research issues for realizing symbiosis were discussed. In order to tackle these issues, several kinds of Robovies have been developed, and field experiments at a school and science museum were conducted. These experimental results suggest that communication robots should be designed to have something in common with their users. Symbiosis with human beings and robots provides a social as well as a technical challenge. The network robot is also very effective for identifying a specified person among several persons.

In the near future, communication robots will likely spend much time talking with persons and sometimes taking care of them. That means symbiosis in the human community will make communication robots ultimate communication media.

Acknowledgements

We are also grateful to the members of our project for several discussions. This research was supported by the Ministry of Internal Affairs and Communications of Japan.

References

1. N. Hagita, "Network Robots Project," *IEEE/RSJ International Conference on Intelligent Robots and Systems (IROS), JST Workshop on Robotics Projects and Future Funding Strategy*, pp. 28–33, Sendai, Japan, October, 2004.
2. T. Kanda, H. Ishiguro, M. Imai, and T. Ono, "Development and Evaluation of Interactive Humanoid Robots," *Proceedings of the IEEE (Special issue on Human Interactive Robot for Psychological Enrichment)*, Vol. 92, No.11, pp. 1839–1850, 2004.
3. T. Kanda, T. Hirano, D. Eaton, and H. Ishiguro, "Interactive Robots as Social Partners and Peer Tutors for Children: a Field Trial," *Human Computer Interaction (Special issues on human-robot interaction)*, Vol. 19, Nos. 1–2, pp. 61–84, 2004.
4. H. Ishiguro, M. Shiomi, T. Kanda, D. Eaton, and N. Hagita, "Field Experiment in a Science Museum with Communication Robots and a Ubiquitous Sensor Network," *IEEE International Conference on Robotics and Automation (ICRA) Workshop on network robot system*, pp. 24–33, Barcelona, Spain, April, 2005.

Wave-based Control of Flexible Mechanical Systems

William J. O'Connor

213 Mechanical Engineering, UCD Belfield, Dublin 4, Ireland
william.oconnor@ucd.ie
Tel +353 1 7161887 Fax + 2830534

Abstract. There are many contexts, from space structures to disk drive heads, from medical mechanisms to long-arm manipulators, from cranes to light robots, in which it is desired to achieve rapid and accurate position control of a system end-point by an actuator working through a flexible system. The system's actuator must then attempt to reconcile two, potentially conflicting, demands: position control and active vibration damping. Somehow each must be achieved while respecting the other's requirements. Wave-based control is a powerful, relatively new strategy for this important problem that has many advantages over most existing techniques. The central idea is to consider the actuator motion as launching mechanical waves into the flexible system while simultaneously absorbing returning waves. This simple, intuitive idea leads to robust, generic, highly efficient, adaptable controllers, allowing rapid and almost vibrationless re-positioning of the remote load (tip mass). For the first time there is a generic, high-performance solution to this important problem that does not depend on an accurate system model or near-ideal actuator behaviour.

Keywords. Flexible mechanical systems, robot analysis and control, slewing of space structures, active vibration control.

1 Introduction

There are many contexts, from space structures to disk drive heads, from medical mechanisms to long-arm manipulators, from cranes to light robots, in which it is desired to achieve rapid and accurate position control of a load (or system end-point) by an actuator that is separated from the load by an intermediate system which is flexible. While all systems are to some extent flexible, issues related to flexibility become decisive as one tries to design lighter mechanisms, or systems that are more dynamically responsive, or softer, or more energy efficient, or simply long in one dimension.

The system's actuator must then attempt to reconcile two, potentially conflicting, demands: position control and active vibration damping. Somehow each must be achieved while respecting the other's requirements.

Previous approaches have included various classical and state feedback control techniques (often using simplified dynamic models); modal control (often considering a rigid-body, or zero frequency mode separately from vibration modes); sliding mode control; input command shaping; time-optimal control leading to bang-bang control; wave-based control; and control based on real-virtual system models [1] [2] [3] [4] [5] [6]. Each method has special characteristics and drawbacks, discussed in the literature. None is completely satisfactory under all headings.

The wave-based control strategy [7] [8] [9] [10] [11] [12] is a powerful, relatively new method of dealing with flexibility that has been shown to be better than existing methods in most respects. The central idea is to consider the actuator motion as launching mechanical waves into the flexible system while simultaneously absorbing returning waves. This simple, intuitive idea leads to robust, generic, highly efficient, adaptable controllers, allowing rapid and almost vibrationless re-positioning of the system and the remote load (tip mass). For the first time there is a generic solution to this important problem that does not depend on an accurate system model and does not demand close to ideal performance by the actuator. Rather than treating the flexibility as a problem, it works with the flexibility to achieve system control in a natural way.

This chapter will investigate the mathematical foundation for a wave-based interpretation of flexible system dynamics, both lumped and continuous, exemplified in Fig. 1. It will then show how this view can be used to interpret the actuator-system interface as a two-way energy flow, leading to the design of controllers that give optimal performance by controlling this energy flow, in ways that are simple, robust, generic, and energy efficient.

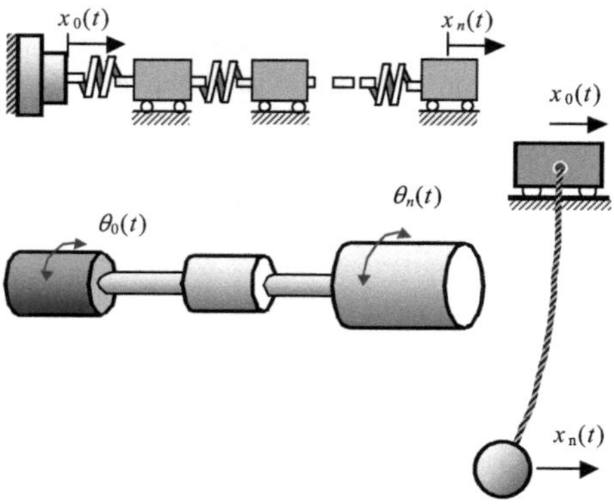

Fig. 1 Typical flexible systems, with actuator position, $x_0(t)$ or $\theta_0(t)$, controlling tip position, $x_n(t)$ or $\theta_n(t)$.

For simplicity, it will be assumed that there is a single actuator, with its own position controller, which is attempting to control the position of the system tip, moving it from rest in one position to rest at a target position. If gravity is active, it is assumed that the initial and final gravitational strains are equal, so that, when the system comes to rest again, the net displacement of the actuator will equal that of the tip. It is further assumed that the actuator position controller has zero steady-state error, so final position accuracy is limited only by the actuator sensor accuracy.

2 Wave Analysis of Flexible Systems

In flexible systems of the above type, the actuator and load are dynamically uncoupled. The interaction between them is mediated by the flexible system dynamics and it involves delays. When the actuator moves it directly affects only the part of the flexible system to which it is attached. A disturbance (or "wave") then travels through the system to the load or tip, and then back towards the actuator, typically dispersing as it goes, in a complex motion. At each end of the system, some of this wave may be reflected and/or absorbed, depending on the instantaneous relationship between the motions of the actuator, or tip, and the motion of the adjacent parts of the flexible system.

The wave-based control strategy depends on (a) understanding, (b) measuring and (c) controlling the (notional) two-way flow of energy and momentum happening at the interface between the actuator and the flexible system. To move the tip from rest to rest the actuator must launch a "wave" into the flexible system and then absorb it, in such a way that when all the energy and momentum of the motion have been extracted, the system is at the target position.

The term "wave" here is very general, and includes not just oscillating motion but also a "step wave" which, after it passes a point, changes the net or DC displacement, implying "rigid body" or "zero frequency" motion. Because the primary focus here is position control, the wave variable is taken as displacement, linear or angular (in meters or radians). In other applications, variables such as force, torque, velocity, or acceleration would be appropriate as wave variables, and the wave control ideas can easily be adapted to suit.

3 Resolving Actuator Motion into 2-Way Waves

The actuator motion, $x_0(t)$ is notionally resolved into two component motions, $a_0(t)$ and $b_0(t)$,

$$x_0(t) = a_0(t) + b_0(t), \tag{1}$$

with a_0 corresponding to an outwards-going, or launch, wave, b_0 corresponding to a return wave, which the actuator attempts to absorb.

For the control application, the resolving in (1) need not be precise: it is necessary to fulfil only certain generic criteria [11]. The simplest definition sets

$$a_0(t) = \tfrac{1}{2}\left(x_0(t) + \int \frac{f(t)}{Z} dt \right) \tag{2}$$

$$b_0(t) = \tfrac{1}{2}\left(x_0(t) - \int \frac{f(t)}{Z} dt \right) \tag{3}$$

where $f(t)$ is the force that the actuator applies to the beginning of the flexible system in the direction of motion, and Z is an impedance term, whose value is not critical. For lumped systems Z can be taken as $\sqrt{(k_1 m_1)}$, corresponding to the first spring stiffness, k_1, and first mass, m_1. For the gantry crane, one can set $Z = \sqrt{(\rho T)}$ where ρ is the linear density of the cable and T the tension at the top, and for a simple pendulum system, $Z = m\sqrt{(g/L)}$, with m the mass and L the length.

For a system of lumped masses and springs in series, a slightly better expression [11] [12] for the s-domain version of the return wave, $B_0(s)$, is given by

$$B_0(s) = G(s)\big(X_1(s) - G(s)A_0(s)\big) \tag{4}$$

with

$$A_0(s) = X_0(s) - G(s)\big(X_1(s) - G(s)A_0(s)\big) \tag{5}$$

where $G(s)$ is a second order mass-spring-damper system, with mass and spring corresponding the beginning of the lumped flexible system, and with viscous damping at half critical [Fig. 2].

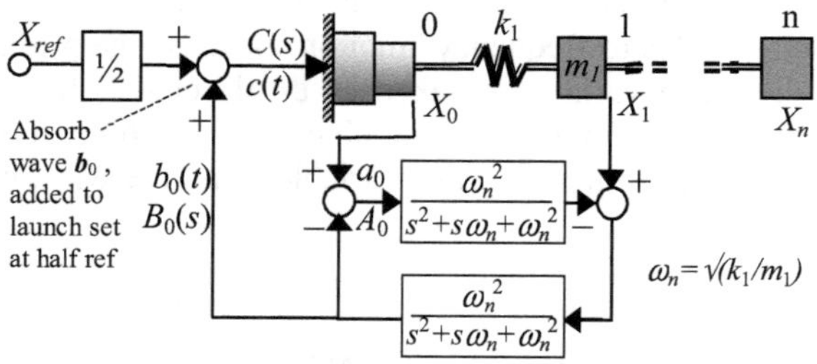

Fig. 2 A wave-based control implementation using (4) and (5) to determine return wave, b_0 from x_0 and x_1, with G as shown.

4 The Control Strategy

For rest-to-rest motion to a target position, the strategy is as follows. The input to the actuator is set as the sum of two components. The first component is set directly by the controlling computer, the only essential requirement being that its time profile ends at half the target displacement and holds there. The second component is the measured return wave, b_0, calculated by measuring two variables, such as x_0 and x_1, or x_0 and f, and calculated using for example (3) or (4).

Adding the second input component, which has the form of a positive feedback signal, provides active vibration damping, by making the actuator appear as a matched viscous impedance to "waves" returning from the flexible system towards the actuator. See for example (3). This causes the actuator to act as a very efficient, one-way, active vibration absorber, yet without impeding the action of simultaneously setting a launch wave.

A second effect of the absorbing component is to cause the total steady-state actuator displacement to become exactly double the launch component. This can be seen most clearly in the case of (2) and (3). For rest to test motion, the force integral terms must become zero, so that at rest a_0 and b_0 must become equal to each other and equal to $\frac{1}{2}x_0$. Thus, if the launch displacement component is set to settle at half the target displacement, adding the absorbing component ensures that (a) the system vibrations are absorbed, and (b) in absorbing them, the system arrives exactly at target. Thus, the main control problem has been solved, in a simple and elegant manner.

The launch component of the actuator motion can be considered as *pushing* the system half the distance to the target, while the absorbing component acts as if the reaction of the system were *pulling* the actuator the other half displacement, but in such a way that all momentum and energy return to zero precisely on completion of the process, just when the system arrives to target.

4.1 Launch Wave Profile

The launch waveform (time profile) is to a great extent arbitrary. It can arrive at the half-target displacement in many ways (step, ramp, constant acceleration, or using a pre-determined motion plan), limited only by the actuator dynamics. The control strategy works very well for all such choices.

There is one choice that is particularly neat [11]. The absorb wave motion is added throughout the entire manoeuvre. If the absorb wave is recorded from the beginning, the initial part of it can be used to determine a very good way to complete the launch wave. The actuator gets half way to target before the launch wave has reached its steady (half-target) value (Fig. 3). At this point, the launch wave can be set to complete its trajectory using an inverted and time-reversed version of the wave that has been absorbed out of the system up to that point. Thus, the "echo" that was

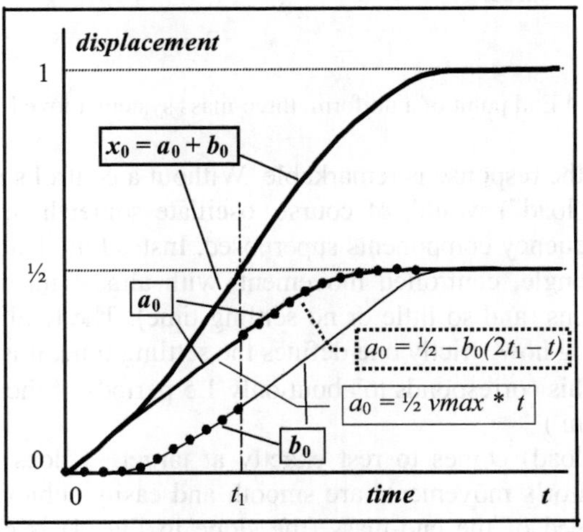

Fig. 3 Wave-echo control to determine actuator motion x_0. The target distance is 1. The a_0 component is set by the controller as a ramp until $t=t_1$ (when $x_0=\frac{1}{2}$), then as a reverse replay of recorded b_0, but inverted (shown dotted). At all stages b_0 is determined from the measured system response.

absorbed from the system in the first half of the manoeuvre is played back into the system in reverse to complete the manoeuvre. This has the effect of bringing the tip to rest in a time-and-space mirror-image of the start-up motion. In other words, the load stops dead, rapidly, precisely at the target, while the actuator continues to move so that the rest of the system then relaxes in just the right way to leave everything motionless in the correct position.

5 Sample Results

As an example, Fig. 4 shows the control of a uniform, lumped three-mass system. The actuator and end-mass positions are shown against time expressed in units of the period, T, or $2\pi/\omega_n$, where $\omega_n = \sqrt{(k/m)}$. The target displacement is 1m. Also shown are $a_0(t)$ and $b_0(t)$.

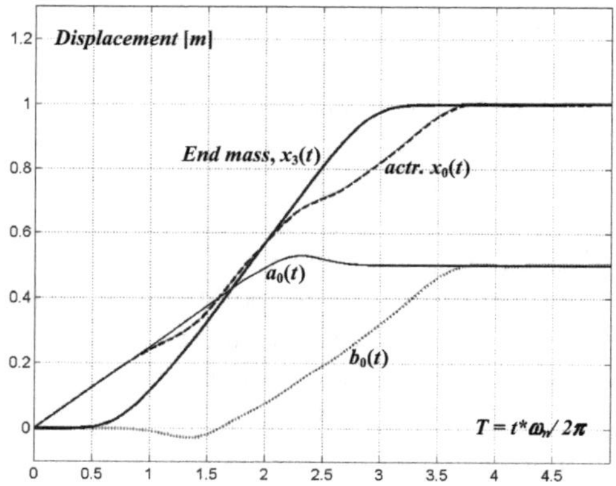

Fig. 4 End point of a uniform, three-mass system, moved 1 m.

As can be seen, the response is remarkable. Without a control strategy, the position of the end mass ("load") would, of course, oscillate somewhere between zero and two, with three frequency components superposed. Instead the load is translated from rest to rest, in a single, controlled movement, with almost no overshoot and with negligible oscillations (and so little or no settling time). The total manoeuvre time is excellent. Depending how strictly one defines the settling time, it is between 3 and 3.5 "periods" of ω_n, (This corresponds to about only 1.5 periods of the fundamental mode of the 3-mass system.)

The end mass (load) comes to rest exactly at target. It does so sooner than its actuator. The actuator's movements are smooth and easily achievable. Around mid-manoeuvre, the speed of the end-mass (the slope in Fig. 4) is close to that of the actuator: the flexible system is then behaving as if rigid, or almost so: vibration is under control.

Similarly impressive results are obtained whether the system is long or short, uniform or not, with linear or hardening or softening springs (other than the first), with or without internal damping, with ideal or realistic actuator, and with or without precise values for all the terms in (3) or (4). Illustrating a mixture of such added complexities, Fig.5 shows the response of a 5 DOF system with non-linear (hardening) springs; variations of the masses of 1, 0.5, 1, 2, 1; damping between the masses of 0, 0.25, 0.1, 0.25, 0, 0.05 times critical damping; an actuator modelled as a first order system of time constant $1/3\omega_n$; and $b_0(t)$ approximated by (3). These parameter values and system size were chosen almost at random: a similar result is obtained for almost arbitrary choices of these system variables.

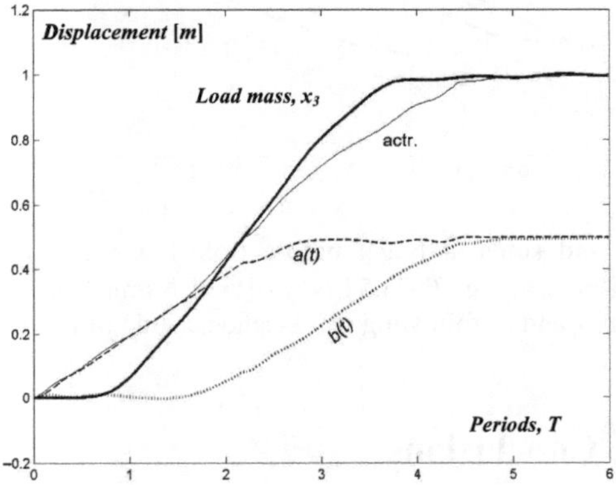

Fig. 5 Response of end mass of 5 DOF system, with non uniform masses, various dampers between the masses, non-linear springs, a first order actuator, and $b_0(t)$ determined simply by (3).

As can be seen, the non-uniformities, actuator limitations, and so on, make things less smooth, but despite everything, the controller gets the load exactly to target, rapidly, and it works very effectively to remove vibratory energy from the system during the motion and on arrival at target.

The presence of system non-uniformities requires no adjustment to the control strategy. The actuator's action is restricted to either launching or absorbing waves. To absorb returning waves the actuator must await their arrival. The non-uniformities will delay, and stretch out, their arrival and therefore delay the absorbing process. Also the non-ideal actuator response will slightly prolong the final tidying up. But even though the strategy and controller settings were not changed, their effectiveness in meeting the much more difficult challenge is almost undiminished.

Figure 6 gives a trolley crane example, moving a load 2 m, with a 4-m cable of significant mass [10]. The launch waveform is set to correspond to half the maximum trolley velocity, to which is added the return wave, b_0. For a long manoeuvre, this causes the trolley velocity to approach the maximum for the middle part of the transit, with the swing angle approaching zero. In other words, the system is then moving at top speed and as if it were rigid, with the load displacement tracking the trolley displacement. After the halfway point for the trolley (1 m), the launch displacement a_0 is based on the previous b_0. This causes the trolley to decelerate in precisely the way needed to get the load to land at target (2 m) and stop dead. The load arrives before the trolley, which continues to move in just the right way to allow the cable to straighten up as all wave energy flows out of the system.

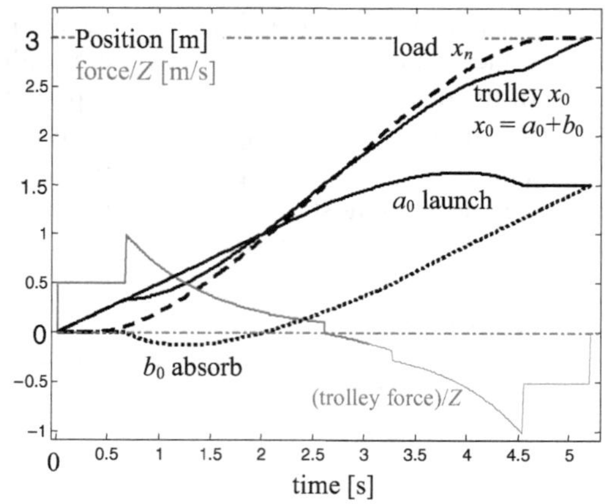

Fig. 6. Load arrives and stops at target before trolley. Crane target distance 3 m. Cable length=4 m, ρ=0.1 kg/m, m=2 kg, T=23.54 N, Z=1.534 Ns/m. Note time symmetry of load & trolley motions, due to a_0 and b_0 following wave-echo scheme of Fig. 3.

6 Discussion & Conclusions

Aspects of the problem of controlling flexible systems have been presented in new and fruitful ways, leading to new control algorithms that perform remarakbly well. They easily move a load from point to point, rapidly, yet with negligible residual vibration and negligible overshoot and zero steady-state error. They move the load at close to the actuator velocity (the ideal), in one controlled motion, without exciting load or system vibrations unnecessarily. The control strategies are very robust; they are applicable to a wide variety of problems; they require minimal system information, little computational overhead, and are very tolerant of limitations in the actuator dynamics. Sensing requirements are also minimal. Other than the actuator's own motion, only one other sensed input is needed, and the second sensor supplying this information is located conveniently close to the actuator, where sensing is generally easiest and safest in practice.

Modelling errors hardly feature. System changes are automatically accommodated. The order of the controller automatically matches that of the system, and explicit information, for example, about locations of poles (or natural frequencies and damping ratios of modes) is not needed.

The control approach can be considered a combination of "command shaping" *and* feedback control, the launch wave being a simple, shaped input, and the absorb part the feedback contribution.

With the wave-echo idea, the returning waveform, b_0, reveals to the controller the entire system dynamics in just the form the controller needs to achieve ideal system deceleration to rest. In a sense, the system itself serves as the system model, which is therefore always accurate, up to date, and of the correct order. The system itself also serves as the model's computer. To put it another way, all the required system identification is done in real time, as part of the controlled motion, with minimal computational overhead. This partly explains the control system's robustness to system changes.

The interface between the actuator and the flexible system is seen as a wave gateway, controlled and managed by the actuator's motion. Energy and momentum enter and leave the flexible system at the interface. They propagate in two directions within the system, from actuator to end-mass, and back again, albeit in ways that are faltering, complex, and highly dynamic. Rest-to-rest motion corresponds to getting the energy and momentum into, and then out of, the system in just the right way to ensure that the entire system comes to rest at the target.

The actuator is the sole agent for all this. But the actuator interacts directly only with the part of the system dynamics to which it is directly connected, namely the interface. All the control can be done, and must be done, at this interface and through this interface. The wave approach shows just how this can be done.

A very comprehensive study of the field [6] has observed that "to date a *general solution* to the control problem [of flexible mechanical systems] has *yet to be found.* One important reason is that computationally efficient (real-time) mathematical methods do not exist for solving the extremely complex sets of partial differential equations and incorporating the associated boundary conditions that most accurately model flexible structures." (*Emphasis added.*)

It is here contended that wave-based methods provide just such a "general solution" for a wide class of problems, with many additional attractive features.

Acknowledgements

This work was funded in part by Enterprise Ireland Basic Research Grant, code SC/2001/319/.

References

1. Book, W.J., 1993, "Controlled Motion in an Elastic World", *ASME Journal Dynamic Systems, Measurement, and Control*, 115, pp. 252–261, June.
2. Jayasuriya, S., Choura, S., 1991, "On the Finite Settling Time and Residual Vibration Control of Flexible Structures", *Journal of Sound and Vibration*, 148, No. 1, pp. 117–136.
3. Meckl, P., Seering, W., 1985 "Active Vibration Damping in a Three-Axis Robotic Manipulator", *ASME Journal of Vibration, Acoustics, Stress and Reliability in Design*, 107, pp. 38–46, January.
4. Meirovitch, L., 1989, *Dynamics and Control of Structures*, Wiley.
5. Pao, L.Y. 1996, "Minimum-time Control Characteristics of Flexible Structures", AIAA Journal of Guidance, Control and Dynamics, 1, pp. 123–129, January–February.
6. Robinett, R.D. III, Dohrmann, C.R., Eisler, G.R., Feddema, J.T., Parker, G.G., Wilson, D.G., Stokes, D., 2002, *Flexible Robot Dynamics and Controls*, Kluwer Academic/Plenum, New York, p. 165 & passim.
7. O'Connor, W.J., Lang D., 1998, "Position Control of Flexible Robot Arms using Mechanical Waves", *ASME Journal Dynamic Systems, Measurement, and Contro*l, 120, No. 3, pp. 334–339, September.
8. O'Connor, W.J., 2002, "Gantry Crane Control: a Novel Solution Explored and Extended", *Proceedings American Control Conference 02*, Alaska, 8 May.

9. O'Connor, W.J., Hu, C. 2002, "A Simple, Effective Position Control Strategy for Flexible Systems", *2nd IFAC Conference on Mechatronic Systems*, Berkeley, California, pp. 153–158, December.
10. O'Connor, W.J. 2003, "A gantry crane problem solved", *ASME Journal Dynamic Systems, Measurement, and Control*, 125, No. 4, pp. 569–576, December.
11. O'Connor, W.J., 2006, "Wave-Echo Control of Lumped Flexible Systems", *Journal of Sound and Vibration*, 298, Nos. 4–5, pp. 1001–1018, December.
12. O'Connor, W.J. 2007, "Wave-Based Analysis and Control of Lump-Modeled Flexible Robots", *Robotics, IEEE Transactions on*, 23, pp. 342–352.

What is Networked Robotics?

Gerard McKee

Active Robotics Laboratory, School of Systems Engineering, The University of Reading, Reading, Berkshire, UK, RG6 6AY
g.t.mckee@reading.ac.uk

Abstract. Networked Robotics is an area that straddles robotics and network technology. A robot system controlled via the WWW exploits the Internet network and hence is one realisation of networked robotics. A set of field robots that exploit wireless networks to share and distribute tasks might also be considered an exemplar of networked robotics. But isn't this just an exemplar of distributed robotics? And if so, what does networked robotics bring to the "robotics" table? These are questions and issues addressed in this chapter. The chapter will propose that networks are at once both enabling and constraining to robotics. They enlarge the scope of the robotics discipline yet introduce challenges that must be overcome if that potential is to be fully realized. In short, when the network becomes a design issue – normally when performance of the system is at a premium – networked robotics is at play.

Keywords. Networked robotics, distributed robotics, robot architectures.

1 Introduction

Computer networks are a pervasive element of everyday life. They allow us to access information at a distance and indeed to link geographically dispersed computing resources into powerful, distributed computing platforms. The World Wide Web (WWW) motivated the first integration of robotics with networking through the creation of online robot systems [12]. Teleoperation, which has a well established record of operating robots remotely, subsequently found a new transport medium in which to explore remote control issues and novel applications [25] [13]. Client-server models, object orientation and multi-agent systems offered robotics the opportunity to compose complex systems from distributed sensor, effector and computational resources [21] [8] [9]. More expansive views envision networked robotics as humans and robots acting together in intelligence spaces [15].

The online robot system, the Internet robot, the distributed robot architecture and the intelligence space are all very different views of networked robotics. What, however, are the underlying principles and purposes of networked robotics? What exactly is networked robotics? What or how does it contribute to robotics and, indeed, to networking, if anything? Or is networked robotics just the latest in a long line of fund-worthy buzzword? These are the questions that this chapter sets out to address. The important message is that networked robotics is an enabler – it allows us to do things that were not previously possible with robots – the online robot is just one example. However, it is also constraining, since in order to do many of these things there are problems that we need to overcome. Ultimately it is a liberator, for in doing these things and overcoming these problems, robot architectures and systems can be liberated from the constraints of fixed wiring.

The remainder of this chapter is organized as follows. The following section presents the component or modular view of robotic systems that is emphasised in the treatment of networked robotics provided in this chapter. Section 3 then makes the distinction between robotics resources, robot systems, robotic agents and tasks that follows from this modular view. Section 4 compares networked robotics with distributed robot systems, introducing robot talk networks and distributed robot architectures. Section 5 widens the discussion to include ambient and pervasive intelligence, and in this context as well sensor and other forms of distributed networks of devices and computational nodes. Section 6 incorporates internet and online robots into the framework developed in the previous sections. Section 7 explores networked robotics within the context of field robotics. Section 8 provides a general framework for exploring important issues in networked robotics. Section 9 concludes the chapter. The chapter does not offer a review of networked robotics but only dips into research that has a bearing on the question posed in the title.

2 Breaking the Robot Apart

Early robot systems incorporated deliberative reasoning architectures in which the functional elements of the architecture, such as vision analysis and path planning, followed sequential programming principles and interacted through procedure calls [22]. The systems were by today's standards monolithic, undifferentiated and unstructured. Indeed, they rapidly lost their appeal when more fine-grained behaviour and hardware-oriented approaches to building robot systems emerged [3]. While the deliberative and the behaviour-based approaches both offered opportunities for modularity, and the latter was heralded as a more modular and hence more flexible approach to systems engineering, they were both slow to follow when the wider discipline of computing embraced object-orientation [7] – the additional effort implied by object orientation was viewed as unnecessary and would only get in the way of engineering the robot system. However, object orientation did appeal to some researchers [10] [23] and the importance of modularity was not lost on the designers of reconfigurable robot systems [1].

Networked robotics is founded on the ability, at different levels of granularity, to model a robot system as a set of components that are glued together to compose robot systems. Object orientation is useful to networked robotics to the extent that it fosters the idea of a software component that encapsulates a certain unit of functionality and provides a well-defined interface by which other objects can avail themselves of that functionality – by way of a set of services on the interface. Therein lies part of the distinctive character of networked robotics and the challenges it faces: What defines a useful component? How are robot architectures composed from these components? Robot systems lend themselves to component-based approaches when we enumerate their sensors and effectors, but it is less obvious how to break apart the rest of the system. This is a key challenge for robotics engineering in general but is specifically addressed by networked robotics in the context of distributed architectures.

Once we tender the notion that a robot system can be modelled as a set of components that interact with each other we can then ask the question as to how these components are represented on a computer system; as the elements of a suite of libraries that are glued together under an application that is compiled and executed; or as active components whose interfaces are made public through a registry service or its equivalent. The latter model fosters the further idea of automated reconfiguration of the system. Reconfiguration is a familiar idea to robotics, but primarily in the guise of modular reconfigurable robot systems [14]. Networked robotics makes the model more accessible to the wider robotics community.

Why would one wish to remap components to create alternative robot architectures? The simple answer is that it is one of the possibilities that an active component-based approach to robot architectures offers. It is not a recognised path to follow in robotics since the standard model for robot systems engineering is to identify a task and to create a robot system to support that and only that task. The task may involve a number of distinct stages that require swapping in or out desired our undesired functionality [4] [3]. In general, however, the one architecture is expected to persist for the duration of the task and indeed indefinitely. The argument is typically offered that this persistence necessarily concedes optimization for specific subtasks in favour of optimization across the whole task. However, reconfiguration may offer a better approach, a least an alternative, and networked robotics makes reconfiguration accessible.

3 Tasks, Agents and Resources

In robotics we can talk of a robot system, typically a mobile robot platform, embodying a set of robotics resources. The resources include sensors, effectors, and computational components. Robot architectures bind together these components in a form suitable to complete some designated task. Under networked robotics the relation between the task and the resources is the architecture. The architecture coordinates the resources to complete the task. It is the embodiment of robotic agency. The role of the robot platform in this model is only incidental: it mounts the physical sensors and actuators and typically as well the computational platform on which the computational components execute. Therefore, in networked robotics we associate tasks with agents and in turn associate agents with resources, not the robot platform. This gives networked robotics it's wider reach to sensor networks and ambient intelligence.

The networking in networked robotics thus eliminates the requirement for the resources that are coordinated to be localised to a single robot platform. The components are enhanced through the awareness of the network and the ability to perform remote procedure calls or adopt process-oriented message-passing models for interaction with other components. In short, the components are network-enabled. The robotic engineer is no longer constrained to map robotic agency one for one to a mobile robot platform, but that option is not conceded either – the engineer now has more options.

4 Robot Talk Networks

We will now draw a distinction between networked robotics (NR) and distributed robotics (DR). Research on multirobot systems covers a very broad area of robotics [2], but this chapter will focus on the specific notion of a team of mobile robots cooperating to complete some task [24]. The distributed in DR then implies multiple robot systems geographically dispersed about the task environment. A network framework is often used in these systems to allow the robots to communicate and indeed to allow one or more human operators to guide the system as a whole or the robot individually. The key difference between NR and DR is that in the former the robot platform is modelled as a physical cluster of robotic resources whereas in the latter the robot platform is a modelled as a single robotic agent.

Given the above interpretation of DR, the problem statement for DR is then: *Given a set of robots and a task that has been broken down into a set of subtasks, how does one allocate these subtasks between the set of robots in order to achieve optimum performance?* NR, on the other hand, takes robotics resources as its primitive architectural unit, not the robot platform, and hence the problem statement is formulated accordingly: *Given a set of robotics resources and a task which can be broken out into subtasks, how does one compose the resources to create robotic agents that will perform the task?*

When the problem statements for NR and DR are defined in this way there is, of course, no explicit reference to networking. The network is really just a part of the implementation. In fact, the NR problem statement is really a statement about distributed robot architectures (DRA). By this we mean that the components of the architecture are distributed across multiple robot platforms and the network provides a means of wiring the components together. This can be contrasted with the role of networking in DR, where it effectively provides a "talk network" which allows robots to talk with each other as and when the need arises.

The basic unit for communication within a robot talk network is a robotic agent. The agent-to-agent communication within a task setting will typically be high-level, for example passing back and forth higher-level observations, negotiating for resources or cooperating on task planning. The basic unit of communication for distributed robot architectures is a resource-to-resource network data connection, which can be referred to as an *arclet*, representing a wired connection. Since the latter is part of the control architecture of the robotic agent, and perhaps requires coordinated action across multiple robot platforms, there may be non-functional performance requirements that need to be satisfied. These requirements may in turn place considerable pressure on the performance of the network, and hence cause the engineer to consider network performance in the design of the networked robotic agent. These requirements are much harder to deal with than the requirements for talk networks, which may simply rest on maintaining agent-agent network connectivity.

5 Embedded Intelligence

Networked robotics embodies the requirement to make explicit the physical, geometrical and functional properties of sensors and effectors in order to allow distributed sensors and effectors to be integrated in networked robotic agents. In doing so it affords as well the opportunity for robotics to exercise a form of outreach into new application areas.

There is no reason, for example, why the resources that are integrated should not include sensors and effector devices embedded in the environment in which a mobile robot operates, or indeed in which humans live and work. In fact, there is no reason why the robotic intelligence should not be based exclusively on sensor and effectors systems embedded within the environment. The contrast can be drawn, therefore, between embodied intelligence (i.e. intelligence embodied in robotic artifacts) and embedded intelligence (i.e. intelligence embedded in the environment).

Of course, what is environment and what is object are relative – a set of mobile robot systems may provide an embedded robotic intelligence to one or more mobile robot systems or humans that they collectively enclose. The latter concept allows us to draw the link, in turn, between network robotics and sensor networks, and ultimately to other forms of device networks through to an all-encompassing ambient and pervasive intelligence. Some of the most innovative work in networked robotics addressed precisely this form of intelligence [15]. These links are important in opening up new opportunities for robotics beyond the traditional industrial robot systems.

6 Internet and Online Robots

The model of networked robotics outlined above is largely one of resources and distributed architectures. Where in this lies Internet and online robots? Online robots have their origin in web-based access to remote laboratory-based robot systems [12], [20]. Internet robotics, on the other hand, has its origins in teleoperation, telemanipulation and telerobotics [25]. It has simply transposed the medium of communication to a setting, the Internet, which is outside the designer's control [13]. In the case of Internet robotics time delay is a major issue, whereas in the case of online robots it is generally not such a major concern. Failure to gain control of time delays in the first instance can result in the failure of the system due to instability, whereas in the second case it can lead to annoyance but generally not complete failure. If the Internet robotic system is part of a remote surgical system, for example, then failure may have severe consequences. Failure of an online robot experiment can lead to frustration, waiting around, but generally nothing catastrophic.

Since the performance of the Internet is a factor that needs to be taken into account in the design of online robot systems and internet robots, though for different reasons, these systems are central to networked robotics. The key challenge, for online robots as well but to a lesser extent than Internet robots, is that fluctuating bandwidth, time delays and network jitters are givens [17]. The obvious approaches to addressing these problems is to reduce the time dependency of the system by incorporating more local intelligence at the robot site (telerobotics versus teleoperation), designing new communication protocols that adapt better to changing bandwidth limitations and requirements, or adapting both the operational speed and performance of the operator interface and remote robot system in tandem with changing bandwidth availability and requirements.

There are two major challenges for online robot systems. The first is to build interesting, reliable, online robot systems that engage the public and students alike, and which can operate 24/7. For use in robotics education they must also incorporate assessment methods within a learning programme. The latter, however, requires the development of a robotics curriculum if the system is to make a substantive contribution to robotics education [18].

The second major challenge is in fact one that requires both Internet and Online Robots techniques. This is precisely the concept of open research laboratories; robot laboratories, that is, that offer a particular research and development challenge, requiring remote contributors to have access to laboratory systems under manual (individual or shared) or mediated control, that provides facilities to allow users to contribute software and hardware components that can be combined to build robot architectures, and offers learning resources introducing laboratory systems and indeed the discipline of robotics. Allied to this is the creation of protocols by which components interact.

7 Field Robot Systems

The Internet is a highway that anyone and everyone can use. There is little scope for robotics to exert an influence on its design in order to better serve the requirements of Internet and online robots. In field robot systems, where a set of robot systems are typically deployed in a natural environment, the possibilities are very different. Such systems, even if networked via wireless Ethernet, can be isolated from the Internet highway. In these setting the range of networking options can be expanded to include radio and even infra-red networks, and the network can be put under tighter control. Indeed, field robot systems are better served by a more networked robotics centred viewpoint that recognises the importance of distributing resources in sensor and actuator networks [6] [16].

Heterogeneous network environments offer new possibilities for networked robotics. For example, radio networks, since they tend to have low protocol overheads can support multi-channel real-time communication requirements between low-level control systems on geographically dispersed robot platforms. Ethernet, on the other hand, can support a form of robot talk network allowing the robot systems to share information and knowledge, distribute tasks and to effect such operations as role-transferral. In order to exploit this opportunity, and to explore distributed robot architectures, as against just robot networks, it is important to gain control not only of the network but also of the processor nodes that support the computational elements of the robot architecture.

Field robotics presents other interesting opportunities and challenges for networked robotics. For example, a human operator may need to interact with the set of robots either individually or as a unit and to a greater or lesser degree. A networked robotics agent may be embedded within the field robot system to mediate between the human operator and the robots. The agent may be distributed across the set of individual robot systems; giving it a presence in every robot system; but nevertheless it has an existence independent of the individual robots.

In a conventional robot team setting the mediating intelligence is typically run on a separate server and instructions or state changes are communicated via a wireless network to the individual robots. Robot soccer teams that participate in the RoboCup competition are one example [11], in fact these represent a case of field robotics in the small. The network robotic agent may in fact not only monitor the state of the robot system but also of the environment and change the state of the robot team accordingly (e.g. from defence mode to attack mode).

These, and other, possibilities are a consequence of fully exploiting networking within robotics. They are afforded by the introduction of networking, but, as stated, they also raise problems that need to be overcome if those possibilities are to be realized. The networking problems for field robot systems in the large, in addition, are very different to those for field robot systems in the small. Indeed, although one might consider that such possibilities are afforded by network technology, the question is whether indeed they really are possible and then as to whether they are useful is a challenge to both networked robotics and to robotics engineering.

8 A Platform Model for Networked Robotics

Networked robotics is concerned with engineering the underlying platform that supports the implementation of robot systems, whether networked, distributed, field, reconfigurable, internet, or online robot system. This section will offer a platform model based on network robotic concepts introduced in previous sections. Figure 1 gives a diagrammatic view of the model.

The model is centred on a modular approach to robotics, whereby a robot system is modelled as a hierarchy of modules. These comprise primitive modules and glue modules. Primitive modules are resource modules – they offer well-defined services. Glue modules (solid circles in Fig. 1) incorporate abstract task requirements and link with resource modules to realise these requirements. They can not only offer the functionality embodied in the module they link to, but also create new functionality. When linking has been completed the glue modules become resource modules. The hierarchy, so crafted, can be constructed to the appropriate level required for the services of an Internet or online robot, or still higher to networked and distributed robotic systems.

The composition of robot architectures can follow traditional manual methods based on compiling and linking code, and process-oriented or remote procedure models for concurrent, decentralised, or distributed applications. Networked robotics, however, offers an alternative model for the composition process. In this model the modules are network enabled, and both self and context aware. Network enablement means that the modules have a unique network address and can listen to and connect with other modules across the network. Self-awareness means that the modules are aware of their own capability, specifically the services they offer and under what

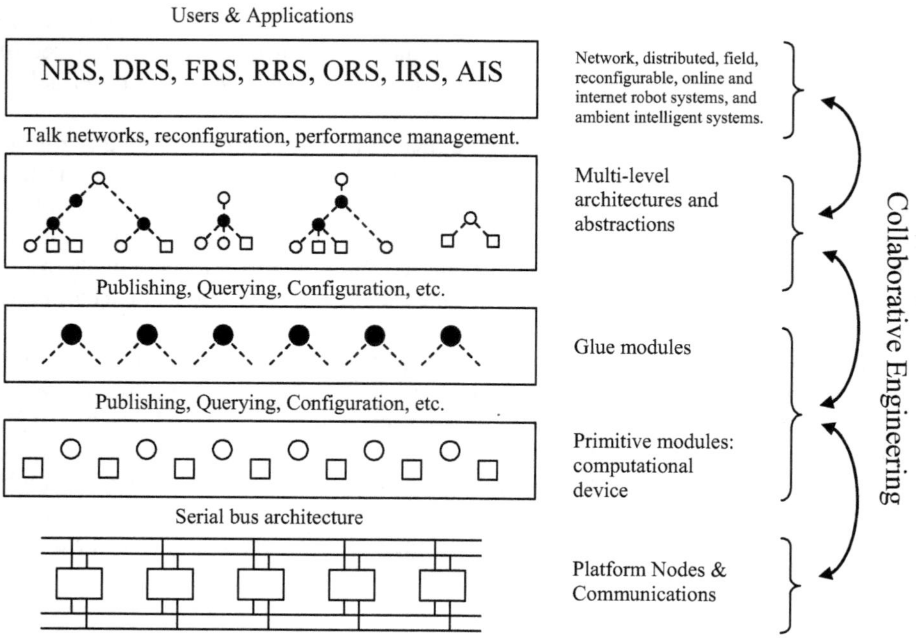

Fig. 1 A platform model for networked robotics.

conditions they can offer these services, and the underlying platform resources (processing and communication) required to provide these services. Context awareness means that the modules are aware of their network and physical location. The latter include, for device modules for example, their physical location on a mobile robot platform.

Primitive modules by definition are assigned a location on the network. So too are glue modules, prior to and following instantiation. In effect, every module is exposed to the network to form a flat serial bus architecture from which application hierarchies can be composed. Additional platform level services are required to support this composition process, including the ability for modules to publish or advertise themselves, for glue modules to query the set of published modules for those that can help them to satisfy their functional and non-functional requirements, and for negotiation to gain access to module services. These services must also be reflected in the design of the modules [5].

At the highest level the architecture forms single or multiple robotic agents in the form of networked, distributed, field, reconfigurable, online, and Internet robot systems (NRS, DRS, FRS, RRS, ORS, and IRS respectively in Fig. 1). These in turn offer application services to users. The modules that form the architecture live within a platform environment comprising process nodes and communication pathways. These platform-level resources need to be properly deployed in the service of the application. This includes the ability for modules to move about in order to satisfy performance requirements. Here one finds typically two opposing forces acting: the first, dispersal of modules in order to balance the load across the processor nodes; the second, clustering of modules on the same processor node in order to satisfy real-time requirements. As modules come to life, move, or die, the available processor and bandwidth resources will change. The modules need to be aware of these changes and react appropriately: knowing when to stay put, to ask for more resources, or to move to other nodes. In this the modules are situated and need situated intelligence to get the best out of the platform for the application.

Platform services must be provided to support the mobility of the modules. These services must include as well a model of the capacity of each processor node and of the communication pathway so that the pressure on resources can be assessed. Reconfiguration of the platform nodes and of the communication pathways as well will provide more scope for optimizing the platform resources available for the application systems. Another set of reconfiguration services are also required to allow the reconfiguration of networked robotic robot systems. Such reconfiguration may be required to meet the requirements of different subtasks. The graceful transitions between these different systems will require additional services.

The modular, networked robotics approach also offers benefits for the process of developing robot systems and applications. The modular, networked approach means that engineers can develop and publish individual modules, create subsystems and offer both of these to others, giving a more open, collaborative engineering environment. This is a natural extension of the online robot and the Internet robot systems concept into an important and challenging area of engineering [19].

Networked Robotics in the model above is first and foremost a medium for wiring together robotics components to form robot architectures. The operation and performance of the platform are important in maintaining the appropriate performance and stability of the architecture. Internet and online robots are special, important, and popular representations of such architectures. However, they are only the surface. The network also offers avenues for integrating robotics technology with other newer network based technologies to create ambient and pervasive robotic intelligence. The platform model above reflects this, but extends the scope beyond hardware-based distributed devices to systems comprising both hardware and software components, and systems offering opportunities for reconfiguration.

9 Conclusions

Network technology is both enabling and constraining for robotics. It offers new opportunities, such as online robots, but it creates challenges that need to be accounted for either in the robot design or in the creation of new network environments. The extent to which the network is taken into account in the design of the robot system determines the extent to which the work contributes to networked robotics. Connecting a set of robots or robot components to a network is only incidentally networked robotics. Incorporating measures to accommodate for network time delays in the controller of a telemanipulation system is hard networked robotics. Gaining control of the underlying network and processor platform, creating new protocols and heterogeneous networking options, is currently only realistic within field robot systems, but is core networked robotics.

Networked robotics allows us to take further the progress that has led from the early structured yet monolithic robot control systems through the object-oriented software paradigm that meshes well with robots at the device level, to active component and ultimately self-aware network enabled modules that can broker their own connectivity. The glue logic that enables primitive robotic resources to be blended into successively higher-level functionality needs to be modelled. A framework must be provided that allows the components to be mapped and remapped between robot architectures. The creation of such a framework is an important challenge for networked robotics.

Networked robotics is little concerned with the content of the components that the robotic engineer defines. It is concerned with content and hence robotic intelligence to the extent that this content needs to be communicated to other modules. This is part of the self-awareness that needs to be built into modules. Networked robotics is only incidentally distributed robotics to the extent that the latter incorporates a network as a medium for the robots to talk to each other. However, a networked robotics environment may be configured as a distributed robotics systems. In this is it is clearly a more general class of system. Networked robotics is more than incidentally cooperative robotics to the extent that a cooperative robotics system incorporates distributed robot architectures and communication resources are explicitly modelled.

Online robots are a popular example of networked robotics, but networked robotics is much more. The purpose of this chapter was to explain how much more. Networked robotics is about modularity, it is about using network technology to wire together distributed, networked enabled, self and context-aware modules that offer well defined sets of services. Networked robotics is about composing robot architectures that express themselves through robotic intelligence in robot-robot and human-robot systems and interactions. Networked robotics offers robotics a timely injection of insight and innovation; we have the opportunity to reflect on the discipline of robotic science and engineering and at the same time engage new technologies as users and enablers. Networked Robotics is an opportunity. Once we have understood better what robotic science is about and what robotics can do, and in particular once we have a new set of tools for composing robot systems, we can set networked robotics aside.

References

1. Anderson, R. J., 1995. SMART: A modular control architecture for telerobotics. In *IEEE Robotics and Automation Magazine*, September, pp. 10–18.
2. Arai, T., Pagello, E., and Parker, L. E., 2002. Guest Editorial, Advances in Multirobot Systems. In *IEEE Transactions on Robotics and Automation*, Vol. 18, No. 5, pp. 655–661.
3. Arkin, R., 1998. *Behaviour-Based Robotics*, MIT Press.
4. Arkin, R. C. and Balch, T., 1997. AuRa: Principles and Practice in Review. In *Journal of Experimental and Theoretical Artificial Intelligence*, Vol. 9, pp. 175–189.
5. Baker, D., 2006. DINERO: Distributed networked robotics. PhD Thesis, The University of Reading, UK.
6. Batalin, M. A. and Sukhatme, G. S., 2003. Using a Sensor Network for Distributed Multi-Robot Task Allocation. In *IEEE International Conference on Robotics and Automation*, New Orleans, LA, April 26–May 1, pp. 158–164.
7. Booch, G., 1993. *Object Oriented Analysis and Design with Applications*, Benjamin Cummings.

8. Brooks, B. G. and McKee, G. T., 1997. Agent-based control architecture for teleoperation. In *SPIE Proceedings Volume 3209, Sensor Fusion and Decentralised Control in Autonomous Systems*, pp. 200–211.
9. Brooks, B. G., McKee, G. T., and Schenker, P. S., 2001. The Visual Acts Model for Automated Camera Placement During Teleoperation. In *Proceedings of the IEEE International Conference on Robotics and Automation*, Korea, pp. 1019–1024.
10. Chochon, H., 1991. Object-oriented design of mobile robot control systems. In *2nd ISER*, Toulouse, France, pp. 317–328.
11. D'Andrea, R., 2005. The Cornell RoboCup Robot Soccer Team: 1999–2003, in D. Hristu-Varsekelis and W. S Levine (eds), *Handbook of Networked and Embedded Control Systems*. Birkhauser, pp. 793–804.
12. Goldberg, K., 2000. *The Robot in the Garden*, Cambridge, MA, MIT Press.
13. Goldberg, K. and Seigwart, R., 2002. *Beyond Webcams: An Introduction to Online Robots*, The MIT Press.
14. Kim, M., Roufas, K., Duff, D., Zhang, Y., Eldershaw C., and Homans, S. B., 2003. Modular Reconfigurable Robots in Space Applications. In *Autonomous Robots* Vol. 14, (Nos. 2–3), pp. 235–237.
15. Lee, J-H. and Hashimoto, H., 2003. Controlling mobile robots in distributed intelligent sensor network. In *IEEE Transactions on Industrial Electronics*, Vol. 50, No. 5, October, pp. 890–902.
16. Makerenko, A., Nettleton, E., Grocholsky, B., Sukkarieh, S., and Durrant-Whyte, H., 2003. Building a decentralized active sensor network. In *Proceedings of ICAR*, pp. 332–337.
17. Liu, P., Meng, M., and Yang, S., 2003. Data Communications for Internet Robotics. In *Autonomous Robots*, Vol. 15, No. 3, pp. 213–223.
18. McKee, G. T., 2006. The maturing discipline of robotics. In Special Issue of the *International Journal of Engineering Education* (IJEE) on Robotics Education, www.ijee.dit.ie.
19. McKee, G. T., 2005. Robots, robots, robots. In *Collaboration@Work, The 2005 Report on New Working Environments and Practices*. European Commission. www.amiatwork.com.
20. McKee, G. T., and Barson, R., 1996. Using the internet to share a robotics laboratory. In *International Journal of Engineering Education*, Vol. 12, No. 2.
21. McKee, G. T., and Schenker, P. S., 2000. Networked robotics. In S*PIE Proceedings on Sensor Fusion and Decentralised Control in Robotics Systems III*, Boston, November, pp. 197–209.
22. Moravec, H. P., 1983. The Stanford Cart and the CMU rover. In *Proceedings of the IEEE*, Vol. 71, No. 7, pp. 872–884.
23. Paredis, C. J. J., and Khosla, P. K., 1993. Synthesis methodology for task based reconfiguration of modular manipulator systems. In *Proceedings of the 6th Int. Symposium on Robotics Research* (ISRR'93).
24. Parker, L. E., 1998. ALLIANCE: An architecture for fault-tolerant multirobot cooperation. In *IEEE Transactions on Robotics and Automation*, Vol. 14, pp. 220–240.
25. Sheridan, T. B., 1992. *Telerobotics, Automation and Human Supervisory Control*, MIT Press.

PART I

Intelligent Control Systems and Optimization

Encoding Fuzzy Diagnosis Rules as Optimisation Problems

Antonio Sala, Alicia Esparza, Carlos Ariño and Jose V. Roig

Systems Engineering and Control Dept., Univ. Politécnica de Valencia
Cno. Vera s/n, 46022 Valencia, Spain
{asala,alespe,jvroig}@isa.upv.es

Abstract. This chapter discusses how to encode fuzzy knowledge bases for diagnostic tasks (i.e., list of symptoms produced by each fault, in linguistic terms described by fuzzy sets) as constrained optimisation problems. The proposed setting allows more flexibility than some fuzzy-logic inference rulebases in the specification of the diagnostic rules in a transparent, user-understandable way (in a first approximation, rules map to zeros and ones in a matrix), using widely-known techniques such as linear and quadratic programming.

Keywords. Fault detection and diagnosis, fuzzy mathematical programming, approximate reasoning, optimisation.

1 Introduction

Many industrial activities depend on the correct operation of complex technological processes. A fault changes the behaviour of a system in such a way that it does no longer satisfy its nominal performance objectives or even the system functionality is lost. The objective of diagnosis (process monitoring) is estimating a vector of fault state parameters, f, from measurements of the outputs of a dynamic system, whose trajectories depend on f and on time, initial conditions, external input variables, physical parameters, etc. Detecting faults with a gradation of *severity* from let's say 0 to 100% may be advantageous in practice: detecting faults in its early stages enables corrective actions to be taken on time, if needed.

Different approaches to the problem appear in literature: data-based, knowledge-based or based on differential-equation analytical models. A description of many of these techniques appear in [1] and [2]. The broadest conception of the problem may be set up in a probabilistic setting. In that situation, a comprehensive approach to the problem would involve estimation under nonlinear stochastic dynamics [3] [4] as well as decision-theoretic criteria [5] apart, of course, of the consideration of robustness of the resulting results when subject to possibly significant modelling errors. The problem, as such, is quite complex and possibly intractable. Hence, simplified assumptions are often stated.

In quite a few practical cases, diagnostic-related knowledge is available from experts, who express it in linguistic terms ("fault F produces a pressure in pipe 3 *lower than normal*"). The meaning of some of those linguistic terms may be understood as rules involving fuzzy concepts defined on the numeric range of the physical variables being measured. This fact inspires the use of fuzzy logic in diagnosis [6] [7]. Other knowledge-based frameworks extend the basic logic reasoning schemes to uncertain reasoning [8] [9], possibilistic reasoning [10] [11], Bayesian networks [12] [13] or uncertain probability [14], or combine the approach with neural networks [15] [16].

This chapter presents an alternative approach: the expert knowledge stemming from some (partially simplified) properties of a nonlinear dynamic system is encoded as a constrained optimisation problem, transcoding fuzzy assertions as approximate linear equations in the linguistic domain. The idea of transforming fuzzy statements into equations also appears in [17] and [18] in a different context of system modelling.

The proposed approach seems to possesses significative advantages with respect to a classical fuzzy IF-THEN rulebase, particularly in multiple fault situations, while keeping the problem readable (reduced number of "rules") and computationally tractable (efficient off-the-shelf linear programming (LP) software exists able to deal with hundreds, even thousands, of constraints and variables, and quadratic programming (QP) routines are also widely available). LP is a widely known tool [19] [20], taught in many undergraduate disciplines so that user understanding of both the rules and the inference tools implies that the approach might be useful in practical applications.

The structure of the chapter is as follows: a preliminary section will justify some approximate additivity properties of systems based on linearisation. Section 3 will discuss how a fuzzy rulebase may be encoded as equations. Section 4 will discuss the available algorithms to solve them and how they should be modified for the problem in consideration. A conclusion section closes the chapter.

2 Preliminaries

Dynamic systems. Let us assume that a system to be diagnosed is governed by equations $\dot{x} = \psi(x, u, f, \theta, t)$ where x is the system's dynamical state, u is a set of known *input* variables, f is the *failure* state to be estimated and θ are a set of system parameters, also assumed to be approximately known, and t is the time variable [4].

First, it will be assumed that the initial conditions $x(t_0)$ and inputs of standarised tests are known beforehand, as well as system-dependent parameters θ. The result of such tests will be a finite collection of measurements at some time instants:

$$y = \psi(x(t_0), \theta, u, t_0, t, f)$$

where, by assumption, only f is unknown. Let us denote as $q = (x(t_0), \theta, u, t_0, t)$ the set of known variables; they will be denoted as the experiment *context variables*. The diagnostic problem may then be cast as estimating f from $y(q, f)$, i.e., obtaining the implicit function, if it exists $f = \gamma(q, y)$. In general, the analytic expression for γ cannot be obtained except in the simplest of the cases. If enough sensors exist,and pre-classified data from the system (y,q,f) were available, a functional approximator such as a neural network might be used in order to try to learn γ by example, or at least to learn the normal behaviour and generate some residuals in case of faults. Applications of the approach are reported in [16] and [21]. However, learning for successful fault isolation in a complex system may require an impractical number of data. The approach will not be pursued further in this chapter.

The diagnosis problem becomes easier if some simplifications and assumptions are made. Taking a Taylor series expansion of ψ on the variable f, around $f = 0$, which will be defined as the "normal" situation, the result is an affine model:

$$y(q, f) = \psi(q, 0) + \frac{\partial \psi}{\partial f}(q, 0) * f + o(f^2) \tag{1}$$

This Taylor series expansion justifies that at least approximately a linearity assumption holds:

$$y(q, f) \approx \psi(q) + C_y(q) * f \tag{2}$$

where C_y is a linear operator (a matrix whose elements depend on the known information q). Usually, a parameterised analytical model of the system is not available so $\psi(q)$ and $C_y(q)$ cannot be calculated. However, human operators possess knowledge that may be encoded in terms of fuzzy sets. Expressing such a knowledge in the form (2) is discussed in next section.

3 Fuzzy Diagnostic Models

In the above discussed context, a collection of *fuzzy* sets mapping y to the interval $[0, 1]$ are assumed to defined by an expert on the system to be diagnosed, $\mu : R^p \rightarrow [0, 1]^k$ where p is the number of measurements and k is the number of fuzzy concepts. Usually, those sets are denoted by user-defined linguistic labels.

Then, the expert also knows which effects each fault has in the measurements. This knowledge is usually expressed in terms of rules:

> If the isolated fault F_i occurs, then abnormal symptoms $S_{i_1}, \ldots, S_{i_p}$ should be observed, being the rest normal.

Those rules, in a fuzzy context, may be basically understood as

> If the *severity* of fault F_i is f_i, $0 \leq f_i \leq 1$, and it is the only fault occurring in the system, then the intensity of symptoms $S_{i_1}, \ldots, S_{i_p}$ is approximately f_i, and the intensity of the rest of them is zero.

where $f_i = 0$ denotes fault not occurring and $f_i = 1$ denotes the fault occurring at a very significative severity level requiring user attention,[1] and the intensity of the symptoms is the membership function of the suitably defined fuzzy concepts.

In a multiple-fault situation, the system is assumed to verify an expression such as (2). If the observed outputs are the fuzzified ones, it will be assumed that the system verifies the following linear equation in the domain $[0, 1]^p$ of logic values:

$$\mu_q(y) \approx C(q)f + D(q) \tag{3}$$

where C and D are a known function of the system's parameters, input variables and initial conditions of the experiment. The notation μ_q indicates that even the definition

[1] It is up to the user to fix a maximum value of f_i (usually 1) in the optimisation procedures to be discussed, or to set $f_i = 1$ as a "landmark" point but considering higher values (more severe) possible.

of the membership function may depend on *a priori* information, such as historical data, system-dependent parameters, etc. $C(q)$ will be similar to $\frac{\partial \mu}{\partial y} C_y(q)$. Note that a differentiable system indeed does verify such an equation[2] for small values of f based in (2). The basic assumption here is that, with suitably defined fuzzy sets, the range in which such an equation fulfills is large enough to be useful for diagnosis.

Fuzzy sets may be defined on the *difference* between the observed readings and those in a normal situation, as an alternative to the *absolute* reference frame for concepts implicit in (3), so that:

$$\mu_q(y - D(q)) \approx C(q) f \tag{4}$$

In a practical situation, both types of fuzzy sets (3) or (4) may be used. In summary, the following linear equation must be solved in the diagnosis process:

$$\mu(y, q) = C(q) f \tag{5}$$

Other authors also pose matrix representations of the relationship between fuzzy faults and symptoms (for instance, Yao and Yao [22] uses a fuzzy relation approach). The assumption of linearity in the logic domain, with suitable definitions of membership functions is also used, in a control context, in [17].

3.1 Rule Encoding

The proposed expert rules are encoded in the format required in (3) in a simple way. Let us discuss several situations which will be clarified by examples.

First, the simplest setting would imply that, under a standarised test (q is fixed for all diagnostic experiments), a fuzzy set denoting an "abnormal" situation is defined for every measured variable. In that case, the knowledge:

F_i causes abnormality in $y_1, y_3, \ldots$

will define a *column* of matrix C, where elements at rows 1, 3, ... will be 1 and the rest of elements not explicitly enumerated at the above assertion will be set as zero. The usual fuzzy negation operator may be used if some variables have a fuzzy set defining the "normal" value of a variable.

Example 1. Let us have an industrial boiler where a fault f_1 causes: no variation on a temperature measurement t_1, and increasing of temperature t_2. Another fault, f_2, causes increases in both temperatures. They both cause an increase of pressure p. Defining abnormally high temperatures with fuzzy sets denoted as "T_1 abnormal", μ_1 and "T_2 abnormal", μ_2, and "normal pressure" with another fuzzy set (such as a triangular one), μ_3, the basic diagnosis equation would be:

$$\begin{pmatrix} \mu_1(t_1) \\ \mu_2(t_2) \\ 1 - \mu_3(p) \end{pmatrix} = \begin{pmatrix} 0 & 1 \\ 1 & 1 \\ 1 & 1 \end{pmatrix} \begin{pmatrix} f_1 \\ f_2 \end{pmatrix}$$

[2] Formally, left and right derivatives may be needed, but details are not relevant.

Alternatively, if the nominal normal pressure depends on a variable q, denoted as $P_0(q)$, (for instance, q might be the load regime of the boiler), a fuzzy set may be defined on the pressure increment so that the diagnosis equation would be written as:

$$\begin{pmatrix} \mu_1(t_1) \\ \mu_2(t_2) \\ 1-\mu_3(p-P_0(q)) \end{pmatrix} = \begin{pmatrix} 0\,1 \\ 1\,1 \\ 1\,1 \end{pmatrix} \begin{pmatrix} f_1 \\ f_2 \end{pmatrix}$$

Model errors and sensor faults. Another situation takes into account the approximate nature of (5) allowing for an instrumental "fault" variable associated to each of the measurements. That instrumental fault variable encompasses both sensor faults and modelling errors (inaccuracy in the definition of the membership functions).

Example 2. In the example being considered, the fault vector may be extended as:

$$\begin{pmatrix} \mu_1(t_1) \\ \mu_2(t_2) \\ 1-\mu_3(p) \end{pmatrix} = \begin{pmatrix} 0\,1\,1\,0\,0 \\ 1\,1\,0\,1\,0 \\ 1\,1\,0\,0\,1 \end{pmatrix} \begin{pmatrix} f_1 \\ f_2 \\ f_1^* \\ f_2^* \\ f_3^* \end{pmatrix} \tag{6}$$

where f^* are three instrumental fault variables and f denotes the "primary" faults.

In other situations, different faults have opposite effects on a particular variable so that its simultaneous occurrence does not deviate its measurements from the normal condition. The following example clarifies how to encode such a knowledge.

Example 3. If, in Example 1 fault 1 decreases T1, being the rest of symptoms the same as previously described, with concepts "abnormally high", μ_h, and "abnormally low", μ_l, defined for T_1, then the rulebase should be encoded as:

$$\begin{pmatrix} \mu_h(t_1) \\ \mu_l(t_1) \\ \mu_2(t_2) \\ 1-\mu_3(p) \end{pmatrix} = \begin{pmatrix} -1 & 1 \\ 1 & -1 \\ 1 & 1 \\ 1 & 1 \end{pmatrix} \begin{pmatrix} f_1 \\ f_2 \end{pmatrix}$$

In the above example, depending on the units and membership definitions, the -1 terms may be a different negative number. Note that some combinations of faults may yield negative membership values; this is not a problem if inequality restrictions are considered, as discussed in Sect. 4.1.

In any of the examples, a coefficient in matrix C lower (higher) than 1 would indicate a milder (stronger) effect of the fault on the symptom, as defined by the membership function. Modifying the coefficients might be needed in a fine-tuning phase because not all the faults influence with the same intensity a particular variable. However, taking into account "sensor" errors f^*, an initial setting with mostly 0 and 1 coefficients in C may be enough in order to achieve a reasonable output.

4 Inference

A fault estimation f is consistent with the observed symptoms if it is a solution of the basic diagnosis equation (5). Hence, fuzzy diagnosis amounts to finding the set of solutions of (5). However, some considerations need to be made: indeed, if sensor faults and modelling errors, f^*, are considered, the solution set includes *any* conceivable primary fault f, as sensor faults can accommodate any reading. For instance, in (6) a value of f^* can be calculated for any sensor reading and any value of f. However, if sensor faults are not considered, with less faults than sensors the set of consistent faults will be usually empty, as (5) will have no solution due to modelling errors. So, the above idea must be refined for practical usability.

Inference as optimisation. As above discussed, sensor faults must be considered in practice. Then, (5) are equality restrictions and a criterion should be used in order to rank all the feasible solutions. Note also that inequality restrictions $f_i > 0$ implicitly apply, unless the user casts a meaning for negative fault severities.

A possible criterion to be chosen is minimising the norm of the "instrumental" fault component, i.e., giving as the diagnosis solution the one that achieves less discrepancy between the measurements and the predictions $C(q)f$. The discrepancy $\mu - C(q)f$ will be denoted as *inference error.*

If the chosen norm is the Euclidean one, fuzzy inference is equivalent to a least squares problem. Let us consider an equation $\mu = Cf$ where sensor faults are also members of f. This is a linear system of equations with more unknowns than equations, which can be solved in the following sense:

The feasible solution f that minimises the squared Euclidean norm of Wf, where W is a diagonal weight matrix, is given by the pseudo-inverse formula [23]:

$$f = W^{-2}C^T(CW^{-2}C^T)^{-1}\mu \tag{7}$$

In a practical setting, a high penalisation in W must be specified for the sensor fault components f^*. For invertibility of W, small positive weights in the primary faults need to be introduced.[3]

If the chosen norm is the 1-norm of Wf (sum of absolute value of the components), then inference can be carried out in a linear programming setting. The LP framework needs to introduce dummy variables for positive and negative sensor errors $f^* = f^+ - f^-$, $f^+ > 0, f^- > 0$, to calculate the 1-norm as the sum of $f^+ + f^-$. This change of variables is standard in LP textbooks.

Note that LS algorithms produce an "intermediate" point as a result (not a vertex of the feasible region), sharing the error between all the equations, as small errors are not significant (because of the squaring) so LS tries to reduce big errors. On the contrary, LP produce a result in a vertex of the feasible solution space, and increments from either small or big errors weight the same.

---

[3] Solving $y = Cf$ by standard least squares, $f = (C^TC)^{-1}C^T\mu$, is equivalent to the proposed approach when the primary-fault weights tend to zero and sensor ones are equal to the same constant.

4.1 Constrained Optimisation

Under the proposed settings, it is implicitly assumed that a reasonable diagnostic should verify $f_i \geq 0$ in all components of the primary faults.

Also, fuzzy concepts saturate in $[0, 1]$. Hence, simultaneous faults yielding the same symptom cannot fulfill, for instance, $1 = f_1 + f_2$ if they are fully active. However, two easy options are available (or a combination of them) in that case:

- When a fuzzified sensor reading is saturated, replace the equality constraint in the diagnosis equation by an inequality ($1 \leq C^i f, 0 \geq C^i f$, where C^i denotes the i-th row of C).
- Translate an ordered fuzzy partition on a domain into numerical values (for instance, {*very low, low, normal, high, very high*} into $\{-2, -1, 0, 1, 2\}$) in the spirit of the so-called linguistic equation [17] [24]. In this way, the basic diagnosis equation (5) may involve sensor values ranging more than $[0, 1]$, but somehow keeping the linguistic meaning.

Algorithms. LP algorithms incorporate linear inequality restrictions seamlessly. However, the least squares formula (7) must then be discarded and quadratic programming (QP) routines used instead. For instance, if the sensor reading in Example 4 had been $\{.2, .18, 0\}$ the output of the LS formula would have been $f_1 = 0.197$, $f_2 = -0.003$, out of the constraint space so QP or non-negative least squares would have been needed.

As the involved restrictions are linear, efficient code exists for both LP and QP settings. For brevity, mostly linear programming settings will be considered in the sequel, although a similar version posed as QP would produce comparable results. For instance, commercially available LP software is able to efficiently deal with hundreds (even thousands) of variables and restrictions, allowing for large-scale implementation of the ideas in this work.

Binary Faults. Some faults may be only either 0 or 1, without intermediate values (for instance, circuit breaker ON vs. OFF). To carry out optimisation, explicit enumeration of all the involved binary (or integer) variables and solving for each of them the optimisation on the remaining real variables may be an approach. Alternatively, mixed linear integer programming or branch-and-bound methodologies may also be applied [19].

Managing Non-unique Solutions. The optimisation routines stop at an approximately optimal point. However, there might be other optimal points or, at least, which may have a very similar value of the cost index. The situation is particularly frequent in the case of missing measurements, which amounts to deleting the corresponding row of C.

In order to choose between possible nonunique solutions, the weights of the different fault components should penalise each fault according to its probability: in that way, the solution would tend to be the most likely fault consistent with the sensor measurements.

Imprecise Measurements. Some sensors may be imprecise, in the sense that a small deviation from the expected values is frequent and acceptable to assume when producing a diagnosis. In a sense, quadratic cost indices naturally take that fact into account, but LP settings need a straightforward change of variable to do that.[4] This is easily carried out, by expressing each "sensor fault" by a sum of two sub-faults, bounding the maximum value of one of them (with a small weight in the inference cost index) and setting a much larger penalisation on the deviations of the non-limited one. If the small sensor errors have zero weight, the setting is equivalent to interval measurements:

$$f^* = f_1^+ + f_2^+ - f_1^- - f_2^- \quad (8)$$

$$0 \le f_1^+ \le l^+,\ 0 \le f_1^- \le l^-,\ 0 \le f_2^+,\ 0 \le f_2^- \quad (9)$$

Heavy cost in f_2^+ and f_2^-, no penalisation in f_1^+, f_1^- results in an interval sensor reading allowing, with no cost, diagnostics involving the actual reading, say σ, plus or minus the desired bound: $[\sigma - l^-, \sigma + l^+]$.

Example 4. Let us consider a knowledge base:

Fault 1 produces S1, S2. Fault 2 produces S2, S3

and a fuzzified sensor reading were $\{0.2, 0.43, 0.15\}$. The setting for minimal squared inference error (weighting by 0.5 the "confidence" on the accuracy of (2), because the addition of individual faults may not result in the exact addition of the results due to possible system nonlinearity) would be carried out by the following Matlab code:

```
C=[1 0;1 1;0 1];C2=[C eye(3)];
j=diag([0.01 0.01 1 .5 1]);
 cw=C2*inv(j);
 f=inv(j)*pinv(cw)*[.2 .43 .15]'
```

The result is $f_1 = 0.21$, $f_2 = 0.16$.

Alternatively, the Matlab code for the above problem minimising the 1-norm of the inference error via linear programming is:

```
 C=[1 0;1 1;0 1];
C2=[C eye(3) -eye(3)];
 j=[0.01 0.01 1 .5 1 1 .5 1];
x=linprog(j,[],[],C2,
      [.2 .43.15],zeros(8,1));
```

which produces $f_1 = 0.2$, $f_2 = 0.15$, apportioning all error to the less reliable sensor 2. Vector j contains the weights for f_1, f_2 and the positive and negative components of the 3 possible sensor/modelling faults. The Matlab manual explains the meaning of the arguments to `linprog`.

5 Conclusions

This paper presents a methodology for approximately translating expert diagnostic knowledge into mathematical programming problems (constrained optimisation). The knowledge is a series of statements about the list of symptoms caused by the occurrence of a particular fault, expressed via linguistic statements involving fuzzy sets.

[4] The same change of variable may be used in QP formulations to fine-tune the cost index formula, if so wished.

The approach deals naturally with multiple faults approximately following a linear equation in the linguistic domain. The diagnostic procedure operates satisfactorily with missing measurements or a limited number of faulty sensors.

The proposed approach can be thought of as an intermediate between IF-THEN rulebases and diagnosis based on a full mathematical model. It keeps a linguistic interpretation while allowing for many combinations of requirements difficult to be taken into account in a pure logic framework. Marginal possible intervals of fault severities are also easily calculated in the case of non-unique solutions of the optimisation problem. The problem is computationally tractable even in a large-scale framework as efficient software exists for the proposed optimisation techniques.

The presented framework has been discussed on a theoretical level with simple academic examples. Detailed comparative analysis and application to realistic diagnostic environments with a large number of "rules" is under research at this moment.

References

1. Chiang, L., Russell, E., Braatz, R.: Fault Detection and Diagnosis in Industrial Systems. Springer-Verlag, London, UK (2001)
2. Blanke, M., Kinnaert, M., Lunze, J., Staroswiecki, M., eds.: Diagnosis and Fault-Tolerant Control. Springer, London (2003)
3. Timmer, J.: Parameter estimation in nonlinear stochastic differential equations. Chaos, Solitons and Fractals **11** (2000) 2571–2578
4. Khalil, H.: Nonlinear Systems. 3rd edn. Prentice Hall, New Jersey, USA (2002)
5. Berger, J.: Statistical Decision Theory and Bayesian Analysis. Springer-Verlag, London (1985)
6. Angeli, C.: Online expert system for fault diagnosis in hydraulic systems. Expert Systems **16** (1999) 115–120
7. Carrasco, E. et. al.: Diagnosis of acidification states in an anaerobic wastewater treatment plant using a fuzzy-based expert system. Control Engineering Practice **12** (2004) 59–64
8. Kruse, R., Schwecke, E., Heinsohn, J., eds.: Uncertainty and vagueness in knowledge based systems: numerical methods (artificial intelligence). Springer-Verlag, Berlin, DE (1991)
9. Shafer, G., Pearl, J., eds.: Readings in uncertain reasoning. Morgan Kauffman, San Mateo (CA), USA (1990)
10. Dubois, D., Prade, H.: Possibilistic logic: a retrospective and prospective view. Fuzzy Sets and Systems **144** (2004) 3–23
11. Yamada, K.: Diagnosis under compound effects and multiple causes by means of the conditional causal possibility approach. Fuzzy Sets and Systems **145** (2004) 183–212
12. Castillo, E., Gutierrez, J., Hadi, A.: Expert Systems and Probabilistic Network Models. Springer, London (1997)
13. Russell, S., Norvig, P.: Artificial Intelligence: A Modern Approach. 2nd edn. Prentice-Hall, New Jersey, USA (2003)
14. Kyburg, H.: Higher order probabilities and intervals. International Journal of Approximate Reasoning **2** (1988) 195–209
15. Ayoubi, M., Isermann, R.: Neuro-fuzzy systems for diagnosis. Fuzzy Sets and Systems **89** (1997) 289–307
16. Jie, Z., Morris, J.: Process modelling and fault diagnosis using fuzzy neural networks. Fuzzy Sets and Systems **79** (1996) 127–140

17. Juuso, E.: Fuzzy control in process industry: the linguistic equation approach. In Verbruggen, H., Zimmermann, H.J., Babuska, R., eds.: Fuzzy Algorithms for Control. Kluwer, Boston (1999) 243–300
18. Sala, A., Albertos, P.: Inference error minimisation: Fuzzy modelling of ambiguous functions. Fuzzy Sets and Systems **121** (2001) 95–111
19. Sierksma, G.: Linear and Integer Programming: Theory and Practice. 2nd edn. Marcel Dekker Pub., New York (2001)
20. Gass, S.: Linear Programming: methods and applications. 5th edn. Dover, Mineola, NY, USA (2003)
21. Chow, M., Sharpe, R., Hung, J.: On the application and design consideration of artificial neural network fault detectors. IEEE Transactions on Industrial Electronics **40** (1993) 181–198
22. Yao, J., Yao, J.: Fuzzy decision making for medical diagnosis based on fuzzy number and compositional rule of inference. Fuzzy Sets and Systems **120** (2001) 351–366
23. Meyer, C.: Matrix Analysis and Applied Linear Algebra. Society for Industrial & Applied Mathematics (SIAM) (2001)
24. Jarvensivu, M., Juuso, E., Ahavac, O.: Intelligent control of a rotary kiln fired with producer gas generated from biomass. Engineering Applications of Artificial Intelligence **14** (2001) 629–653

A Multi-agent Home Automation System for Power Management

Shadi Abras[1], Stéphane Ploix[2], Sylvie Pesty[1] and Mireille Jacomino[2]

[1] Laboratoire Leibniz-Institut IMAG, CNRS, UMR5552, 46, Avenue Félix Viallet, France
{Shadi.Abras, Sylvie.Pesty}@imag.fr
[2] Laboratoire d'Automatique de Grenoble, CNRS, UMR5528, BP 46, France
{Stephane.Ploix, Mireille.Jacomino}@inpg.fr

Abstract. This chapter presents the principles of a Home Automation system dedicated to power management that adapts power consumption to available power resources according to user comfort and cost criteria. The system relies on a multi-agent paradigm. Each agent is embedded into a power resource or an equipment, which may be an environment (thermal-air, thermal-water, ventilation, luminous) or a service (washing, cooking), and cooperates and coordinates its action with others in order to find acceptable near-optimal solution. The control algorithm is decomposed into two complementary mechanisms: an emergency mechanism, which protects from constraint violations, and an anticipation mechanism, which computes the best future set-points according to predicted consumptions and productions and to user criteria. The chapter details a negotiation protocol used by the both mechanisms and presents some preliminary simulation results.

Keywords. Home automation system, multi-agent systems, automatic control, negotiation and cooperation, power management.

1 Introduction

For the next decades, the two major problems concerning energy are the greenhouse effect and the depletion of petrol resources especially the energy provided by oil and gas. Therefore, by conscience or by necessity, the resort to renewable resources of energy such as wind or solar radiations, arrives in the buildings knowing that the building represents 47% of the energy consumption and it is responsible for 25% of the greenhouse effect [1]. Moreover, undoubtedly, the user will be confronted by variable tariffs of energy according to the hour and the days and to the energy producers. It is in this varied and dynamic context of production and consumption of energy that a building, equipped with a Home Automation system to control the energy, takes its importance. The role of a Home Automation system dedicated to power management is to adapt the power consumption to the available power resources taking into account user comfort criteria: it permits to limit the use of supplementary resources which require additional investment and to avoid the expensive need of storage. A Home Automation system has to reach a compromise between the priorities of the user in term of comfort and in term of cost while satisfying technological constraints of equipment and user's comfort constraints.

This problem can be formulated as a scheduling problem. In [2], a solution based on a Resource Constrained Project Scheduling Problem (RCPSP), to improve the management of thermal-air equipments, is presented. Its aim is to satisfy resource constraints by coordinating the control of thermal-air equipment. Nevertheless, this approach requires precise predictive models and RCPSP techniques are hardly adaptable to the context of multi energy resources and multi equipments. In [3], an anticipation mechanism using Bellman-Ford's algorithm [4] is presented for solving the problem of managing predicted events in a Home Automation system. The principal advantage of Bellman-Ford's approach is that the optimal solution is guaranteed (if exist) but the major disadvantage is the high order of complexity.

An alternative approach is to use Multi-Agent techniques. Algorithms based on Multi-Agent Systems are nowadays used in several areas such as Computer science or Automatic Control. The first MAS approach for energy distribution have been presented in [5] and [6]. Kok et al. [7] put forward a market-based control concept for the supply and demand matching (SDM) in electricity networks. It aims to propose a Multi-Agent system for the electronic market. Its purpose is to control tasks in future electricity network which is expected to develop into a network of networks in which a vast number of system parts communicate and coordinate with each other.

The developments of solutions based on Multi-Agent Systems, well suited to solve spatially distributed and opened problems, permit to imagine an intelligent Multi-Agent Home Automation system. This chapter presents a Multi-Agent Home Automation System (MAHAS). It focuses on the definition of a negotiation protocol between agents embedded into equipments as well as in energy resources. The chapter is organized as follows: Section 2 describes, in a general view point, the Multi-Agent Home Automation System. Section 3 presents the two main mechanisms of this system: *the emergency and anticipation mechanisms*. Section 4 presents, in detail, the principle of the negotiation protocol for emergency and anticipation mechanisms. Then, the chapter presents some preliminary results and highlights the future work which will be done.

2 Multi-Agent Home Automation System

The three main features of the Multi-Agent Home Automation System (MAHAS) (Fig. 1), which consists of agents embedded into energy resources and into the different equipments, are the following:

- Distributed: the energy resources and equipments are distributed spatially and their control systems are independent.
- Flexible: the energy resources are few but also some equipments can accumulate energy (thermal-air, thermal-water) or satisfy with delay to demands of services (washing service, cooking service).
- Opened: the number of connected resources and equipments may vary with time (equipments or resources can be connected or disconnected) without having to completely redefine the control mechanism.

In Multi-Agent Systems, the notion of control involves operations such as coordination and negotiation among agents, elimination of agents that are no longer present and adding new agents when needed.

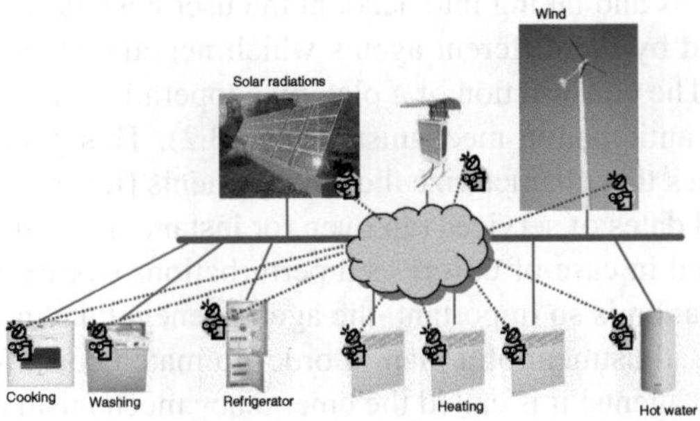

Fig. 1 Energy network and communication between embedded agents housing.

2.1 Agent Architecture

The main functionalities of an agent in MAHAS are shown in Fig. 2.

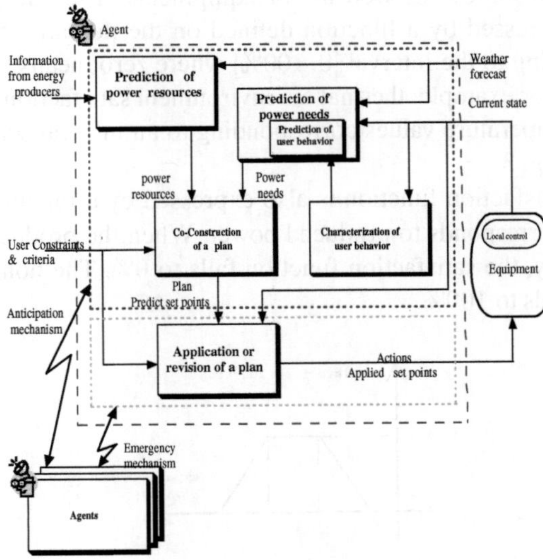

Fig. 2 Structure of an agent in MAHAS.

Depending on weather forecast, energy resource information and user habits:

– Resource agent calculates the available power resources: to determine what is and what will be the available power. For the moment, the energy resources are represented by a virtual energy resource which manages operations between the different resources.
– Equipment agent calculates the prediction of power consumption: to determine what are the future power needs taking into account the usual behaviour of users.

From these predictions and taking into account the user constraints and criteria, a plan is jointly constructed by the different agents which negotiate their future power consumption (Sect. 4). The construction of a plan by cooperating and negotiation between agents is called the anticipation mechanism (Sect. 3.2). This plan includes predicted values of the variables that characterize the environments (for example: the room temperatures) or the end dates of services (an oven for instance). Then, this plan is applied but it can be modified in case of unforeseen perturbations (for example: consumption peak). If the perturbation is so important, the agents renegotiate in order to recalculate plans. The real time adjustment of a plan in order to match constraints is achieved by cooperation between agents: it is called the emergency mechanism (Sect. 3.2).

A third mechanism may exist: the local control mechanism i.e. the controllers endowed into equipments by the manufacturers. It's time response is very fast. This mechanism receives set points from the agents. Besides, some information on its current state (power needs) are sent back to the agents so that they can be taken into account in the future plans. This mechanism is not mentioned in this chapter because other mechanisms are slower and local controls are assumed to be transparent.

One of the objectives of the MAHAS is to fulfil user comfort. A notion bound directly to the comfort is the satisfaction function [8]. Satisfaction functions have been defined for energy resources as well as for equipments. The equipment satisfaction function will be expressed by a function defined on the domain of the characteristic variable corresponding to the interval $[0, 100\%]$ where zero means "inadmissible" and 100% is "perfect". For example: thermal air environment satisfaction function, which is defined on room temperature values corresponding to an interval selected by user, can be represented by Fig. 3.

The resource satisfaction function is also expressed by a function where the characteristic variable corresponds to produced power. When the produced power exceeds the resource capacity, the satisfaction function falls to 0%. The nominal power of the resources corresponds to 100%.

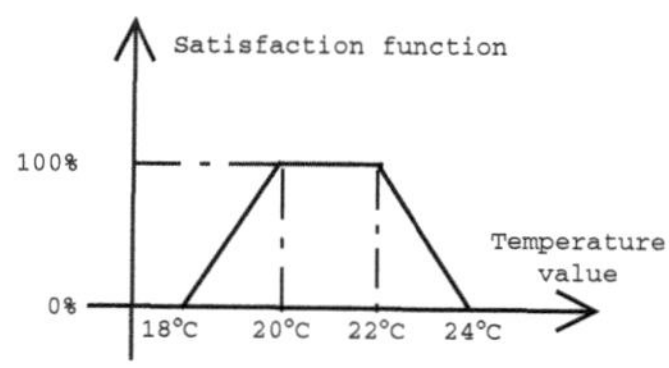

Fig. 3 Thermal air environment satisfaction function.

3 Agent Mechanisms

3.1 Emergency Mechanism

The emergency mechanism is a real time adjustment mechanism which is triggered out when the level of satisfaction of an agent falls below the weak values (10% for example). This mechanism, which relies on the negotiation protocol (Sect. 4), permits to react quickly to avoid violations of energy constraints and to guarantee a good level of user satisfaction. It is considered as transparent for anticipation mechanism because emergency adjustments have very small impact on the period considered by anticipation.

Therefore, the emergency mechanism adjusts, in real time, set points coming from the predicted plan, equipment's current state (equipment satisfaction value) and constraints and user criteria. The predicted set points can be directly transmitted to the local control mechanism or modified in case of emergency.

When the emergency mechanism is triggered, each agent has multiple roles:

- It evaluates, at predefined intervals, its current satisfaction. Therefore, it uses an infinite internal loop. This interval of time is called checking period.
- It can request help from other agents, by sending messages, when its satisfaction falls below a level of emergency.
- It analyzes the other agent demand and makes some propositions.
- When it receives some answers to its demands, it chooses and accepts the interesting propositions (to have a maximum value of satisfaction).
- It can allow, according to received messages, to activate or inactivate its associated equipment.

If an equipment agent satisfaction decreases, it sends messages requesting *help* from resource agents to initiate a negotiation. Other agent answers are collected during a fixed delay and are sorted out according to their satisfaction values. Then a solution which maximizes the satisfactions of equipments and resources is chosen.

3.2 Anticipation Mechanism

The emergency mechanism is sufficient to avoid constraint violations but a MAHAS can be improved in order to avoid emergency situations. This improvement is obtained thanks to the anticipation mechanism. The objective of this mechanism is to compute the predicted set points depending on predictions of consumptions and on predictions of energy resources. The anticipation mechanism relies on the fact that there is on the one hand, some electric equipments which are capable of accumulating energy and on the other hand, some services that have a variable date as for their execution: some services can both be delayed or advanced. From these preliminary observations, it is possible to imagine that if the equipment consumption can be anticipated, there is a way to organize it better.

The anticipation mechanism relies on learning algorithms which are not explained in this chapter. As for the emergency mechanism, the anticipation mechanism relies also on a negotiation protocol (Sect. 4). It works on a time window (anticipation period) larger than the checking period and works with average values of energy, because it is difficult to make precise predictions, in order to keep emergency mechanism transparent for it.

During anticipation mechanism, each agent has multiple roles:

- When requested, it predicts future needs or resources over a given number of anticipation period. This period is a multiple of the checking period.

- It analyzes the other agent demands and makes some propositions.
- When it receives some answers to its demands, it chooses and accepts the best propositions (to have a maximum value of satisfaction for all).
- It calculates, according to received messages, its predicted set points.

The message exchanges between agents during emergency and anticipation negotiations are defined by a protocol which is presented in the next section.

4 Negotiation Protocol

The negotiation protocol has been defined on the basis of the contract negotiation model [9], CNP protocol [10] [11] and algorithms of distributed constraint satisfaction problems [12]. This protocol can be used for agent mechanisms according to the checking period for emergency mechanism and anticipation period for anticipation mechanism. The negotiation protocol is characterized by successive messages exchanged between resource and equipment agents. Agents exchange messages for two objectives:

- To avoid to overpass the maximum available energy.
- To keep the satisfactions over a certain value: acceptable characteristic variable for environments which accumulate energy and acceptable shifts for services.

The agreements issued from negotiations are based on satisfactions of equipments (representing user comfort criteria) and on satisfactions of resources (representing the ideal power production).

4.1 Phases of Negotiation Protocol

The negotiation protocol (Fig. 4) may be decomposed into three phases:

- Energy demand phase: During this phase, the resource agents request equipment agents for propositions that lead to satisfactions greater or equal to an attempting satisfaction value and wait for equipment agent answers.
- Proposition phase: A conversation between resource agents and equipment agents takes place during which new propositions are exchanged. Then, resource agents analyse these propositions and can either accept them or request for equipment agents to send all the solutions for a new attempting satisfaction.
- Final decision phase: The resource agents take the decision, so equipment agent demands can either be accepted or refused.

The global success of negotiation is reached when all the equipments have reached quite similar satisfactions. When an event is under negotiation and no solution is possible, a negotiation with the user starts to modify user constraints.

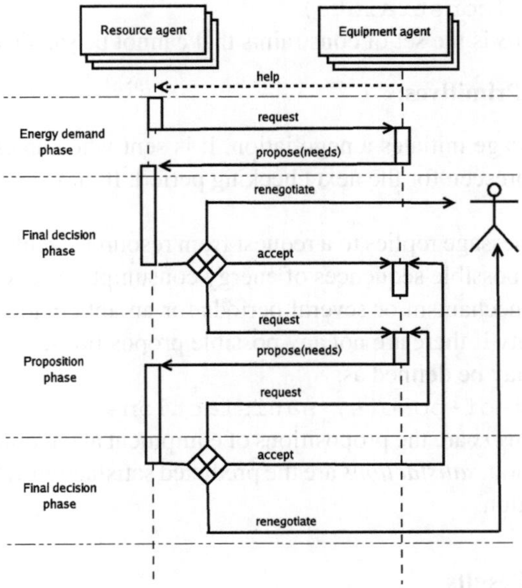

Fig. 4 Negotiation protocol.

4.2 Primitives of Negotiation Protocol

The primitives of negotiation protocol are decomposed into two groups.

Energy Resource Agent Primitives:

- Request: The resource agents initiate a negotiation by asking equipment agents to send them their power needs in order to reach a satisfaction greater or equal than an attempting satisfaction provided by resource agent. It collects the answers, it verifies if there is a global solution. Next request indicates to the equipment agents that there is no solution for the attempting satisfaction because the energy asked by equipments exceeds the maximum available energy provided by resources, so resource agents request equipment agents to send them other propositions about their needs for a smaller attempting satisfaction. A request may be defined as:
  `request(mechanism-name, period, satisfaction)`
  where *mechanism-name* has two values "emergency" or "anticipation". *period* value may be equal to the checking period or to the anticipation period. *satisfaction* is the attempting satisfaction value provided by resource agent.
- Accept: This message indicates to equipment agents that one of the proposed solutions has been accepted by a resource agent. This message may be defined as:
  `accept(proposition)`
  where *proposition* is one of the solutions proposed by the equipment agents.
- Renegotiate: This message indicates that there is no solution that satisfies the constraints defined by the user. A negotiation with the user starts. This message may be defined as:

```
renegotiate(constraints)
```
where *constraints* is the set of constraints that cannot be satisfied.

Equipment Agent Primitives:

- Help: This message initiates a negotiation. It is sent when an emergency situation is detected or foreseen for the next checking period. It may be defined as: `help()`.
- Propose: This message replies to a request from resource agents. It contains a set of propositions of possible sequences of energy consumption covering one period for an emergency mechanism or several periods for an anticipation mechanism. This list may be empty if there are not any possible propositions.
  This message may be defined as:
  `propose(set-of-powers, satisfactions)`
  where *set-of-powers* are the propositions of equipment agent during the checking or anticipation period. *satisfactions* are the predicted satisfaction values corresponding to each proposition.

4.3 Preliminary Results

In this subsection, an illustrative example is presented for Home Automation system which consists only of thermal air environments which are the largest part of consumption of electricity in buildings in winter. This system consists of three electrical heaters of 1 kW each and a 2100 W energy resource knowing that the initial temperatures in rooms are fixed to 18°C and the desired value of temperature is 20°C (satisfaction function takes its values between 0% for 18°C and 100% for 20°C). The temperature values

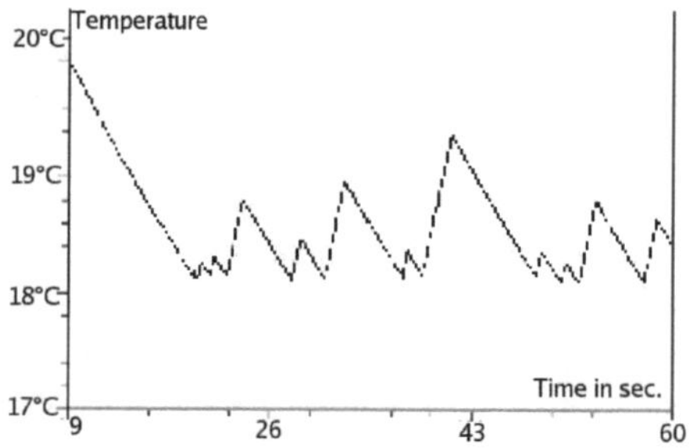

Fig. 5 Simulated temperature in room 1.

for other rooms are quite similar to room 1 (Fig. 5). The control system, in this example, is capable of maintaining temperature values for each environment above 18°C: because of the lack of power, the temperatures remain close to the minimum acceptable value. An example of exchanged messages between the energy resource agent and the equipment agents is presented below:

```
Heater2: help(heater2)
Resource: request ("emergency",15s,90%)
```

```
Heater2: propose (900W,90%)
Heater1: propose (900W,90%)
Heater3: propose (900W,90%)
Resource: request ("emergency",15s,80%)
Heater3: propose (800W,80%)
Heater1: propose (700W,70%)
Heater2: propose (750W,75%)
Resource: request ("emergency",15s,70%)
Heater1: propose (650W,65%)
Heater3: propose (700W,70%)
Heater2: propose (600W,60%)
Resource: accept(650W, 600W, 700W)
Heater1: help(heater1)
Resource: request ("emergency",15s,90%)
Heater2: propose (900W,99%)
Heater1: propose (900W,90%)
Resource: accept(900W, 900W, 0W)
```

Heater2 agent has requested help from the resource agent to start the negotiation. Then a conversation between the agents takes place during which the resource agent requests the equipment agents to send their propositions for an attempting satisfaction value, and during which the equipment agents send their propositions, which may be empty, to the resource agent.

In the absence of MAHAS but with an unbalancing system, always the same heater is penalized when all heaters simultaneously consume energy according to the user's predefined priorities. Contrary to MAHAS, the maximum user satisfaction cannot be guaranteed.

5 Conclusions and Perspectives

This chapter has presented a Multi-Agent Home Automation system allowing the agents to cooperate and coordinate their actions in order to find the accepted near-optimal solution for power management. Negotiation protocol has been detailed. The experimental results have showed the performance of the negotiation algorithm. This chapter have provided evidence that cooperation and negotiation capabilities of Multi-Agent systems can be advantageously used in automatic control systems for spatially distributed and opened systems.

The implementation of a simulator for the emergency and anticipation mechanisms is not finished yet. This simulator will be tested on a reduced-scale model of an apartment composed of two thermal environments and several services (washing machine,...). Each environment contains a reduced-scale electric heater, a temperature sensor and a micro-controller card with an embedded Java Virtual Machine.

References

1. Fontaine, N.: Livre blanc sur les énergies. débat national sur les énergies, http://www.industrie.gouv.fr/energie/politiqu/ploe.htm (2003)
2. Ha, D.L., Ploix, S., Zamai, E., Jacomino, M.: Control of energy consumption in home automation by ressource constraint scheduling. In: The 15th International Conference on Control System and ComputerScience, Bucharest, Romania (2005)
3. Ha, D.L., Ploix, S., Zamai, E., Jacomino, M.: A home automation system to improve household energy control. In: The 12th IFAC Symposium on Information Control Problems in Manufacturing (2006)
4. Cherkassky, B.V., Goldberg, A.V., Radzik, T.: Shortest paths algorithms: theory and experimental evaluation. In: SODA '94: Proceedings of the Fifth Annual ACM-SIAM Symposium on Discrete Algorithms, Philadelphia, USA, Society for Industrial and Applied Mathematics (1994) 516–525
5. Jennings, N.R.: The ARCHON system and its applications. In: Second International Working Conference on Cooperating Knowledge Based Systems (CKBS-94), Keele, UK (1994) 13–29
6. Hägg, S., Ygge, F.: Agent-oriented programming in power distribution automation. PhD thesis, University of Karlskrona/Ronneby, Ronneby, Sweden (1995)
7. Kok, J.K., Warmer, C.J., Kamphuis, I.G.: Powermatcher: multiagent control in the electricity infrastructure. In: AAMAS '05: Proceedings of the Fourth International Joint Conference on Autonomous Agents and Multiagent Systems, New York, USA, ACM Press (2005) 75–82
8. Simonin, O.: Le modèle satisfaction-altruisme: coopération et résolution de conflits entre agent situés réactifs, application à la robotique. PhD thesis, Université Montpellier II (2001)
9. Mathieu, P., Verrons, M.H.: Three different kinds of negotiation applications achieved with GeNCA. In: Proceedings of the International Conference on Advances in Intelligent Systems – Theory and Applications (AISTA) In Cooperation with the IEEE Computer Society, Centre de Recherche Public Henri Tudor, Luxembourg-Kirchberg, Luxembourg (2004)
10. Smith, R.G.: The contract net protocol: high-level communication and control in a distributed problem solver. IEEE Transaction on Computers **C-29** (1980) 1104–1113
11. Yang, J., Havaldar, R., Honavar, V., Miller, L., Wong, J.: Coordination of distributed knowledge networks using contract net protocol. IEEE Information Technology Conference, Syracuse, NY (1998)
12. Makoto, Y., Hirayama, K.: Algorithms for distributed constraint satisfaction: a review. Autonomous Agents and Multi-Agent Systems **3** (2000) 185–207

Feature Selection for Identification of Spot Welding Processes

Eija Haapalainen, Perttu Laurinen, Heli Junno, Lauri Tuovinen and Juha Röning

Intelligent Systems Group, Department of Electrical and Information Engineering
P.O. Box 4500, FIN-90014 University of Oulu, Finland
{Eija.Haapalainen, Perttu.Laurinen, Heli.Junno
Lauri.Tuovinen, Juha.Roning}@ee.oulu.fi

Abstract. Process identification in the field of resistance spot welding can be used to improve welding quality and to speed up the set-up of a new welding process. Previously, good classification results of welding processes have been obtained using a feature set consisting of 54 features extracted from current and voltage signals recorded during welding. In this study, the usability of the individual features is evaluated and various feature selection methods are tested to find an optimal feature subset to be used in classification. Ways are sought to further improve classification accuracy by discarding features containing less classification-relevant information. The use of a small feature set is profitable in that it facilitates both feature extraction and classification. It is discovered that the classification of welding processes can be performed using a substantially reduced feature set. In addition, careful selection of the features used also improves classification accuracy. In conclusion, selection of the feature subset to be used in classification notably improves the performance of the spot welding process identification system.

Keywords. Feature selection, k-nearest neighbour classifier, process identification, resistance spot welding.

1 Introduction

Resistance spot welding is one of the most important methods for joining metal objects. It is in widespread use in, for example, the automotive and electrical industries, where more than 100 million spot welding joints are produced daily in European vehicle industry only [1].

In resistance spot welding, two or more metal sheets are joined together by passing an electrical current through them. The current is conducted through two electrodes pressed against the metal surfaces to hold the parts to be welded tightly together. The heat produced by the flowing current melts the metals, and a welding spot is formed. The amounts of current, pressure and time are all carefully controlled and matched to the type and thickness of the material.

After cooling, the quality of the welding joint can be estimated by measuring its diameter. In general, the bigger the diameter is, the firmer is the welding joint. Some other factors, such as faults and embrittlement in the welding joint, also affect its strength.

The most reliable and commonly used method to verify the quality of a welding joint is to tear the welded parts apart and to measure the spot diameter. However, the welding joint is thereby destroyed. Some nondestructive methods for estimating the spot diameter also exist, but so far, no real-time, nondestructive method for online use on production lines has been developed. The two most common methods of nondestructive testing are radiographic and ultrasonic weld inspection [2]. These methods can also be used to detect discontinuities within the internal structure of a weld. Another example of nondestructive quality control methods of spot welding is the method based on primary circuit dynamic resistance monitoring by Cho and Rhee [3].

Different combinations of welding machines used and materials welded constitute distinctive welding processes. In other words, welding processes could also be called production batches. In this study, the properties of welding experiments that distinguish different processes are the type of welding machine used, the materials welded, the thicknesses of the materials and the welding time. However, changes in current, electrode force and electrode wear are thought to be internal changes of processes. Recognition of the most similar process from a pool of previously stored processes is called process identification.

This study is a follow-up on a previous article by the authors, in which different classification methods were evaluated for use in the identification of spot welding processes [4]. That study showed that welding processes can be reliably identified by extracting certain statistical and geometrical features of current and voltage signals measured during welding and performing classification based on these features. The k-nearest neighbour classifier (kNN) with the parameter value $k=3$ was found to be the most suitable method for the classification of spot welding processes.

Process identification is needed to be able to utilise information collected from previously run processes to produce new welding spots of good quality. The characteristics of a sample from a new welding process can be compared to information collected from previously run processes to find a similar process. After that, the process parameters of the previous process already proven to lead to high-quality welding joints can be applied to the new process. With this approach, good welding results are achieved right from the beginning, and the time needed for the set-up of a new process can be significantly reduced. In addition, if a similar process is found, the quality control methods that proved viable for that process can also be used for the new process.

In the previous study by the authors, classification of welding processes was performed using altogether 54 distinctive features extracted from the signal curves recorded during welding. The aim of this study was to reduce the dimension of the feature space by eliminating features with less classification-relevant information and to consider the usefulness of the individual features. This was expected to cut down the time needed for classification and, most importantly, to further improve the classification accuracy. Various feature selection methods were tested to find the minimal feature set yielding good classification results.

Previously, feature selection in the field of spot welding has been studied by Stoppiglia et al. [5]. In that study, however, only one process was considered at a time, and the study concentrated exclusively on nondestructive estimation of the diameter of the welding spot. The existing feature selection methods have been extensively reviewed in the studies [6] and [7]. Especially, in [7] and [8], the Sequental Floating Feature Selection methods used in this study have been shown effective and suitable for feature selection problems with the dimension of the feature data of the same magnitude as the data used in this study.

2 The Data

The data used in this study were supplied by two welding equipment manufacturers. There were altogether 20 processes, of which 11 had been welded at Harms+Wende GmbH & Co.KG and 9 at Stanzbiegetechnik. A total of 3879 welding experiments were covered. The experiments were done by welding two metal objects together using a resistance spot welding machine. Each of the observations contained measurements of current and voltage signals recorded during welding.

The raw signal curves contained plenty of oscillatory motion and a pre-heating section, and they were therefore pre-processed. The pre-heating parts of the curves were cut off, so that all that remained was the signal curves recorded during the actual welding phase. In addition, the curves were smoothened using the Reinsch algorithm [9]. An example of a signal curve before and after pre-processing is shown in Fig. 1a,b.

3 The Features

Altogether 54 geometrical and statistical features were extracted from the two signal curves relating to a single welding experiment. The geometrical features were chosen to locate the transition points of the curves as precisely as possible. The statistical features included the median of the signal and the arithmetic means of the signal values calculated on four intervals based on the transition points. In addition, the signal curve was divided into ten intervals of equal length, and the means of the signal values within these intervals were used as features. There were altogether 12 geometrical and 15 statistical features extracted from both signal curves. The features are demonstrated in Figs. 2 a,b.

In practice, it often happens that some of the geometrical features overlap, and that the overlapping features vary from one curve to another. However, this can also be regarded as a characteristic of the curve. In Fig. 2a, all the geometrical features are demonstrated on an artificial curve simulating the real data. On this curve, the features do not overlap, but the curve is otherwise notably similar to genuine signal curves. Figure 2b shows an example of the features calculated on a real signal curve.

The ten means of a signal curve have earlier been used as features in the articles [4][10] [11] and [12]. It has been discovered that they present the main characteristics and differences of the curves very well and are therefore suitable to be used in the quality control and identification of welding processes.

4 Feature Selection

Of the classification methods evaluated in the previous article by the authors [4], the k-nearest neighbour classifier proved to be optimal for the identification of spot welding

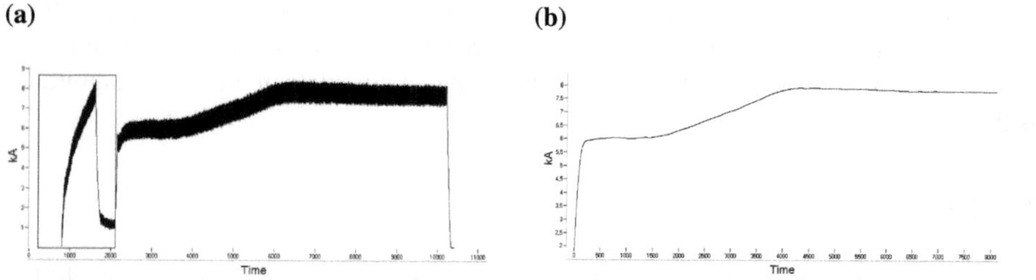

Fig. 1 (a) A raw signal curve. The pre-heating section is outlined with a rectangle. **(b)** The same curve after pre-processing.

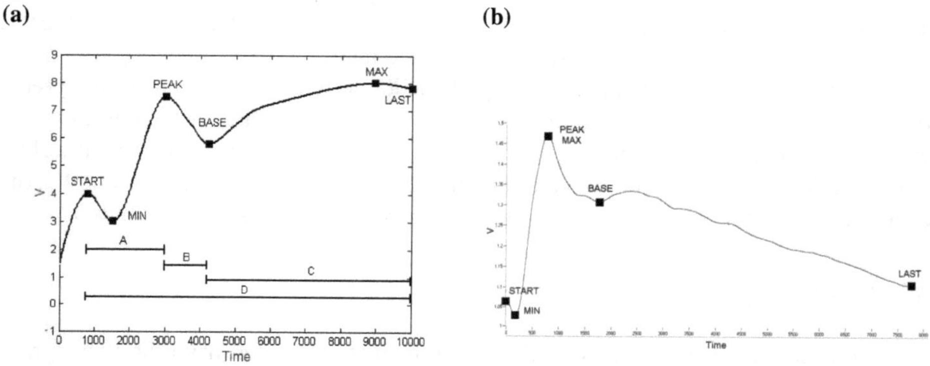

Fig. 2 (a) The geometrical features on an artificial voltage curve. The line segments A–D below the curve demonstrate the intervals based on the transition points on which means were calculated. **(b)** An example of how the geometrical features often partially overlap in practice. On this voltage curve, the features named "peak" and "max" overlap.

processes. The parameter value $k = 3$ was selected based on the comparative study. The classification was performed using alternatively either all the 54 features extracted from the curves or only the ten mean values calculated on both signal curves (altogether 20 features). The best results were obtained using the ten means. In that case, a classification accuracy of 98.53 percent was obtained, while the classification based on all the 54 features yielded notably inferior results with a classification accuracy of only 84.13 percent.

Although good classification results were obtained using the ten means as features, there was still an interest to study the usability of the other features. Also, the amount of classification-related information carried by each of the ten means was unknown. Therefore, various feature selection methods were applied to the entire feature set to find the optimal feature subset to be used in classification. It was studied whether classification accuracy could be further improved by discarding redundant features. Since dimension reduction of the feature set also reduces the computational time required for classification, a minimal subset was searched for.

Five different feature selection methods were tested: *Sequential Forward Selection* (SFS), *Sequential Backward Selection* (SBS), *Sequential Forward Floating Selection* (SFFS), *Sequential Backward Floating Selection* (SBFS) and *n Best Features Selection*.

SFS is a simple bottom-up search procedure in which one feature at a time is added to the current feature set. At each stage, the feature to be included is selected from the set of remaining available features, so that the new extended feature set yields a maximum value of the criterion function used [13]. The SBS method is the top-down counterpart of the SFS algorithm.

The Floating Forward and Backward Feature Selection methods, SFFS and SBFS, introduced in [14], are based on the *plus l-take away r* method [15], in which the feature set is alternately enlarged by l features using the SFS method and reduced by discarding r features applying the SBS algorithm. In the Floating Feature Selection methods, however, the number of forward and backward steps is dynamically controlled instead of being fixed in advance. The conditional inclusion and exclusion of features is controlled by the value of the criterion function. In the bottom-up algorithm, SFFS, after each forward step, a number of backward steps are applied as long as the resulting subsets yield better values of the criterion function than the previously evaluated ones of the same dimension. In the top-down counterpart, SBFS, an exclusion of a feature is followed by a series of successive conditional inclusions if an improvement to the previous sets can be made. The feature to be included into the current feature set or excluded from it is always the one that improves the set most or degrades the value of the criterion function least [14].

The n Best Features Selection method simply means selection of the n individually best features in the sense of maximizing the criterion function. It is the simplest alternative for feature subset selection, but also the most unreliable since the features may correlate with each other. Therefore, it was only used for comparison in this study.

The best possible way to design the process identification system would have been to select the feature set and the classification method used simultaneously. However, because the kNN classifier had previously been found suitable for the process identification task [4], the effectiveness of the different feature subsets produced by the feature selection methods were evaluated using the classification accuracy of the 3NN classifier as the criterion function. In addition, the k-nearest neighbour method has been used to measure the goodness of a feature set also in the studies [7] and [8].

The SFS, SBS and n Best Features methods were selected to be used in this study because of their easy application and relatively short calculation time. Compared to the basic sequential feature selection methods, the main advantage of the floating methods is that the resulting feature sets of different dimensions are not necessarily nested, as in the case of the SFS and SBS methods. This is because the floating methods are able to correct the erroneous decisions made at the previous steps of the algorithm. Therefore, these methods provide a close to optimal solution to the problem of feature subset selection [14]. Because of this characteristic, they are also highly applicable to problems involving nonmonotonic feature selection criterion functions, which was the case in this study. In addition, even though the floating feature selection methods are only nearly optimal, they are much faster than the optimal but computationally prohibitive *Branch and Bound* algorithm [16].

In order to evaluate classification accuracy when using different feature sets, the data were divided into training and test data sets, which consisted of 2/3 and 1/3 of the data, respectively. The training data set was used to train the 3NN classifier, and the test data set was used to evaluate the classification accuracy.

5 Results

The best possible feature subsets for maximizing the 3NN classifiers classification accuracy were searched for. The feature selection methods were applied to both the original and a normalized feature set. The latter was formed by normalizing the feature values of the original feature set to have an average of zero and a standard deviation of one. The results of the classification using feature subsets constructed by the various feature selection methods are presented in Table 1a,b.

The tables show the best classification accuracy obtained using feature sets formed by each of the feature selection methods. The feature subsets of all dimensions between 1 and 54 (dimension of the original feature set) were formed with each of the feature selection methods. Classification using each of these sets was performed, and the best classification accuracy obtained was recorded in the tables. The percentages in the middle row indicate the ratios of correctly classified processes and the numbers in the bottom row stand for the dimension of the feature set used in classification. It should be noted that the best feature subsets produced by the different feature selection methods are composed of unequal numbers of features.

The classification results of feature subsets formed from the unnormalized features are presented in Table 1a, and the results of the subsets constructed from the normalized features are shown in Table 1b. It can be seen that the subsets of the set of normalized features are notably larger than the subsets of the set of original features. However, the dimensions and classification accuracies of the different feature subsets are difficult to compare since only the subsets yielding the best classification results were considered at this point. Only the backward methods, SBS and SBFS, seem to yield better feature subsets when applied to normalized data. Nevertheless, the dimensions of these sets, 17 and 29, are much larger than those of the subsets formed from the unnormalized feature data, which are both of dimension 7. For comparison, it can be studied what the classification accuracies would be for smaller subsets of the set of normalized features formed with the backward methods. These results are presented in Table 2. It can be seen that quite good classification results are also obtained by using the smaller feature sets. However, these results do not compare with the classification results of the subsets produced from the unnormalized feature set by the forward methods, SFS and SFFS.

From the point of view of this study, it was considered more important to find a moderately small feature set that yields excellent classification results than to reduce the dimension of the feature set used to the absolute minimum. It can be stated, however, that the best classification results are obtained with small feature subsets formed from the original feature set. The best of all feature subset with a classification accuracy of 99.30 percent was produced by the SFFS method. The dimension of this set was 11. The SFS method also yielded a feature subset with almost equally good classification accuracy of 98.92 percent. And what is remarkable about this result is that this set consists of only 6 features.

Table 1 Classification results of the 3NN classifier using different feature subsets formed by the feature selection methods together with the number of features included in each feature set. (a) Subsets formed from the original feature set. (b) Subsets formed from normalized feature data

(a)

Feature selection method	**SFS**	**SBS**	**SFFS**	**SBFS**	**nBEST**
Classification accuracy	98.92	84.67	99.30	84.75	95.28
Number of features used	6	7	11	7	10

(b)

Feature selection method	**SFS**	**SBS**	**SFFS**	**SBFS**	**nBEST**
Classification accuracy	98.30	97.14	98.45	97.45	95.12
Number of features used	18	17	19	29	52

Table 2 Classification accuracies of subsets of different dimensions formed by the SBS and SBFS methods from the normalized feature set

Method/Dim.	**6**	**7**	**8**	**10**	**15**	**20**
SBS	95.74	96.28	96.44	96.83	96.75	97.06
SBFS	95.67	96.21	96.52	96.90	97.14	97.21

The observation that, in general, the subsets generated from the set of original features yield better classification results than the subsets formed from the normalized data alludes that the measurements of current contain more information related to process identification than the measurements made of voltage. This is because the original range of current measurement values was wider than the range of voltage values. Therefore, in the case of the original data, the features extracted from the current signals affect the classification more than the features calculated from the voltage signals. Since the influence of the two signals is equalized in normalized data, this implies that the features calculated from current signals are more significant to process identification than the features extracted from the voltage signals. From experience, however, it is known that only a rough classification of welding processes can be made based on the current signals alone, and the information carried by the voltage signals is also needed to get more precise results.

The significance of the features extracted from the current signals can also be established by comparing the amount of classification-related information contained in each of the features individually. Table 3a presents the 20 individually best features of the normalized feature set. The column in the middle shows the accuracy of classification based on only one feature. The classification results of the feature subsets formed with the n Best Features Selection method are presented in the column on the right. It can be seen that all the 20 individually best features are extracted from the current curves. (The first feature extracted from voltage signals would be the 22nd on this list.) However, the individual goodness of the features is not a sufficient criterion for feature selection since the features may correlate with each other. Because of this, the feature subsets produced by the n Best Features Selection method are generally inferior to the subsets formed by the other methods. The best subset formed from the normalized feature set yielded a classification accuracy of 95.12 percent, and the subset consisting of unnormalized features yielded an accuracy of 95.28 percent as seen in Table 1.

Table 3 The 20 individually best features of the normalized feature set. A *c* at the end of a feature name means that the feature has been extracted from the current signal, while a *v* stands for voltage signal

Feature	%	*n*
median_c	72.45	72.45
start_value_c	72.29	89.47
mean_C_c	71.67	90.40
interval_mean_5_c	71.36	92.57
mean_D_c	69.51	92.57
interval_mean_6_c	69.20	92.26
interval_mean_4_c	68.65	92.57
interval_mean_2_c	68.50	94.58
interval_mean_7_c	68.19	94.74
mean_A_c	66.33	94.97
interval_mean_8_c	65.87	94.58
interval_mean_9_c	65.71	94.50
interval_mean_10_c	64.94	94.35
interval_mean_3_c	64.63	94.43
peak_value_c	62.38	94.50
mean_B_c	61.84	94.50
max_value_c	61.77	93.96
interval_mean_1_c	61.53	94.43
last_value_c	60.06	94.81
min_value_c	59.68	94.66

It should also be noted in Table 3 that all of the ten means calculated on the current signal (called interval means) are among the 18 individually best features. After this, it only seems logical that remarkably better classification results were obtained in the previous study using a feature subset consisting of the ten means than using the entire feature set in classification.

In Table 4, the features selected by the SFS, SFFS and n Best Features Selection methods are presented. These feature subsets of the dimensions 6, 7 and 10 were formed from the unnormalized feature set. It can be seen that several of the features have been selected by two or all three of the methods. Again, the ten means are well represented in the set of features selected. It should also be noted that the SFS and SFFS methods have selected approximately evenly features extracted from the current and the voltage signals. As the best classification results are obtained using feature subsets formed by these methods, it can be concluded that the use of features extracted from both the signals is necessary to obtain excellent classification results.

6 Conclusions

In this study, various feature selection methods were discussed with an aim to improve the performance of a spot welding process identification system. The methods were applied to a set of features extracted from current and voltage signals recorded during welding. The classification accuracy of the 3NN classifier found suitable for the process

Table 4 The features selected by the SFS, SFFS and n Best Features Selection method from the original feature set

Feature	SFS	SFFS	nB
start_value_c	x	x	x
interval_mean_2_c	x	x	x
interval_mean_1_v	x	x	
interval_mean_5_v	x	x	
median_c	x		x
interval_mean_5_c		x	x
interval_mean_6_c		x	x
interval_mean_10_v	x		
last_value_v		x	
peak_value_v		x	
median_v		x	
interval_mean_3_v		x	
interval_mean_6_v		x	
mean_A_c			x
mean_C_c			x
mean_D_c			x
interval_mean_4_c			x
interval_mean_7_c			x

identification task in a previous study was used as the criterion function for the feature subsets produced by the methods. Altogether five different feature selection methods were tested on both the original and a normalized feature set. These methods were the *Sequential Forward Selection* (SFS), the *Sequential Backward Selection* (SBS), the *Sequential Forward Floating Selection* (SFFS), the *Sequential Backward Floating Selection* (SBFS) and the *n Best Features Selection*. In general, the subsets generated from the set of original features yielded better classification results than the subsets formed from the normalized data.

It was discovered that classification accuracy can be improved from the 84.13 percent obtained previously to 99.30 percent simultaneously reducing the dimension of the feature set from 54 to 11. This reduction in the feature set size facilitates both the extraction of features and the actual process identification. The best feature subset (classification accuracy of 99.30 percent) was obtained using the SFFS method. The feature subset produced by the SFS method also yielded a very good classification result. The classification accuracy obtained using this set was 98.92 percent. And what is most important about this result is that the dimension of this set was notably smaller than the dimension of the set produced by the SFFS method being only 6. Hence, it was shown, that the dimension of the feature subset used in classification can be significantly reduced and the performance of the spot welding process identification system notably improved.

By considering the amount of classification-related information contained in each of the features individually, it was confirmed that the ten means calculated on intervals of equal length of a signal curve are indeed very good features to be used in classification. In addition, it was discovered that the features extracted from current signals contain more information related to process identification than those calculated from voltage signals. Nevertheless, the use of both of these is necessary to obtain excellent classification results.

Acknowledgements

We would like to thank our partners in cooperation Fachochschule Karlsruhe, Stanzbiegetechnik and Harms + Wende GmbH & Co.KG. Furthermore, we would like to express our gratitude for the financial support provided by the Commission of the European Communities, specific RTD programme "Competitive and Sustainable Growth", G1ST-CT-2002-50245, "SIOUX", and Infotech Oulu, an umbrella organization for information technology research in Oulu region.

References

1. TWI: World centre for materials joining technology. Resistance Spot Welding, www document (2004) Retrieved September 29, 2005, from `www.twi.co.uk/j32k/protected/band_3/kssaw001.html`.
2. Anderson, T.: Radiographic and ultrasonic weld inspection: establishing weld integrity without destroying the component. Practical Welding Today, December 13 (2001)
3. Cho, Y., Rhee, S.: Primary circuit dynamic resistance monitoring and its application to quality estimation during resistance spot welding. Welding Journal **81** (2002) 104–111
4. Haapalainen, E., Laurinen, P., Junno, H., Tuovinen, L., Röning, J.: Methods for classifying spot welding processes: a comparative study of performance. In: Proc. 18th International Conference on Industrial & Engineering Applications of Artificial Intelligence & Expert Systems (2005) 412–421
5. Stoppiglia, H., Dreyfus, G., Dubois, R., Oussar, Y.: Ranking a random feature for variable and feature selection. Journal of Machine Learning Research **3** (2003) 1399–1414
6. Dash, M., Liu, H.: Feature selection for classification. International Journal of Intelligent Data Analysis **1** (1997) 131–156
7. Kudo, M., Sklansky, J.: Comparison of algorithms that select features for pattern classifiers. Pattern Recognition **33** (2000) 25–41
8. Jain, A., Zongker, D.: Feature selection: evaluation, application, and small sample performance. IEEE Transactions on Pattern Analysis and Machine Intelligence **19** (1997) 153–158
9. Reinsch, C.: Smoothing by spline functions, II. Numerische Matematik **16** (1971) 451–454
10. Junno, H., Laurinen, P., Haapalainen, E., Tuovinen, L., Röning, J., Zettel, D., Sampaio, D., Link, N., Peschl, M.: Resistance spot welding process identification and initialization based on self-organizing maps. In: Proc. First International Conference on Informatics in Control, Automation and Robotics, Volume 1 (2004) 296–299
11. Junno, H., Laurinen, P., Tuovinen, L., Röning, J.: Studying the quality of resistance spot welding joints using self-organising maps. In: Proc. 4th International ICSC Symposium on Engineering of Intelligent Systems (2004)
12. Junno, H., Laurinen, P., Haapalainen, E., Tuovinen, L., Röning, J.: Resistance spot welding process identification using an extended knn method. In: Proc. IEEE International Symposium on Industrial Electronics, Volume 1 (2005) 7–12
13. Devijver, P., Kittler, J.: Pattern Recognition – A Statistical Approach. Prentice-Hall, Englewood Cliffs, NJ (1982)

14. Pudil, P., Ferri, F., Novovičová, J., Kittler, J.: Floating search methods for feature selection with nonmonotonic criterion functions. In: Proc. 12th International Conference on Pattern Recognition (1994) 279–283
15. Stearns, S.: On selecting features for pattern classifiers. In: Proc. Third International Conference on Pattern Recognition (1976) 71–75
16. Narendra, P., Fukunaga, K.: A branch and bound algorithm for feature selection. IEEE Transactions on Computers **C-26** (1977) 917–922

Fuzzy Logic Based UAV Allocation and Coordination

James F. Smith III and ThanhVu H. Nguyen

Code 5741, Naval Research Laboratory, Washington, DC, 20375-5320, USA
james.smith@nrl.navy.mil, thanhvu.nguyen@nrl.navy.mil

Abstract. A fuzzy logic resource allocation algorithm that enables a collection of unmanned aerial vehicles (UAVs) to automatically cooperate will be discussed. The goal of the UAVs' coordinated effort is to measure the atmospheric index of refraction. Once in flight no human intervention is required. A fuzzy logic based planning algorithm determines the optimal trajectory and points each UAV will sample, while taking into account the UAVs' risk, risk tolerance, reliability, and mission priority for sampling in certain regions. It also considers fuel limitations, mission cost, and related uncertainties. The real-time fuzzy control algorithm running on each UAV renders the UAVs autonomous allowing them to change course immediately without consulting with any commander, requests other UAVs to help, and change the points that will be sampled when observing interesting phenomena. Simulations show the ability of the control algorithm to allow UAVs to effectively cooperate to increase the UAV team's likelihood of success.

Keywords. Decision support systems, distributed control systems, fuzzy control, knowledge-based systems applications, software agents for intelligent control systems.

1 Introduction

Knowledge of meteorological properties is fundamental to many decision processes. Due to personnel limitations and risks, it is useful if related measurement processes can be conducted in a fully automated fashion. Recently developed fuzzy logic based algorithms that allow a collection of unmanned aerial vehicles (UAVs) and an interferometer platform (IP) [9] to automatically collaborate will be discussed. The UAVs measure the index of refraction in real-time to help determine the position of an electromagnetic source (EMS). The IP is actually an airplane with an interferometer onboard that measures emissions from the electromagnetic source whose position is to be estimated. Each UAV has onboard its own fuzzy logic based real-time control algorithm. The control algorithm renders each UAV fully autonomous; no human intervention is necessary. The control algorithm aboard each UAV will allow it to determine its own course, change course to avoid danger, sample phenomena of interest that were not preplanned, and cooperate with other UAVs.

Section 2 provides an overview of the meteorological sampling problem and a high level description of the planning and control algorithms that render the UAV team fully autonomous. Section 3 discusses the electromagnetic measurement space,

UAV risk, and the planning algorithm. Section 3 also discusses the UAV path construction algorithm that determines the minimum number of UAVs required to complete the task, a fuzzy logic based approach for assigning paths to UAVs and which UAVs should be assigned to the overall mission. Section 4 describes the control algorithm that renders the UAVs autonomous. Section 4 also describes the priority for helping (PH) algorithm, a part of the control algorithm based on fuzzy logic that determines which UAV should help another UAV requesting help. The three subclasses of help requests are also discussed in this section. Section 5 discusses experimental results including UAV path determination, UAV path assignment, determination of which UAVs should fly the mission and the result of a request for help during the mission. Finally, Sect. 6 provides a summary.

2 Meteorological Sampling and Cooperative Autonomous Platforms

For many applications it is useful to be able to make meteorological measurements in real-time. Examples include determining the index of refraction of the atmosphere to facilitate geo-location [9]; determination of the presence and extent of such phenomena as radio holes and ducts, which may interfere with communications; tracking atmospheric contaminants [10]; and sand suspended in the atmosphere that can interfere with sensors.

The fuzzy logic based planning and control algorithms that have been developed allow a collection of UAVs making up the UAV team to engage in cooperative sampling of the atmosphere in real-time without human intervention. Each UAV will have its own control algorithm allowing it to determine new optimal trajectories in real-time subject to changing conditions. Also, the control algorithm on the UAVs will allow them to cooperate to increase the probability of mission success. There will be two different types of cooperation allowed by the control algorithm and three classes of help requests which are discussed in Sect. 4.

3 Planning and Risk

The measurement space consists of the electromagnetic propagation environment where UAVs and the IP make their measurements. This environment includes sample points and the desirable neighborhoods that surround them. The sample points or the desirable neighborhoods are where the UAVs will make measurements. The method of determining the sample points and desirable neighborhoods is described below.

The measurement space also includes taboo points and the undesirable neighborhoods that surround them. The taboo points are points of turbulence and other phenomena that could threaten the UAVs. The undesirable neighborhoods surrounding them also represent various degrees of risk. The method of specifying taboo points and quantifying the degree of risk associated with their undesirable neighborhoods employs fuzzy logic and is discussed in this section.

The planning algorithm allows the determination of the minimum number of UAVs needed for the mission subject to fuel constraints, risk, UAV cost, and importance of various points for sampling. Risk refers to turbulent regions or regions undesirable for other reasons, e.g., the presence of enemy observers or physical obstructions. The planning algorithm automatically establishes the order in which to send the UAVs taking into account the UAV's value; onboard sensor payload; onboard resources such as fuel, computer CPU and memory; etc. The priority of sample points and their desirable neighborhoods are taken into account. The planning algorithm also calculates the optimal path around undesirable regions routing the UAVs to or at least near the points to be sampled.

In the planning phase, the location of the EMS is unknown. Some positions are more likely than others for the EMS's location. When establishing likely positions for the EMS, human experts are consulted. The experts provide subjective probabilities of the EMS being located at a number of positions. These likely EMS locations are referred as *hypothesis positions*. Ray-theoretic electromagnetic propagation [1] is conducted from each hypothesis position to each interferometer element on the IP. The points on the sampling grid nearest the points of each ray's passage are the sample points. The priority of a sample point is related to the subjective probability of the hypothesis position from which the associated ray emerges. Sample points arising from the highest probability hypothesis positions have priority one; sample points associated with lower probability hypothesis positions, priority two; etc.

Each sample point is surrounded by what are referred to as *desirable neighborhoods*. Depending on local weather, topography, etc., the desirable neighborhoods are generally concentric closed balls with a degree of desirability assigned to each ball. The degree of desirability characterizes the anticipated variation in the index of refraction.

A point may be labeled taboo for a variety of reasons. A taboo point and the undesirable neighborhoods containing the point generally represent a threat to the UAV. The threat may take the form of high winds, turbulence, icing conditions, mountains, etc. The undesirable neighborhoods around the taboo point relate to how spatially extensive the threat is.

When determining the optimal path for the UAVs to follow both the planning algorithm and the control algorithm running on each UAV take into account taboo points and the undesirable neighborhood around each taboo point. The path planning algorithm and control algorithm will not allow a UAV to pass through a taboo point. Both the concepts of risk and risk tolerance are based on human expertise and employ rules each of which carry a degree of uncertainty. This uncertainty is born of linguistic imprecision [11], the inability of human experts to specify a crisp assignment for risk.

Risk is represented as a fuzzy decision tree [2] [4] [5] [6] [7] [8]. The risk subtree defined below is a subtree of the larger risk tree that was actually used. The risk tree is used to define taboo points and the undesirable neighborhoods surrounding the taboo points.

The root concepts on the risk tree use the membership function defined in (1–3),

$$\mu_{\alpha}(\vec{q}_{taboo},\vec{x})=\begin{cases} 1, & if \quad r=0 \\ 3/4, & if \quad 0<r\leq 1\cdot\Delta l \\ 1/2, & if \quad 1\cdot\Delta l<r\leq\sqrt{2}\cdot\Delta l \\ 1/4, & if \quad \sqrt{2}\cdot\Delta l<r\leq\sqrt{3}\cdot\Delta l \\ 0, & if \quad r>\sqrt{3}\cdot\Delta l \end{cases} \tag{1}$$

$$r=\left\|\vec{x}-\vec{q}_{taboo}\right\| \tag{2}$$

$$\vec{q}_{taboo}=\text{position of taboo point} \tag{3}$$

where the "taboo point," $\vec{q}_{taboo}$ is the point at which the risk phenomenon has been observed. The root concepts used on the risk subtree are given in (4), and the subscript α is an element of the root concept set , *RC*, i.e.,

$\alpha\in$ ***RC***=*{Mountains, High Tension Wires, Buildings, Trees, Smoke Plumes, Suspended Sand, Birds/Insects, Other UAVs, Air Pollution, Civilian, Own Military, Allied Military, Neutral Military, Cold, Heat, Icing, Rain, Fog, Sleet, Snow, Hail, Air Pocket, Wind, Wind Shear, Hostile Action/Observation}* (4)

The norm in (2) is typically taken as an Euclidean distance. The values taken by the quantity Δl will be discussed in a future publication.

The fuzzy membership function for the composite concept "risk" is defined as

$$\mu_{risk}(\vec{q}_{taboo},\vec{x})=\max_{\alpha\in RC}\mu_{\alpha}(\vec{q}_{taboo},\vec{x}) \tag{5}$$

The best path algorithm is actually an optimization algorithm that attempts to minimize a cost function to determine the optimal trajectory for each UAV to follow, given a priori knowledge. The cost function for the optimization algorithm takes into account various factors associated with the UAV's properties, mission and measurement space. Two significant quantities that contribute to the cost are the effective distance between the initial and final proposed positions of the UAV and the risk associated with travel.

For purposes of determining the optimal path, the UAV is assumed to follow a rectilinear path consisting of connected lines segments, where the beginning and ending points of each line segment reside on the UAV's sampling lattice. Let A and B be two grid points on the UAV's sampling grid with corresponding position vectors, $\vec{r}_A$ and $\vec{r}_B$, respectively. Denote the Euclidean distance between A and B as $d(\vec{r}_A,\vec{r}_B)$. Let $v(\vec{r}_A,\vec{r}_B)$ be the speed at which the UAV travels in going from $\vec{r}_A$ to $\vec{r}_B$. If both $\vec{r}_A$ and $\vec{r}_B$ are sample points then the UAV travels at sampling velocity, otherwise it travels at non-sampling velocity. The path cost is given by

$$path\_cost(\vec{r}_A,\vec{r}_B)=\frac{d(\vec{r}_A,\vec{r}_B)+\beta\cdot\sum_{i=1}^{n_{taboo}}\mu_{risk}(\vec{t}_i,\vec{r}_B)}{v(\vec{r}_A,\vec{r}_B)} \tag{6}$$

where n_{taboo} is the number of taboo points, i.e., columns in the taboo point matrix

$$Taboo \equiv \left[\vec{t}_1, \vec{t}_2, \ldots, \vec{t}_{n_{taboo}}\right] \tag{7}$$

and $\vec{t}_i, i = 1, 2, \ldots, n_{taboo}$ are the taboo points determined to exists in the measurement space when $path\_cost(\vec{r}_A, \vec{r}_B)$ is calculated. The quantity, β, is an expert assigned parameter. Note that $path\_cost(\vec{r}_A, \vec{r}_B)$ is an effective time. When risk is not present, i.e., $\beta \cdot \sum_{i=1}^{n_{taboo}} \mu_{risk}(\vec{t}_i, \vec{r}_B)$ is zero, then $path\_cost(\vec{r}_A, \vec{r}_B)$ is the actual travel time. When risk is present then the travel time is increased. The time increase will be significant if the risk is high.

If the candidate path for the mission consists of the following points on the UAV lattice given by the path matrix in (8),

$$Path_i = \left[\vec{r}_1, \quad \vec{r}_2, \quad \ldots, \quad \vec{r}_n\right] \tag{8}$$

then the total path cost is defined to be

$$total\_cost(Path_i) \equiv \sum_{j=1}^{n-1} path\_cost(\vec{r}_j, \vec{r}_{j+1}) \tag{9}$$

Determining the optimal path for the *i*th UAV consists of minimizing the total path cost given by (9) such that there is enough fuel left to complete the path.

The planning algorithm determines the path each UAV will pursue, which points will be sampled, the minimum number of UAVs required for sampling the points and makes assignments of UAVs for measurements at particular points. UAVs are assigned as a function of their abilities to sample high priority points first. The planning algorithm determines flight paths by assigning as many high priority points to a path as possible taking into account relative distances including sampling and non-sampling velocity, risk from taboo points, and UAV fuel limitations. Once flight paths are determined, the planning algorithm assigns the best UAV to each path using the fuzzy logic decision rule for path assignment described in this section.

The planning algorithm must assign UAVs to the flight paths determined by the optimization procedure described below in this section. This is referred to as the UAV path assignment problem (UPAP). The planning algorithm makes this assignment using the following fuzzy logic based procedure. To describe the decision rule it is necessary to develop some preliminary concepts and notation.

Each UAV will fly from lattice point to lattice point, i.e., grid point to grid point, let one such route be given by the matrix of points,

$$Path = \left[\vec{P}_1, \vec{P}_2, \ldots, \vec{P}_{n_{path}}, \vec{P}_1\right] \tag{10}$$

where the ordering of points gives the direction of the route, i.e., starting at $\vec{P}_1$ and ending at $\vec{P}_1$. Let the taboo points be those given in (7). Let the degree of undesirability of the neighborhood associated with taboo points, $\vec{t}_i, i = 1, 2, \ldots, n_{taboo}$ be denoted $\mu_{risk}(\vec{t}_i, \vec{P}_j)$ for the route points $\vec{P}_j, j = 1, 2, \ldots, n_{path}$. The definition of the mission risk is

$$mission\_risk(Path) \equiv \sum_{i=1}^{n_{taboo}} \sum_{j=1}^{n_{path}} \mu_{risk}(\vec{t}_i, \vec{P}_j) \tag{11}$$

Within the path specified by (10), let there be the following sample points to be measured, $\vec{S}_j, j = 1,2,\ldots,n_{sp}$. Let the function *prio* assign priorities to the sample points, i.e, $prio(\vec{S}_j)$ is the priority of the *j*th sample point. The values that $prio(\vec{S}_j)$ can take are positive integers with one representing the highest priority, two the next highest priority, etc. The mission priority for *Path* is defined to be

$$mission\_prio(Path) \equiv \sum_{i=1}^{n_{sp}} \frac{1}{prio(\vec{S}_i)} \tag{12}$$

Furthermore, let the $T(\mathrm{UAV}(i), Path)$ be the amount of time it will take UAV(*i*) to fly and make measurements along $Path$.

The fuzzy degree of reliability experts assign to the sensors of UAV(*i*) is denoted as $\mu_{sr}(\mathrm{UAV}(i))$. This is a real number between zero and one with one implying the sensors are very reliable and zero that they are totally unreliable. Likewise, $\mu_{nsr}(\mathrm{UAV}(i))$ is the fuzzy degree of reliability of other non-sensor systems onboard the UAV(i). This fuzzy concept relates to any non-sensor system, e.g., propulsion, computers, hard disk, deicing systems, etc. The value of UAV(*i*) in units of \$1000.00 is denoted as $V(\mathrm{UAV}(i))$. The amount of fuel that UAV(*i*) has at time *t* is denoted $fuel(\mathrm{UAV}(i), t)$. All the UAVs participating in a mission are assumed to leave base at time, $t = t_o$.

Let UAV(*i*)'s fuzzy grade of membership in the fuzzy concept "risk tolerance" be denoted as $\mu_{risk-tol}(\mathrm{UAV}(i))$. The quantity, $\mu_{risk-tol}(\mathrm{UAV}(i))$, is a number between zero and one and will be simply referred to as UAV(i)'s risk-tolerance. If the risk tolerance is near zero then the UAV should not be sent on very risky missions. If the UAV's risk tolerance is near one then it can be sent on very risky missions. It seems natural to compare risk-tolerance to value. So the comparison can be carried out on the same footing, a fuzzy concept of value should be defined.

The fuzzy grade of membership in the fuzzy concept "Value" of each UAV that can be assigned to the mission is defined as

$$\mu_V(\mathrm{UAV}(i)) \equiv \frac{Value(\mathrm{UAV}(i))}{\max_j \{Value(\mathrm{UAV}(j))\}} \tag{13}$$

The "*max*" operation in (13) is taken over the set of all possible UAVs that can be assigned to the mission.

The advantage of the concept of "risk-tolerance" is that it gives the user an extra concept to exploit. If the UAV is not of great relative value, but it still might be needed for a crucial mission after the current one, it might be useful to give it a low risk tolerance so that it is not lost on the current mission. This may allow it to be used on the following mission.

The final concept and related fuzzy membership function that must be defined is "slow". A UAV is said to be slow if it takes a long time to travel a particular path. The fuzzy membership function for the concept "slow" is defined as follows:

$$\mu_{slow}\left(\text{UAV}\,(i), Path\right) \equiv \frac{T\left(\text{UAV}\,(i), Path\right)}{\max\limits_{j}\left\{T\left(\text{UAV}\,(j), Path\right)\right\}} \tag{14}$$

A "slow" UAV experiences a higher relative mission risk since it is in the field longer and may be exposed to risk longer.

To construct the fuzzy membership function for the fuzzy concept "assign UAV to Path" (AUP) make the following definitions:

$$f_1(UAV(i), Path) \equiv \chi\left(fuel(UAV(i), t_o) + \varepsilon_{fuel} - T(UAV(i), Path)\right). \tag{15}$$

$$f_2(UAV(i), Path) \equiv \frac{mission\_prio}{denom(UAV(i), Path)} \min\left[\mu_{sr}(UAV(i)), \mu_{nsr}(UAV(i))\right] \tag{16}$$

$$\begin{aligned} &denom(UAV(i), Path) \equiv 1 + \\ &\min\left[1 - \mu_{risk-tol}(UAV(i)), \mu_V(UAV(i))\right]\mu_{slow}(UAV(i), Path) \cdot mission\_risk(Path) \end{aligned} \tag{17}$$

$$num(UAV(i), Path) \equiv f_1(UAV(i), Path) f_2(UAV(i), Path). \tag{18}$$

The Heaviside step function denoted as χ in (15) takes the value one when its argument is greater than or equal to zero and is zero otherwise. The quantity ε_{fuel} is added to the fuel term to make sure the UAV selected has more than enough fuel. Given the definition of $num(\text{UAV}(i), Path)$ the fuzzy membership function that gives the grade of membership of UAV(i) in the fuzzy concept "assign UAV to Path" is defined as

$$\mu_{AUP}(UAV(i), Path) \equiv \frac{num(UAV(i), Path)}{\max\limits_{j} num(UAV(j), Path)} \tag{19}$$

where the "*max*" operation in the denominator of (19) is taken over the set of all UAVs that can be assigned to the path.

4 Control Algorithm

Each UAV has a real-time algorithm onboard it that allows recalculation of paths during flight due to changes in environmental conditions or mission priorities. These changes typically become apparent after the planning algorithm has run during the pre-flight stage. As in the case of the planning algorithm the control algorithm uses an A-star algorithm [3] to do the best path calculation, employs fuzzy logic and solves a constrained optimization problem. Although this can require a number of minutes of computation on a two to three gigahertz computer, this is considered adequate given the required UAV flight time between points.

The control algorithms' recalculation of flight paths can be triggered by a number of events such as weather broadcasts that indicate new taboo regions or changes of priority of sample points. For those changes that do not require UAVs supporting each other, the control algorithm does not differ from the planning algorithm. The control algorithm is faster by virtue that it only need process those parts of the measurement space where there have been changes relative to sample or taboo points.

A UAV may requests help if it discovers a potential elevated system like a radio hole, malfunctions or suspected malfunctions. All of these conditions can result in help messages being transmitted between the UAVs. These help messages can result in interactions between the UAVs based on transmission of the results of priority calculations for rendering support to the requesting UAVs.

Currently in the control stage, when a UAV discovers an interesting physical phenomenon, is malfunctioning, or suspects due to internal readings that it is malfunctioning, it sends out a request for help. Each UAV receiving this message calculates its priorities for providing assistance to the UAV in need. This priority calculation gives rise to a number between zero and one, inclusive, which is subsequently transmitted to the original UAV desiring support. The requesting UAV sends out an omni-directional message with the ID of the UAV with highest priority for contributing support. The high priority UAV then flies into the necessary neighborhood of the requesting UAV to provide help.

There are three classes of help request. The first occurs when a UAV, the requester, determines it may have discovered an interesting physical phenomenon. This phenomenon may be an elevated duct, radio hole, rain system or some other type of system with physical extent. The requester desires to determine if the phenomenon has significant extent. It will request that a helping UAV or UAVs sample likely distant points within this phenomenon.

The second class of help request relates to a UAV that according to internal diagnostics may be experiencing a sensor malfunction. This UAV will requests that another UAV or UAVs measure some of the points that the requesting UAV measured. This will help determine if the UAV is actually malfunctioning. If the requesting UAV is determined to be malfunctioning, then it will fly back to base, if it is capable. The determination of whether it is actually malfunctioning requires some consideration. Since the second UAV will probably be measuring a distant point at a time different than the original requesting UAV made its measurements, potential variation in the index of refraction over time must be taken into account.

When a UAV sends out an omni-directional request for help, those UAVs receiving the message will calculate their fuzzy priority for helping, denoted as "PH."

The UAV that will ultimately help the requester is the one with the highest fuzzy priority for helping. The fuzzy priority for helping takes into account a variety of properties of the potential helper. The set of UAVs that receive the request for help from UAV(*i*) at time t is denoted as $help(i,t)$. If UAV(*i*) request help at time t and UAV(*j*) receives the message then UAV(*j*) will take into account the amount of time, denoted, $help\_time(\mathrm{UAV}(j))$, it will take it to fly from the point where it received the request to the point where it would provide support. It also takes into account the amount of fuel UAV(*j*) has left at the time of the request, denoted $fuel(\mathrm{UAV}(j),t)$; UAV(*j*)'s fuzzy concept of price denoted as "*price*", and UAV(*j*)'s fuzzy concept of "*mission priority*" at time, t. Let the set of relevant UAV properties be denoted as UAV_prop and be defined as

$$UAV\_prop = \{help\_time, fuel, mission\_prio, price\} \tag{20}$$

The fuzzy priority for helping denoted as μ_{PH} takes the form

$$\mu_{\mathrm{PH}}(\mathrm{UAV}(i),\mathrm{UAV}(j)) = \sum_{\alpha \in \mathrm{UAV\_prop}} w_\alpha \cdot \mu_\alpha(\mathrm{UAV}(j)) \tag{21}$$

The quantities w_α and μ_α for $\alpha \in UAV\_prop$ are expert defined weights and fuzzy membership functions, respectively. The fuzzy membership functions are defined in (22–25) and given below,

$$\mu_{help\_time}(\mathrm{UAV}(i),\mathrm{UAV}(j)) = \left[\frac{help\_time(\mathrm{UAV}(j))}{\max\limits_{k \in help(i,t)} \{help\_time(\mathrm{UAV}(k))\}} + 1\right]^{-1} \tag{22}$$

$$\mu_{fuel}(\mathrm{UAV}(i),\mathrm{UAV}(j)) = \frac{fuel(\mathrm{UAV}(j))}{\max\limits_{k \in help(i,t)} \{fuel(\mathrm{UAV}(k))\}} \tag{23}$$

$$\mu_{mission\_prio}(\mathrm{UAV}(i),\mathrm{UAV}(j)) = \left[\frac{mission\_prio(\mathrm{UAV}(j))}{\max\limits_{k \in help(i,t)} \{mission\_prio(\mathrm{UAV}(k))\}} + 1\right]^{-1} \tag{24}$$

$$\mu_{price}(\mathrm{UAV}(i),\mathrm{UAV}(j)) = \left[\frac{Value(\mathrm{UAV}(j))}{\max\limits_{k \in help(i,t)} \{Value(\mathrm{UAV}(k))\}} + 1\right]^{-1} \tag{25}$$

It is assumed that all evaluations are processed at time, t, so time dependence is suppressed in (21–25) for notational convenience. A more sophisticated version of the control logic that takes path risk, changes in risk, UAV reliability, UAV risk-tolerance and missed sample points into account will be the subject of a future publication.

5 Computational Experiments

The planning and control algorithms described in the previous sections have been the subject of a large number of experiments. This section provides a description of a small subset of these experiments. They serve to illustrate how the algorithms were tested. Due to space limitations only experiments involving up to three UAVs are discussed.

UAV experiments using only one UAV demonstrate how the planning and control algorithm will determine the route the UAV flies so that it is successful in making measurements at sample points in space, while the UAV avoids taboo points, that is points in space that could damage or destroy the UAV. Experiments using two UAVs illustrate how the control algorithm allows the UAVs to automatically support each other to increase the probability their joint mission is successful.

Figures 1–4 use the same labeling conventions. Sample points are labeled by concentric circular regions colored in different shades of gray. The lighter the shade of gray used to color a point, the lower the point's grade of membership in the fuzzy concept "desirable neighborhood." The legend provides numerical values for the fuzzy grade of membership in the fuzzy concept "desirable neighborhoods". If the fuzzy degree of desirability is high then the index of refraction is considered to be close to the index of refraction of the sample point at the center of the desirable neighborhood. This allows the UAV to make significant measurements while avoiding undesirable neighborhoods.

Each sample point is labeled with an ordered pair. The first member of the ordered pair provides the index of the sample point. The second member of the ordered pair provides the point's priority. For example, if there are n_{sp} sample points and the qth sample point is of priority p, then that point will be labeled with the ordered pair (q,p).

Points surrounded by star-shaped neighborhoods varying from dark grey to white in color are taboo points. As with the sample points, neighborhoods with darker shades of gray have a higher grade of membership in the fuzzy concept "undesirable neighborhood." The legend provides numerical values for the fuzzy grade of membership in the fuzzy concept "undesirable neighborhood." UAVs with high risk tolerance may fly through darker grey regions than those with low risk tolerance. When comparing planning and associated control pictures, if a point ceases to be taboo, the neighborhood where it resides is marked by a very dim gray star as well as being labeled by a dialog box as being an "old taboo point." New taboo points and their associated undesirable neighborhoods are labeled with dialog boxes indicating that they are "new."

UAVs start their mission at the UAV base which is labeled with a diamond-shaped marker. They fly in the direction of the arrows labeling the various curves in Figs. 1–4.

Figure 1 provides the sample points, taboo points and sample path for one UAV as determined by the planning algorithm. It is important to notice that the UAV's path passes directly through each sample point, i.e., through the center of the concentric circular regions representing the fuzzy degree of desirability of neighborhoods. Fortuitously, the taboo points and their neighborhoods are so positioned that they do not interfere with the UAV's measurement process or its return to base.

Figure 2 depicts the actual path the UAV flies as determined by the UAV's real-time control algorithm. The path determined by the control algorithm differs from the one created by the planning algorithm due to real-time changes in taboo points. After leaving the UAV base new weather data was acquired informing the UAVs that the exact position of the third sample point, i.e., the one labeled (3,1) actually resides within an undesirable neighborhood. Due to the high priority of the sample point and the UAV's risk-tolerance, the UAV flies into the taboo points' undesirable neighborhood as indicated in Fig. 2.

In both the planning and control algorithms the UAV measures sample points of two different priorities, with the direction of the flight path selected so that the higher priority points are measured first. By measuring high priority points first, the likelihood of an important measurement not being made is diminished, if the UAV can not complete its mission due to a malfunction, change in weather, etc.

Also, due to movement of old taboo points or the emergence of new taboo points which are marked "New," the path determined for the UAV using the control algorithm is significantly different than the one created by the planning algorithm. The path change represents the control algorithm's ability to reduce UAV risk.

Figure 3 depicts the sampling path determined by the planning algorithm for an experiment involving two UAVs. The first, UAV(1) follows the dashed curve; the second, UAV(2), the solid curve. The UAVs were assigned to the different paths by the fuzzy path assignment decision rule described in Sect. 3. UAV(1) is assigned to sample all the highest priority points, i.e., the priority one points. UAV(2) samples the lower priority points, i.e., those with priority two. Due to the greedy nature of the point-path assignment algorithm, the highest priority points are assigned for sampling first.

Figure 4 depicts the actual flight path the UAVs take during real-time. Initially, UAV(1) is successful in measuring sample points one and two as assigned it by the planning algorithm. Just beyond sample point two, UAV(1) experiences a malfunction. UAV(1)'s real-time control algorithm subsequently sends out a help request informing the only other UAV in the field, UAV(2) of the malfunction. UAV(2)'s control algorithm determines a new path for UAV(2) to fly so that the priority one points, labeled (3,1) and (4,1), that UAV(1) was not able to sample are subsequently measured. After UAV(2) measures sample point five, its new flight path allows it to measure sample points three and four. UAV(2)'s control algorithm determined it was very important that these priority one points be measured. Unfortunately, due to the extra fuel expended in reassigning sample points three and four to UAV(2), UAV(2) did not have enough fuel to measure sample points seven and eight which were of priority two. UAV(2)'s real-time control algorithm determined the best possible solution in the face of changing circumstances and limited resources.

It is important to note that the control algorithms running on UAV(1) and UAV(2) direct both UAVs to alter their return paths to the base due to the emergence of new taboo points making the planning algorithm determined flight paths too dangerous. The control algorithm uses each UAV's fuzzy risk-tolerance to determine how near each UAV may approach a taboo point.

Figure 5 provides an example of the AUP decision tree's assignment of three UAVs to three paths. The highest priority locations are assigned to UAV(1) as it has the greatest fuel capacity, i.e., 90 minutes. UAV(1) however does not have enough fuel to handle the high priority points located at positions six and seven and therefore UAV(2) is assigned these points along with the second degree high priority locations.

Table 1 provides numerical details of the tasks depicted in Fig. 5. The column labels have the following interpretation: "Location," the UAV coordinates on the map; "Fly mode," whether the UAV sampled from its previous location to its current position. If the UAV sampled then a "S" was entered. "NS" was entered if sampling did not occur. "Fuel Time" refers to how much fuel remained by the time the UAV reached the associated location.

Table 1 Details of three UAV mission depicted in Fig. 5

UAV 1 MISSION			UAV 2 MISSION			UAV 3 MISSION		
Locations	**Fly Mode**	**Fuel Time Remain (minutes)**	**Locations**	**Fly Mode**	**Fuel Time Remain (minutes)**	**Locations**	**Fly Mode**	**Fuel Time Remain (minutes)**
Base		90.0	Base		85.0	Base		85.0
(1,1)	NS	76.5088	(6,1)	NS	67.9691	(11,3)	NS	64.2839
(2,1)	S	61.5088	(7,2)	S	55.2412	(12,3)	S	51.0412
(3,1)	S	54.2662	(8,2)	S	47.9986	(13,3)	S	39.5559
(4,1)	S	42.7809	(9,2)	S	39.5133	(14,3)	S	31.0706
(5,1)	S	28.2956	(10,2)	S	22.028	Base	NS	6.2574
Base	NS	6.7113	Base	NS	11.7854			

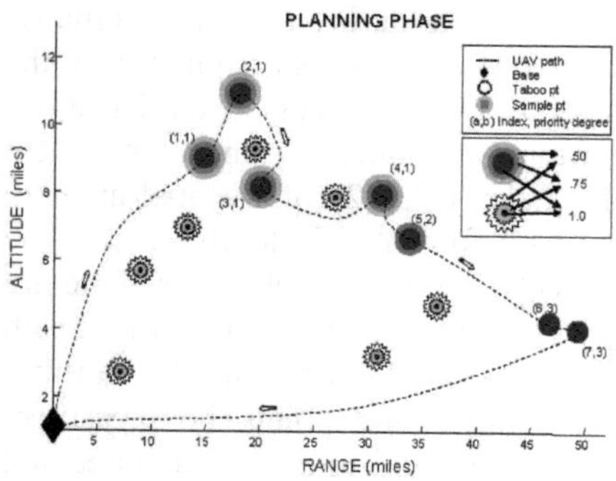

Fig. 1 One UAV trajectory as determined by the planning algorithm.

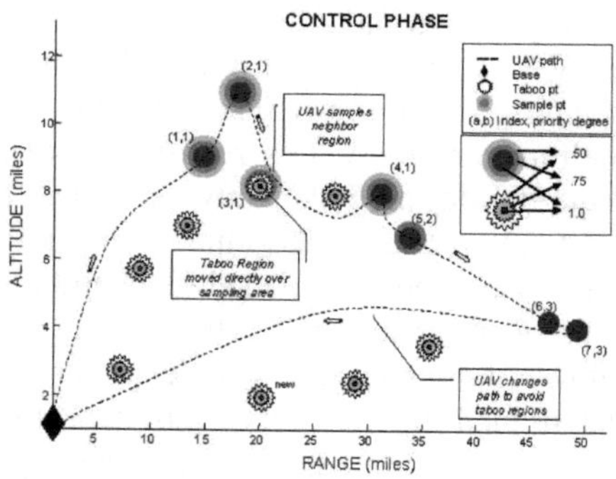

Fig. 2 One UAV trajectory as determined by the real-time algorithm.

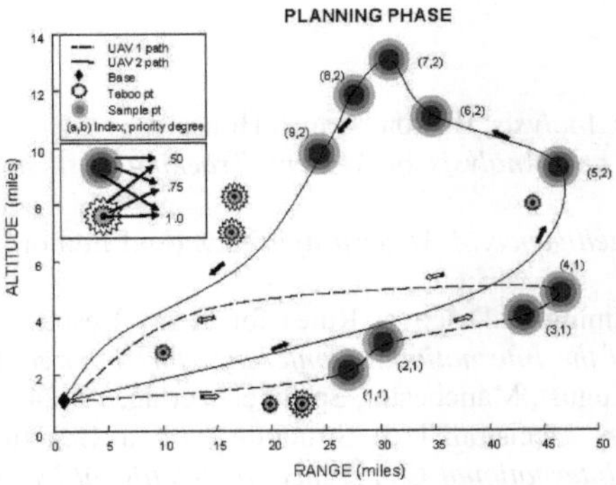

Fig. 3 Trajectory of two UAV as determined by the planning algorithm.

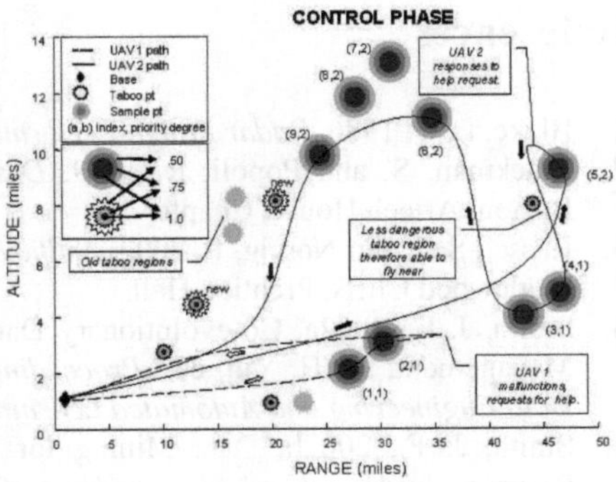

Fig. 4 During flight, updates about environmental changes cause real-time control algorithms on the two UAVs to change their trajectories.

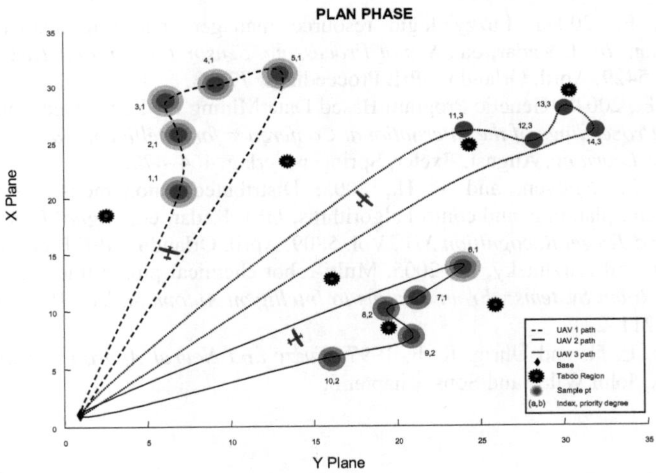

Fig. 5 Three UAV mission described in Table 1, an example of the AUP decision tree's assignments.

6 Summary

Fuzzy logic based planning and control algorithms that allow a team of cooperating unmanned aerial vehicles (UAVs) to make meteorological measurements have been developed. The planning algorithm including the fuzzy logic based optimization algorithm for flight path determination and the UAV path assignment algorithm are discussed. The control algorithm also uses these fuzzy logic algorithms, but also allows three types of automatic cooperation between UAVs. The fuzzy logic algorithm for automatic cooperation is examined in detail. Methods of incorporating environmental risk measures as well as expert measures of UAV reliability are discussed as they relate to both the planning and control algorithms. Experimental results are provided. The experiments show the algorithms' effectiveness.

References

1. Blake, L. V. 1986. *Radar Range-Performance Analysis*, Boston, Artech House.
2. Blackman, S. and Popoli, R. 1999. *Design and Analysis of Modern Tracking Systems*, Boston, Artech House, Chapter 11.
3. Russel, S. J. and Norvig, P. 2002. *Artificial Intelligence: A Modern Approach* (2nd Edition), Englewood Cliffs, Prentice-Hall.
4. Smith, J. F., 2002a. Co-evolutionary Data Mining to Discover Rules for Fuzzy Resource Management, *In:* H. Yin, ed., *Proceedings of the International Conference for Intelligent Data Engineering and Automated Learning*, August, Manchester, Springer-Verlag, 19–24.
5. Smith, J. F., 2002b. Data Mining for Fuzzy Decision Tree Structure with a Genetic Program, *In:* H. Yin, ed., *Proceedings of the International Conference for Intelligent Data Engineering and Automated Learning*, August, Manchester, Springer-Verlag, 13–18.
6. Smith, J. F., 2003. Fuzzy logic resource manager: decision tree topology, combined admissible regions and the self-morphing property, *In:* I. Kadar ed., *Signal Processing, Sensor Fusion, and Target Recognition XII*, Vol. 5096, April, Orlando, SPIE Proceedings, 104–114.
7. Smith, J. F., 2004a. Fuzzy logic resource manager: real-time adaptation and self organization, *In:* I. Kadar, ed., *Signal Processing, Sensor Fusion, and Target Recognition XIII*, Vol. 5429, April, Orlando, SPIE Proceedings, 77–88.
8. Smith, J. F., 2004b. Genetic Program Based Data Mining for Fuzzy Decision Trees, *In:* H. Yin, ed., *Proceedings of the International Conference for Intelligent Data Engineering and Automated Learning,* August, Exeter, Springer-Verlag, 464–470.
9. Smith, J. F., Nguyen, and T. H., 2005. Distributed autonomous systems: resource management, planning, and control algorithms, *In:* I. Kadar, ed., *Signal Processing, Sensor Fusion, and Target Recognition XIV*, Vol. 5809, April, Orlando, SPIE Proceedings, 65–76.
10. Spears, D. and Zarzhitsky, D., 2005. Multi-robot chemical plume tracing, *In:* A. Schultz, ed. *Multi-Robot Systems: From Swarms to Intelligent Automata*, Vol. III, May, New York, Springer, 211–221.
11. Tsoukalas, L. H. and Uhrig, R. E. 1997. *Fuzzy and Neural Approaches in Engineering*, New York, John Wiley and Sons, Chapter 5.

Neural Network Model Based on Fuzzy ARTMAP for Forecasting of Highway Traffic Data

D. Boto-Giralda, M. Antón-Rodríguez, F.J. Díaz-Pernas and J.F. Díez-Higuera

Departamento de Teoría de la Señal, Comunicaciones e Ingeniería Telemática
ETSIT Universidad de Valladolid, Campus Miguel Delibes s/n, 47011 Valladolid, España
danbot@tel.uva.es, mirant@tel.uva.es, pacper@tel.uva.es, josdie@tel.uva.es

Abstract. In this chapter, a neural network model is presented for forecasting the average speed values at highway traffic detectors locations using the Fuzzy ARTMAP theory. The performance of the model is measured by the deviation between the speed values provided by the loop detectors and the predicted speed values. Different Fuzzy ARTMAP configuration cases are analysed in their training and testing phases. Some ad-hoc mechanisms added to the basic Fuzzy ARTMAP structure are also described to improve the entire model performance. The achieved results make this model suitable for being implemented on advanced traffic management systems (ATMS) and advanced traveller information system (ATIS).

Keywords. Fuzzy ARTMAP, travel cost estimates, ATIS.

1 Introduction

Traditional models of traffic congestion and management lack the adaptability and sophistication needed to effectively and reliably deal with increasing traffic volume on certain road stretches. A realistic estimate of planned routes travel cost with reasonable accuracy is essential for successful implementation on an advanced traveller information system (ATIS) for use in an intelligent transportation system (ITS). An ATIS consists of a route guiding system (RGS) that recommend the most suitable route based on the traveller's requirements, using the information gathered from various sources as loop detectors and probe vehicles. The success of an RGS will depend on its ability to predict the anticipatory travel cost in addition to the historical and real-time travel cost [4].

Several aspects should be taken into account to evaluate the travel cost such as distance, time, economy, danger or personal preferences. From the distance point of view, the travel cost quantification will be strictly static, only dependent on the sum of the stretches length. A time based estimate will be dynamic and dependent on multiple factors. It could be directly measured or by the distance-speed relationship. For a economic estimate, toll fares, vehicle consumption and wear will be considered. Road accident risks as well as driving easiness at some particular stretches might be a decisive factor to rule out a route. Finally the traveller's preferences for route services or particular scenarios such as mountain or landscape roads could affect the decision eventually. This chapter will focus on speed estimate in road stretches with traffic detectors using a Fuzzy ARTMAP neural network structure. As said before, speed may be used to calculated the travel time cost as long as distance is known.

Neural network computing applied to travel cost forecast appeared to overcome the shortcomings of preceding methods whose forecasts deteriorate over multiple time steps [5]. A neural network provides a mapping between a set of inputs and corresponding outputs [1]. The network is trained to learn this mapping using a number of training examples. Backpropagation (BP) is the most widely used neural network model in civil engineering applications, primarily due to its simplicity. However, backpropagation has shortcomings, including a very slow rate of convergence and arbitrary and problem-dependent selection of the learning and momentum ratios [2].

A neural model for forecasting the freeway link travel time using counter propagation neural (CPN) network is presented in [4]. There, it was showed that CPN model was nearly two orders of magnitudes faster than BP training algorithm for the same level of accuracy. In this chapter, a neural network model based on Fuzzy ARTMAP is presented for forecasting the average speed values at highway traffic detectors locations. Faster as the aforesaid CPN model, the presented model gives lightly better average errors in forecasting values in more realistic both training and testing scenarios.

2 Fuzzy ARTMAP Basis

The Fuzzy ARTMAP, introduced by Carpenter et al. [3] is a supervised network composed of two Fuzzy ARTs (ART_a and ART_b) interconnected by a series of connections between their output layers. Each connection has an associated weight value (w_{ij}) between 0 and 1, and may be considered as the membership function value in the fuzzy sets theory of the corresponding network category.

These connections form what is called the map field F_{ab}. The weights of the map field are all initialised to 1. The map field has two parameters: the learning rate β_{ab}, and vigilance criterion ρ_{ab}, and an output vector x_{ab}. Figure 1 shows a graphic sample representation of a Fuzzy ARTMAP network.

The input data of both ART_a and ART_b are normalized values, between 0 and 1 (minimum and maximum expected input values respectively), and form the network input vectors a and b. This normalization ensures a proportional response of the network from the input data. Input vector of ART_a is put in complement coding form, resulting in vector A. Complement coding is not necessary in ART_b so the input vector B directly presented to the network.

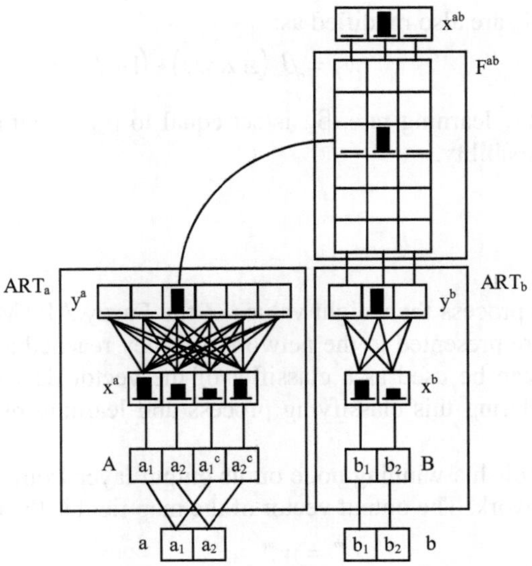

Fig. 1 Sample Fuzzy ARTMAP network.

2.1 Training

Fuzzy ARTMAP networks usage requires a training process before being able to classify input data. In this process, a vector representing a data pattern is presented to ART_a, and a vector which is the desired output corresponding to this pattern is presented to ART_b. The relationship between these two vectors is learned through the weight values of the map field. The vigilance criterion of ART_a, ρ_a, varies during learning from a initial value called the baseline vigilance $\overline{\rho}_a$. The vigilance parameter of ART_b, ρ_b, is set to 1 to perfectly distinguish the desired output vectors.

When vectors A and B are presented to ART_a and ART_b, both networks soon enter resonance. The map field vigilance criterion is then evaluated to verify if the winning neuron of ART_a corresponds to the desired output vector presented to ART_b. This criterion is:

$$\frac{\left|y^b \wedge w_J^{ab}\right|}{\left|y^b\right|} \geq \rho_{ab} \tag{1}$$

where y^b is the output vector of ART_b, J is the index of the winning neuron on the output layer of ART_a, w_{ab}^{J} corresponds to the weights of the connections to the *J*th neuron of the output layer of ART_a and $\rho_{ab} \in [0,1]$ is the vigilance criterion of the map field. If the criterion is not respected, the vigilance of ART_a is increased just enough to select another winning neuron ($\rho_a > |A \wedge w_J| / |A|$) and the vector A is repropagated in ART_a.

When the vigilance criterion is respected, the vigilance value of ART_a is set to its initial baseline value $\overline{\rho}_a$ and the map field learns the association between vectors A and B by modifying its weights as follows:

$$w_J^{ab} = \beta_{ab} x^{ab} + \left(1 - \beta_{ab}\right) w_J^{ab} \tag{2}$$

The weights in ART_a are also modified as:

$$w_J = \beta_a (A \wedge w_J) + (1 - \beta_a) w_J \tag{3}$$

In practice, the ART_a learning rate, β_a, is set equal to β_{ab}, or simply β, defining the learning network capability.

2.2 Classifying

During the training process the weight values of the Fuzzy ARTMAP were updating, as new patterns were presented to the network, till they reached a final value. At this point the network can be used as a classifier of the vector data presented to ART_a. ART_b is not used during this classifying process and learning network capability is deactivated (β=0).

ART_a will establish a winning node on its output layer from each input vector A presented to the network. The output vector of the map field is then set to:

$$x^{ab} = w_J^{ab} \tag{4}$$

where J is the index of the winning node on output layer of ART_a, and $w_{ab}{}^J$ is the corresponding weight values vector on the map field . The index J of this component is the number of the category in which the input vector A has been classified. The use of the map field is thus to associate a category number to each neuron of ART_a's output layer. However, not just the category that best fits an input is the only result of the classifying process. The weight values associated with this category may also be useful to get ulterior information about the relationship between the input vector and the categories learned by the network in the training process.

3 Working Model

3.1 Training the Network

The data for this experiment were collected through the Freeway Performance Measurement System (PeMS) project, and could be obtained thanks to the Next Generation Simulation (NGSIM) and Federal Highway Administration (FHWA) web page at http://ngsim.fhwa.dot.gov. PeMS project was conducted by the Department of Electrical Engineering and Computer Sciences at the University of California, at Berkeley, with the cooperation of California Department of Transportation. Available data from 5 detector stations on US 101 South for 11 days, from June 8 to June 22, excluding the weekends, are provided in this data set. Speed, volume and occupancy at each detector for the 5-min time step are presented at each detector in each lane.

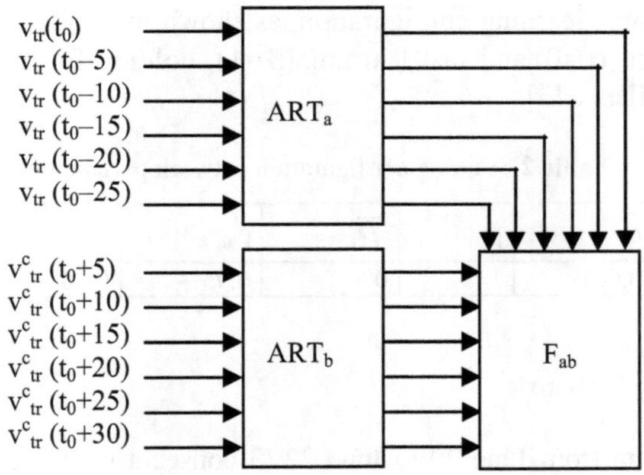

Fig. 2 Fuzzy ARTMAP training structure. $v_{tr}(t)$, training speed value at instant t; $v^c_{tr}(t)$, training speed categorized value at instant t.

Average speed data from June 13 to June 17 (5 consecutive weekdays, making a total of 1440 samples) collected from two stations with different traffic congestion levels (717486, light; 717489, heavy), were employed to train the net. Figure 2 shows the structure of the Fuzzy ARTMAP during this process.

Table 1 Training cases

Case	Input training subsets time step (min)	ART_a input nodes	ART_b input nodes	No. of Categories
A	6*5	6	1	9
B	6*5	6	1	81
C	6*5	6	6	81
D	6*5	4	4	81
E	6*5	8	8	81
F	1*5	6	6	81

Six different structure model cases, shown in Table 1, were considered with different number of input and output nodes, categories and time step between consecutive input training subsets. In Case A, six ART_a input nodes and one ART_b input node were used to associate six consecutive 5-min time step normalized past speed values to one categorized future speed value in the map field of the Fuzzy ARTMAP, F_{ab}. This categorized speed value is calculated as the average of the six 5-min time step speed values following the normalized speed values presented to ART_a, and categorized into one of 9 possible categories. Numerically, these categories are linearly spaced and normalized speed values in the range of 0–80 mph. In Case B, and following ones, the number of categories were increased to 81. In Case C, each six consecutive 5-min time step normalized past speed values presented to ART_a were associated to six consecutive 5-min time step normalized future speed values presented to ART_b. In Case D and E the number of input nodes were changed to 4 and 8 respectively. Finally in Case F, with six input nodes anew, the consecutive sets of values presented to the ART_a and ART_b were 5 min ahead of the former, instead of 30 min.

A one-shot stable learning configuration, as shown in Table 2, has been adopted: conservative limit ($\alpha \cong 0$) and fast learning ($\beta = 1$), holding for fuzzy ART modules with constant vigilance [3].

Table 2 Training configuration network parameters

α	β	$\overline{\rho}_a$	ρ_{ab}	ε
0.001	1	0	0.95	0.001

3.2 Testing the Network

Average speed data from June 20 to June 22 (3 consecutive weekdays following the training ones, making a total of 864 samples) collected from the same two stations that in the training process and shown in station forecasting speed error. Figure 4a and 5a, were employed to make a test of the speed forecasting net capability. A test performance was made for each training case.

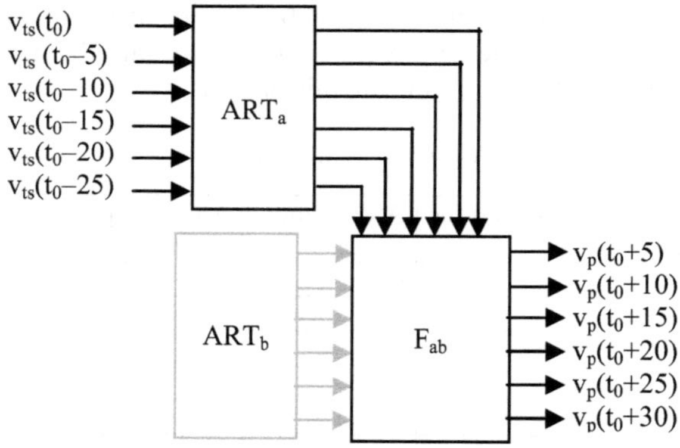

Fig. 3 Fuzzy ARTMAP testing structure. $v_{ts}(t)$, testing speed value at instant *t*; $v_p(t)$, forecasting speed value for instant *t*.

Figure 3 shows the structure of the Fuzzy ARTMAP during this process. Sets of consecutive 5-min time step normalized speed testing values were presented to ART_a. ART_b in testing phase is not used. A number of forecasting speed values, equals to the number of input nodes in ART_b in the training phase, were obtained from each set. These forecasting speed values were calculated from the weight vectors of the map field F_{ab}, w_{ab}, multiplying the selected weight vector, w_{ab}^J, by the speed value associated to the higher training category.

The fuzzy ARTMAP configuration for the testing phase was similar to the one adopted in the training phase but with the learning capability deactivated ($\beta = 0$), as shown in Table 3.

Table 3 Testing configuration parameters

α	β	$\overline{\rho}_a$	ρ_{ab}	ε
0.001	0	0	0.95	0.001

3.3 Forecasting Results

The Fuzzy ARTMAP model have been implemented in MATLAB® Release 12 technical language on a mobile AMD Athlon™ XP 2000+ computer. In order to measure the forecasting accuracy, an average error term was defined in the following form:

$$E(\%) = \frac{100}{N} \sum_{t=1}^{N} \frac{\left| v_t[t] - v_p[t] \right|}{v_t[t]} \tag{5}$$

where N is the number of predicted speed values; $v_p[t]$, the predicted speed value for moment t; $v_t[t]$, the testing speed value measured by the station detector at moment t.

Figures 4 and 5 show the forecasting speed (b) and the forecasting error (c) values over the testing days time. The maximum error values occur close to high traffic congestion situations in 717489 station, when vehicles speed changes too fast in the 5-min step time. This maximum error values are dramatically high but are quickly reduced as the next forecasting speed values are available. Hence, global error performance keeps a satisfactory level. Table 4 shows the average error in forecasting speed for the considered cases.

Table 4 Average error in forecasting speed

Station	E(%)					
	Case A	Case B	Case C	Case D	Case E	Case F
717486	4.12	2.91	3.16	3.77	3.12	2.64
717489	14.78	13.95	10.96	9.67	15.84	7.78

Case B improves Case A forecasting precision by simply increasing the number of categories, particularly with the 717486 station in which all speed values concentrate in a range of 30 mph.

Forecasting precision for 717489 station, in which speed values change quickly close to high congestion situations, strongly improves in Case C as more predicted speed values are given (six instead of one) for the 30-min forecasting time interval considered. Applying no interpolation rule, six is the highest number of predicted values since detectors present new data each 5 min.

The increase in the number of nodes in Case E implies that both the time interval of past speed values presented to the net and the time interval of forecasting speed values are longer, since the number of nodes in ART_a and ART_b were equal in the training process. For the 717489 station, with speed values changing quickly close to high traffic congestion situations, the longer the forecasting time interval , the bigger the error will be. In Case D, the opposite situation occurs but the processing time increases substantially for the same forecasting time interval. No significantly error variation for 717486 station in Cases D and E.

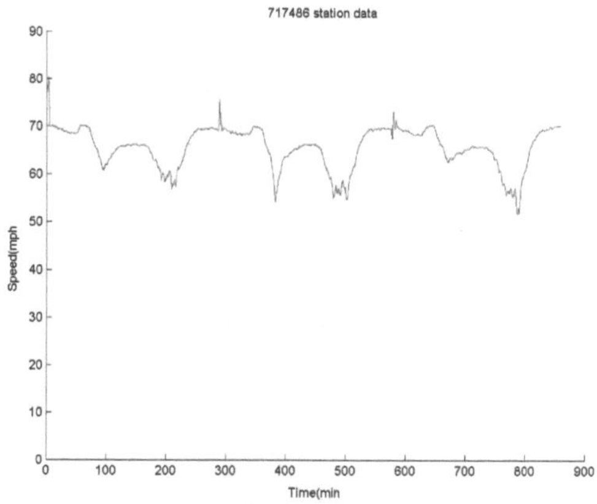

(**a**) 717486 station detector speed data

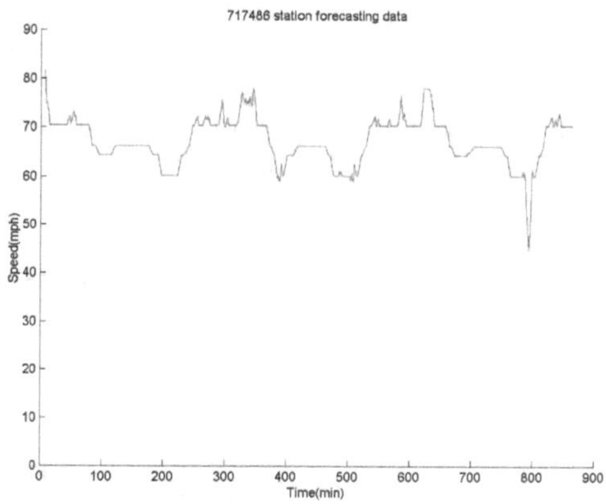

(**b**) 717486 station forecasting speed data

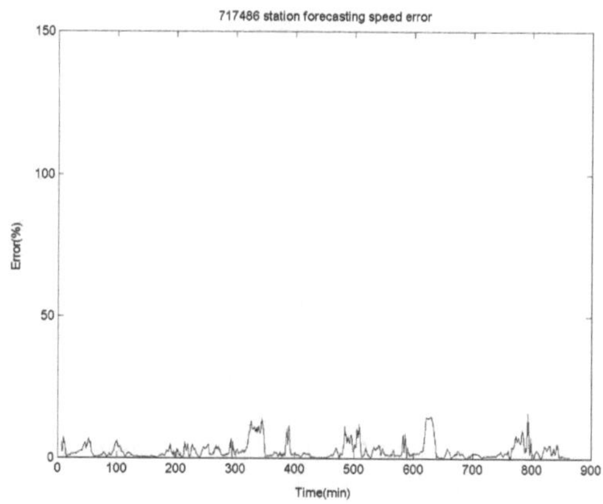

(**c**) 717486 station forecasting speed error

Fig. 4 717486 station data and forecasting results.

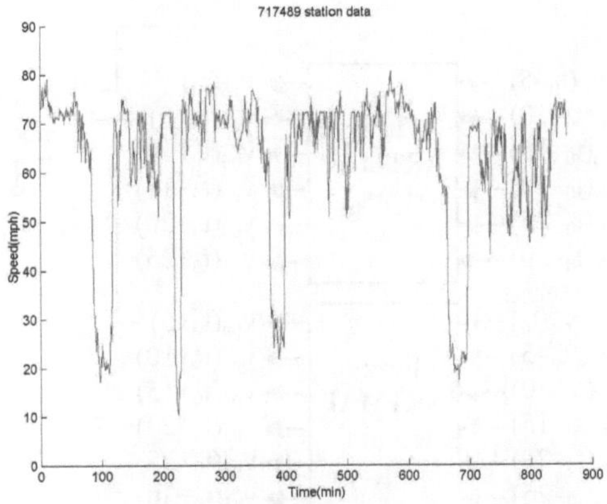

(a) 717489 station detector speed data

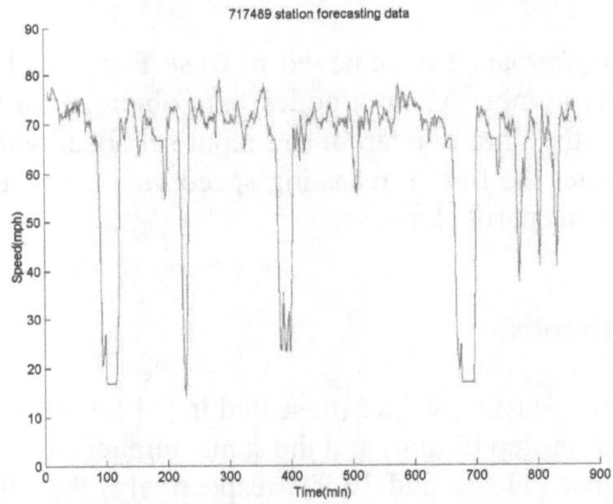

(b) 717489 station forecasting speed data

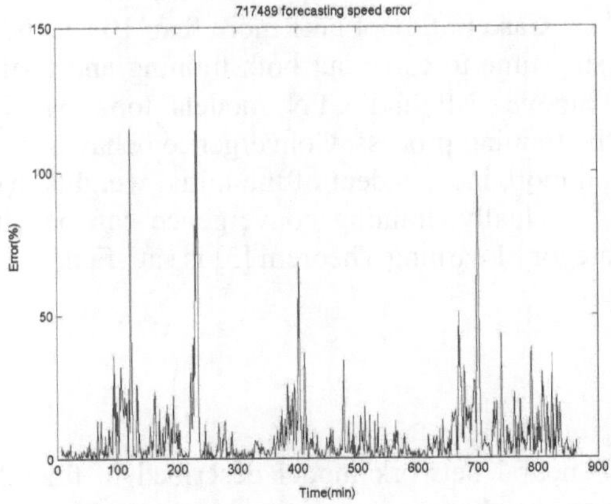

(c) 717489 station forecasting speed error

Fig. 5 717489 station data and forecasting results.

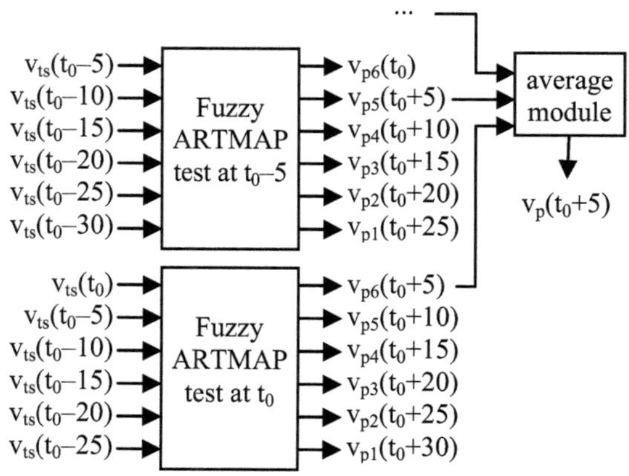

Fig. 6 Case F model structure. $v_{ts}(t)$, testing speed value at instant t; $v_{pi}(t)$, ith forecasting speed value for instant t; $v_p(t)$, averaged forecasting speed value for instant t.

The best error performance is achieved in Case F in which several forecasting speed values (up to the number of input nodes) are obtained for a particular moment into the future (due to the time overlap of the input set speed values). An average of them is then made to get the final forecasting speed value. Figure 6 shows the model structure designed for this particular case.

3.4 Comparative Results

The average errors in forecasting values presented in [4] for BP and CPN models with the same duration of time step (5 min) and the same number of input and output nodes (6), were slightly higher (11.5% and 10.9% respectively) than the one achieved for the Case F (7.8% for the station with the most congested traffic level) in the Fuzzy ARTMAP model. However, speed travel values, instead of travel time values, were predicted in the model presented in this chapter and real traffic data were used instead of simulated traffic data. Case F did not take more than 10 seconds (a rough measure was made) of processing time to carry out both training and testing phases with the conditions described above. BP and CPN models took 312.7 and 3.8 seconds respectively just for the training process. Convergence behaviour of Fuzzy ARTMAP networks are faster and more independent of the initial weights than Back or Counter propagation networks. Actually, training convergence can be guaranteed as far as Fuzzy ART Stable Category Learning Theorem [3] is satisfied.

4 Conclusions

The Fuzzy ARTMAP neural network model described in this chapter provides an appropriated forecasting travel cost mechanism, in terms of average speed values for being integrated in travel cost estimates systems supplied with traffic dynamic parameters such as speed, occupancy or volume data. Multiple training and working configurations for the network are possible in order to match host system requirements, all of them with a remarkable time processing and forecasting error performance. Forecasting test results obtained accuracy levels under the 8% of precision from real congested highway traffic data. A figure slightly lower than previous neural network models developed for highway traffic predictions. So it represents a promising challenge in the evolution of neural networks appliance to intelligent transportation system (ITS).

Acknowledgements

NGSIM Website – Home of the Next Generation Simulation Community, at http://ngsim.camsys.com – for the traffic data set.

References

1. Adeli, H. and Hung, S. L., 1995. Machine Learning-Neural Networks, Genetic Algorithms, and Fuzzy Systems. Wiley, New York.
2. Adeli, H. and Hung, S. L., 1994. "An adaptive conjugate gradient learning algorithm for efficient training of neural networks". Applied Mathematics and Computation 62 (1), 81–100.
3. Carpenter, G. A. et al., 1992. "Fuzzy ARTMAP: A neural network architecture for incremental supervised learning of analog multidimensional maps". Neural Networks, IEEE Transactions on Volume 3, Issue 5, September, pp. 698–713.
4. Dharia, A. and Adeli, H., 2003. "Neural network model for rapid forecasting of freeway link travel time". Engineering Applications of Artificial Intelligence, Volume 16, Issues 7–8, October–December, pp. 607–613.
5. Park, D. and Rilett, L.R., 1999. "Forecasting freeway link travel times with a multilayer feedforward neural network". Computer-Aided Civil and Infrastructure Engineering 14 (5), 357–367.

Automated Generation of Optimal Controllers through Model Checking Techniques

Giuseppe Della Penna[1], Daniele Magazzeni[1], Alberto Tofani[1], Benedetto Intrigila[2]
Igor Melatti[3] and Enrico Tronci[3]

[1]Dipartimento di Informatica, Università di L'Aquila, Italy
{dellapenna,magazzeni,tofani}@di.univaq.it
[2]Dipartimento di Matematica Pura ed Applicata, Università di Roma "Tor Vergata", Italy
intrigil@mat.uniroma2.it
[3]Dipartimento di Informatica, Università di Roma "La Sapienza", Italy
{melatti,tronci}@di.uniroma1.it

Abstract. We present a methodology for the synthesis of controllers, which exploits *(explicit) model checking techniques*. That is, we can cope with the systematic exploration of a very large state space. This methodology can be applied to systems where other approaches fail. In particular, we can consider systems with an *highly non-linear dynamics* and *lacking a uniform mathematical description (model)*. We can also consider situations where the required control action cannot be specified as a local action, and rather a kind of *planning* is required. Our methodology individuates first a raw optimal controller, then extends it to obtain a more robust one. A case study is presented which considers the well known *truck-trailer obstacle avoidance parking problem*, in a parking lot with *obstacles* on it. The complex non-linear dynamics of the truck-trailer system, within the presence of obstacles, makes the parking problem extremely hard. We show how, by our methodology, we can obtain optimal controllers with different degrees of robustness.

Keywords. Controller synthesis, controller optimization, model checking, nonlinear systems.

1 Introduction

Control systems (or, shortly, *controllers*) are small hardware/software components that control the behavior of larger systems, the *plants*. A controller continuously analyzes the plant state (looking at its *state variables*) and possibly adjusts some of its parameters (called *control variables*) to keep the system in a condition called *setpoint*, which usually represents the *normal or correct* behavior of the system.

In the last years, the use of sophisticated controllers has become very common in robotics, critical systems and, in general, in the hardware/software *embedded systems* contained in a growing number of everyday products and appliances.

However, since the primary aim of a controller is to ensure the correct behavior of the controlled plant, we have to guarantee the *efficiency* and the *robustness* of controllers. By *efficiency*, we mean the capability of the controller to bring the system to the set point in the shortest possible time (also called *time optimality*). By *robustness*, we mean the capability of the controller to perform well when the state variables vary outside the design range.

Therefore, the verification of these properties is a crucial task that is being addressed using different formal methodologies (e.g., model checking and theorem proving) developed in different research communities (e.g., automata theory and artificial intelligence [1]). In particular, much work is being done to provide a methodology for the automatic (or semi–automatic) synthesis of correct controllers directly from the plant specifications.

However, efficiency and robustness can hardly be simultaneously fulfilled, especially in critical systems, where the system dynamics is difficult to understand and to control.

Therefore a possible approach for approximate a correct solution, can be to start by synthesizing a very efficient or even optimal controller and – as a second step – to make it robust. (Observe, however, that in case *safety* is involved, a better approach could be to choose the best controller in the class of the *safe* ones [2]).

To this aim, in this chapter we describe an automatic methodology composed by:

1. a procedure for the synthesis of an optimal controller;
2. a procedure for the transformation of the optimal controller into a robust one.

Our methodology exploits explicit model checking in an innovative way w.r.t the approaches adopted so far, especially in the AI planning area. In particular, the first procedure – looking for an *optimal* solution – actually can also be considered as a *planner*. Indeed, this procedure does not simply individuate a good *local* move, but searches for *the best possible sequence of actions* to bring the plant to the setpoint. Therefore, it can be also used as a planner, though we will not pursue this point in this chapter.

2 Controller Synthesis Techniques

In this section we recall some basic notions about controller synthesis and describe the main results presented in this field by the recent research.

2.1 Controller Synthesis

There are a number of well-established techniques for the synthesis of controller. For short, we mention only three of them:

1. PID controllers;
2. fuzzy controllers;
3. dynamic programming techniques.

As it well-known [3], PID-based techniques are very effective for linear systems, while they badly perform w.r.t. non-linear ones. On the contrary our technique is able to cope with such systems, as shown in the case study.

Fuzzy control is well known as a powerful technique for designing and realizing control systems, especially suitable when a mathematical model is lacking or is too complex to allow an analytical treatment [4] [5]. However fuzzy rules correspond to *local* actions, so that, in general, they do not result in an *optimal* controller. Moreover there are situations where local actions are not viable at all, and rather a kind of *planning* is required. For an example of such a situation see the case study.

Dynamic programming techniques are very suitable for the generation of optimal controllers [6] [7]. Although our methodology has a *dynamic programming flavor*, it can cope with (and it is especially suitable for) very rough plant descriptions, whose mathematical definition cannot be adapted to the dynamic programming preconditions, when the cost function cannot be decomposed or the system dynamics function cannot be inverted. Again, see the case study, where a backward decomposition of the cost function (in this case, the length of the path) is hard to perform, due to the complexity of the system dynamics function and to the presence of obstacles. Indeed, we performed a direct systematic analysis of the trajectories, using model checking techniques to support the required computational effort.

3 Optimal Controller Generation through Model Checking

Our objective is to build an optimal controller for a system (or plant) $\mathcal{S}$ which, at every state, has a limited number of allowed actions. Moreover, we suppose that $\mathcal{S}$ starts at a given interesting initial state s_0, and that the final goal is to bring $\mathcal{S}$ in a *goal* state (however, we can easily generalize to the case with n initial states). We recall that our optimality criterion is essentially the *time optimality*: that is, we want to bring $\mathcal{S}$ in a goal state in the smallest possible number of steps.

Thus, our controller has to be able to decide, for every state of $\mathcal{S}$ which is reachable from s_0, which is the action that brings to the nearest (w.r.t. the number of steps) goal state. The optimality of the action chosen implies the optimality of the generated controller. Note that forcing the controller to consider all the states reachable from s_0, instead of controlling only the states in the unique optimal path from s_0 to a goal state, allows us to handle the cases in which a bad move is made as a consequence of a given action.

In order to build such a controller, we consider the *transition graph* G of $\mathcal{S}$, where the nodes are the reachable states and a transition between two nodes models an allowed action between the corresponding states. In this setting, the problem of designing the optimal controller reduces to finding the minimum path in G between each state and the nearest goal state.

Unfortunately a transition graph for complex, real-world systems could be often huge, due to the well-known *state explosion* problem. Thus it is likely that G would not fit into the available RAM memory, and then the minimum path finding process could be highly time-consuming.

However, *Model Checking* techniques [8] [9] [10] developed in the last decades have shown to be able to deal with very huge state spaces. Thus, our idea is to reuse such model checking algorithms, reshaping them to be a controller generator. Note that in this chapter we focus on protocol-based hybrid systems, so we use model checking techniques based on an *on-the-fly explicit enumeration* of the system under analysis, since for such kind of systems these algorithms often outperform the *symbolic* ones [11].

More in detail, in our technique we have two phases, which we describe in the following.

3.1 Optimal Raw Controller Synthesis Phase

In the first phase, an explicit model checking algorithm is used, which performs a *Depth First* (DF) visit of all the reachable states of $\mathcal{S}$, starting from s_0. As usual, a *hash table* `HT` is used in order to store already visited states. Moreover, the *stack* holds, together with states, also the next action to be explored.

However, the DF visit is enriched in order to generate the controller `C`. To this aim, `HT` also stores, for each visited state, a flag `toGoal`, initially set to 0. When a goal state g is reached, then the states in the current path from s_0 to g (that is to say, the states currently on the stack) will have this flag set to 1, as soon as the visit backtracks to them. This is to signify that such states indeed reach a goal, and may be put in `C` – together with the action taken and the number of steps they need to reach the goal itself. In this way, when a state s with the `toGoal` flag set to 1 is reached, then we can analogously set the `toGoal` flag on all the states currently on the stack, and put them on `C`.

However, this scheme may fail in the following case. Suppose that a cycle $s_1 r_1 \ldots r_h s_2 t_1 \ldots t_k s_1$ is found, were s_1 and s_2 are on the stack. When analyzing t_k, the `toGoal` flag of s_1 is not set to 1, since we have not backtracked from s_2 yet. However, the visit is truncated, since s_1 is already visited, thus t_k will not be inserted in `C` (unless it was already present, or it reaches a goal through another path which does not intersect the stack).

To avoid this, a *predecessor table* `PT` is maintained for each state which is visited again while it is on the stack. We have that `PT` stores all the paths leading from a state on the stack and another state previously on the stack. Thus, in the situation described above, the path $r_1 \ldots r_h s_2 t_1 \ldots t_k$ is added to the predecessor table of s_1. Thus, when the DF visit of s_1 is finished, all the states in its predecessor table are added on `C` by using a backward visit, provided that s_1 indeed reaches a goal.

Finally, in order to preserve optimality, each insertion on `C` is effectively performed only if the number of steps to be inserted is less than the already stored one.

3.2 Controller Strengthening Phase

The second phase of our approach performs a *strengthening* of the controller `C` generated by the first phase. In fact, `C` only contains an *optimal plan* that can be used to drive $\mathcal{S}$ from s_0 to the goal. That is, `C` does not take into account any state outside the optimal plan. The final controller should be aware of a larger set of states: indeed, the dynamics of $\mathcal{S}$ can be very complex, and a particular setting of the control variables may not always drive $\mathcal{S}$ to the expected state. That is, all the state variables usually have a specific *tolerance*, and the reactions to controls are subject to these tolerances. For this reason, we refer to the controller `C` output by the first phase as a *raw controller*.

Therefore, to ensure the robustness of the controller, in the second *strengthening* phase we explore a larger number of states obtained by randomly perturbing the raw controller states. That is, for each state s in the raw controller table C, we apply a set of small random changes, bounded by the state variables tolerances, and obtain a new state s'. Then, from each new state s', we start a breadth first visit of the state space of $\mathcal{S}$ stopping as soon as we reach a state s'' that is already in C. The path from s' to s'' is stored in C and the process is restarted.

After some iterations of this process, we have that C is now able to drive $\mathcal{S}$ from any reasonable system state to the nearest state of the optimal controller and, from there, reach a goal. That is, C is now our final optimal controller.

4 The Controller Generation Process

The CGMurφ tool is an extended version of the CMurφ [12] [13] model checker. It is based on an explicit enumeration of the state space, originally developed to verify protocol-like systems. We choose CMurφ as a base to develop our controller generator since it already implements the most common state space compression techniques, such as *bit compression* [14] and *hash compaction* [15] [16], useful to decrease the memory requirements of the controller generation process when dealing with large-dimensional control systems. In particular, when bit compression is enabled, CMurφ saves memory by using every bit of the *state descriptor*, the memory structure maintaining the state variables, instead of aligning the state variables on byte boundaries (this saves on average 300% of memory). When using hash compaction, compressed values, also called *state signatures*, are used to remember visited states instead of full state descriptors. The compression ratio can be set to obtain an arbitrary state site (CMurφ default is 40 bits), but is *lossy*, so there is a certain probability that some states will have the same signature after compression.

Moreover, the CMurφ code is very easy to modify: indeed, in order to generate controllers for complex and hybrid systems we added to CGMurφ some important extensions, i.e., finite precision real numbers handling (to model systems with continuous variables) and external linking to C/C++ functions (to easily model the most complex aspects of the plant, or even interface the model with a complete plant simulator).

The behavioral part of the plant is modeled in CGMurφ through a collection of guarded *transition rules*, whereas the `goal` construct is used to define the *goal properties*, that is "normal" or "safe" states of the plant, i.e. the states that the controller should to bring (or maintain) the plant to.

In the following sections we describe the controller generation algorithm that is the core of CGMurφ. In particular, first we show the data structures used, then we illustrate the procedure for the synthesis and strengthening of the controller table.

4.1 Data Structures

The controller generation algorithm of CGMurφ uses the following data structures:

- the **stack** ST contains pairs (s, r), where s is a state and r is the index of last transition (i.e., CMurφ *rule*) fired from s.

```
DFS(state p) { //p is the start state
 //initialization (start state)
 if (isGoal(p)) return;
 stack_push(ST,(p,first_enabled_rule(p)));
 hashtable_store(HT,p);
 HT[p].toGoal = false; HT[p].inPT = false;
 //main DFS loop
 while (!stack_empty()) {
  (p,r) = stack_top(ST);
  if (r is not null) {
   s = apply_rule(p,r);
   stack_top(ST) = (p,next_enabled_rule(p,r));
   if (Insert(s,p,r)) UpdatePaths(s,p,r)
  } else { //r is null, no more rules for p
   UpdatePathsPt(p); stack_pop(ST);
 }} //while
 UpdatePathsFl();
}
```

Fig. 1 Extended CGMurφ depth first search.

- each slot of the **hash table** `HT` contains a (visited) state and two special flags: the *toGoal* flag indicates that a goal can be reached from this state in one or more steps (transitions), whereas the *inPT* flag is true if the state has been saved in the predecessors table.
- the **predecessors table** `PT` is an hash table storing (s,l) pairs where s is a state on the DFS branch being currently explored, and l is the list of paths leading from an initial state to s. Each step of the path contains a state and the action that leads to that state from the previous step.
- the **final transitions list** `FL` stores paths to visited states (similarly to the predecessors table) that are discovered when such states are *outside* the current DFS branch. These paths are merged with the ones in the predecessors table to compute the shortest path to the corresponding states at the end of the state space exploration.
- the **controller table** `CTRL` contains, for each reachable system state s that leads (in one or more steps) to a goal, a pair $< r, c >$ indicating that the shortest path leading from s to a goal state has c steps, where the first step is the action given by rule r.

4.2 Optimal Raw Controller Synthesis Algorithm

The optimal raw controller synthesis algorithm, as shown in Fig. 1, consists of an extended depth-first visit of the plant state space. As in a standard DF visit, each state s to be visited is generated by applying a particular rule r to the current state p. In Fig. 1, function `first_enabled_rule(p)` returns the first rule that can be applied on a particular state `p`, whereas function `next_enabled_rule(p,r)` returns the next rule that can be applied on `p` after rule `r`. Both functions return null if such transition does not exist.

In addition, during the DF visit our algorithm updates the controller table when a goal is encountered (in function `Insert`), when an already visited state is encountered (in function `UpdatePaths`), when all the children of a state have been explored (in function `UpdatePathsPt()`) and when the state space exploration ends (in function `UpdatePathsFl`).

The function `Insert`, given a new state s reached from state p by firing rule r, checks if s is a goal state and, if so, it creates an entry in the controller table for the state

```
UpdatePaths(state s, state p, rule r) {
 if (HT[s].toGoal==true) {
  if (p is not in CTRL or CTRL[p].count >= CTRL[s].count+1) {
   HT[p].toGoal = true; CTRL[p].rule = r;
   CTRL[p].count = CTRL[s].count+1;
 }} else if (s is on the stack ST) {
  //s may reach a goal
  foreach ((p',r') on the stack ST) {
   save (p',r') in PT[s]; HT[p'].inPT = true;
 }} else if (HT[s].inPT == true) {
  //s was on the stack
  insert (p,r,s) in FL;
}}
```

Fig. 2 Function `UpdatePaths`.

p using the rule r. When s is not a goal, `Insert` behaves as in a standard DFS: if s is non visited the function pushes it on the stack and stores it in the hash table; otherwise, the function simply returns true to indicate that s is a visited state.

Figure 2 shows the details of function `UpdatePaths` that is called when the DFS reaches an already visited state s by applying a rule r on a state p. In this case, we may have to update the controller table `CTRL`:

- if s reaches a goal, then also p does. Thus, if p is not in the controller table, we insert it together with r. Otherwise, if p is already present in the controller table, we update its rule with r if the goal path through s is shorter than the path previously set for p in the controller table. This update ensures the optimality of generated controller.
- if s is in the stack, then it may still reach a goal. Thus we remember all the states on the path leading to s that is represented by the current stack content by saving them in the predecessors table `PT`.
- finally, if s is in the predecessors table, but not on the stack, we save it in the final list `FL`, together with its parent p and the transition r. This information will be later used to resolve cyclic paths in the predecessor table.

The function `UpdatePathsPt` is called when a state s has been completely expanded by the DFS algorithm. If s reached a goal, then for each state p in the predecessors table of s, we add to `CTRL` a rule that allows p to reach the goal through s.

Finally, the function `UpdatePathsFl`, called at the end of the visit, completes the controller table by adding rules for states in the final list `FL`. This is similar to what is done by `UpdatePathsPt`, but is applied at the end of the state space exploration and on a separate set of states. Such states belong to intersecting paths of the transition graph, so their shortest path to the goal can be computed only when all the goal paths have been generated.

4.3 Controller Strengthening Algorithm

The controller strengthening is implemented by the `exploreNeighborhood` function shown in Fig. 3. For each state p in the controller table, the function generates `MAX_VARS_PER_STATE` variations by applying small changes to the state variables. Then, the algorithm checks if each of the new states is in turn in the controller table. If

```
ExploreNeighborhood() {
 repeat {
  complete = true;
  foreach (p in CTRL)
   for vars = 1 to MAX_VARS_PER_STATE {
    s = add_random_variations_to(p);
    if (s is not in CTRL) {
     complete = false;
     //get a path from s to a state in CTRL
     path = BFS_lookup(s,CTRL);
     //store new path in CTRL
     foreach ((s', r') in path)
      CTRL[s'].rule = r';
 }}} until (complete)
}
```

Fig. 3 Function `exploreNeighborhood`.

any generated state s is not yet handled by the controller, the function performs a BFS search from s until it reaches a controlled state, and inserts the path from s to such state in `CTRL`. The process is repeated until all the generated variations are found in `CTRL`. At this point, `CTRL` knows how to drive the plant on the optimal plan and how to bring the plant on the nearest optimal plan state from a reasonable number of states outside the optimal plan.

5 Truck-and-Trailer Obstacles Avoiding Controller

To show the effectiveness of our approach, we show how it can be applied to the *truck and trailer with obstacles avoidance* problem.

The goal of a truck and trailer controller is to back a truck with a trailer up to a parking place starting from any initial position in the parking lot. This is a non trivial problem due to the dynamics of the truck-trailer pair (see the mathematical model in Sect. 5.1).

Moreover, we added to the parking lot some obstacles, which have to be avoided by the truck while maneuvering to reach the parking place. In this setting, also finding a suitable maneuver to reach the goal for any starting position may be an hard task. On the other hand, finding an *optimal* maneuver is a *very* complex problem, that cannot be modeled and resolved using common mathematical or programming strategies, e.g., using a dynamic programming approach.

Indeed, in the truck-and-trailer-with-obstacles problem, a backward decomposition of the cost function (e.g., the length of the path) is hard to perform, due to the complexity of the system dynamics function and to the presence of obstacles, whereas a forward decomposition does not satisfy the optimality principle, since the presence of obstacles may make an optimal *local* maneuver not optimal w.r.t. the final goal. This also makes fuzzy controllers not suitable for this problem, since fuzzy rules have a local character.

In the following sections we give details of the truck and trailer model and show the results obtained by applying the controller generation process described in Sect. 3 to perform a systematic analysis of the truck trajectories, discretized as a sequence of forward steps.

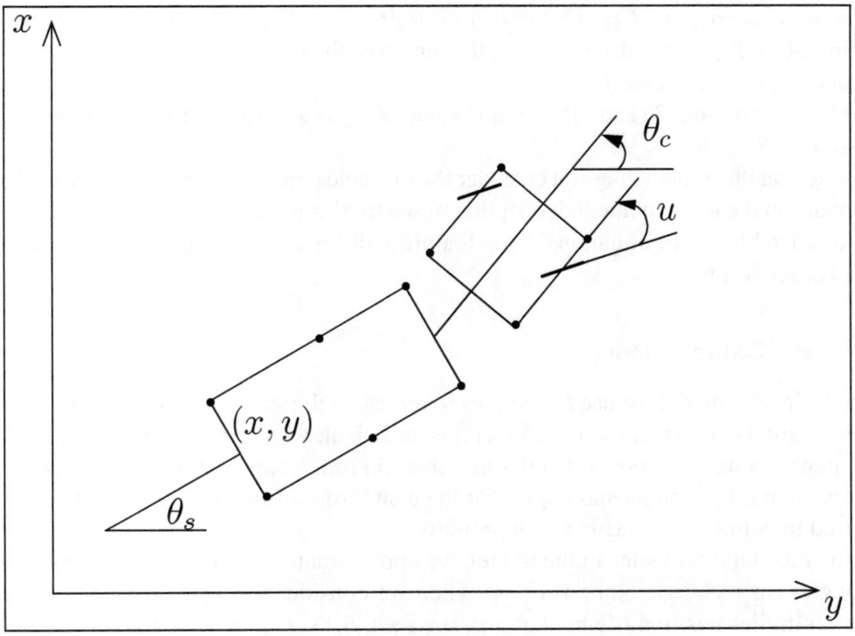

Fig. 4 Truck and trailer system description.

5.1 Model Description

Our model of the truck and trailer is based on the set of equations presented in [17]. The system has four state variables, that is the coordinates of the center rear of the trailer ($x, y \in [0, 50]$), the angle of the trailer w.r.t. the x-axis ($\theta_S \in [-90°, 270°]$) and the angle of the cab w.r.t the x-axis ($\theta_C \in [-90°, 270°]$). We assume that the truck moves backward with constant speed of 2m/s, so the only control variable is the steering angle $u \in [-70°, 70°]$. Figure 4 shows a schematic view of the truck and trailer system with its state and control variable. Moreover we single out 10 points in the truck and trailer border (displayed in the Fig. 4 by bold points) *representative* of the truck and trailer position.

If the values of the state variables at time t are $x[t]$, $y[t]$, $\theta_S[t]$ and $\theta_C[t]$, and the steering angle is u, then the new values of state variables at time $t+1$ are determined by following equations:

$$x[t+1] = x[t] - B * \cos(\theta_S[t]) \tag{1}$$

$$y[t+1] = y[t] - B * \sin(\theta_S[t]) \tag{2}$$

$$\theta_S[t+1] = \theta_S[t] - \arcsin\left(\frac{A * \sin(\theta_C[t] - \theta_S[t])}{L_S}\right) \tag{3}$$

$$\theta_C[t+1] = \theta_C[t] + \arcsin\left(\frac{r * \sin(u)}{L_S + L_C}\right) \tag{4}$$

where $A = r * \cos(u)$, $B = A * \cos(\theta_C[t] - \theta_S[t])$, $r = 1$ is the truck movement length per time step, $L_S = 4$ and $L_C = 2$ are the length of the trailer and cab, respectively (all the measures are in meters).

After computing (3) and (4), the new value of θ_C is adjusted to respect the *jackknife* constraint: $|\theta_S - \theta_C| \leq 90°$.

Note that this model does not consider the obstacles: indeed, embedding the obstacle avoidance in the mathematical description of the truck and trailer dynamics would result in a untractable set of equations. This feature will be added directly in the CGMurφ model described below.

5.2 The CGMurφ Model

In the CGMurφ model we use real values to represent the state variables x and y, whilst for the angle values (i.e., θ_S, θ_C and u) it is sufficient, w.r.t. the system dimensions, to use integer values. Moreover, we define some *tolerance* constants to set up a range of admissible final positions and angles for the center rear of the trailer. These tolerances are used to define the CGMurφ goal property.

To embed the obstacles in the model, we approximate them through their bounding rectangles (or rectangle compositions). Then we consider the *representative* points of the truck-trailer position (defined above, see Sect. 5.1) and, each time a new truck position is computed, we use a function to check if any of these points has hit the parking lot obstacles or borders. Therefore, our controller synthesis algorithm considers only feasible maneuvers to the goal state.

Moreover, in order to obtain a more *robust* controller we also considered the maneuvering errors due to the truck-trailer complex dynamic properties (e.g., friction, brakes response time, etc.) that cannot be easily embedded in the mathematic model. We used such errors to draw a *security* border around each obstacle and used these augmented obstacles in the collision check described above.

To estimate maximum maneuvering error we applied a *Monte Carlo's method* described as follows. We consider a large set of valid parking lot positions $S = \{s_k | 1 \leq k \leq 500000\}$. Given a position $s_k \in S$, (1) we apply a random maneuver m_k obtaining the new position $\bar{s}_k$. Then (2) we randomly perturb s_k generating the position s_k^p and apply the same maneuver m_k on s_k^p obtaining the position $\bar{s}_k^p$. Finally, (3) we compute the distance of the selected truck points P_i between the positions s_k^p and $\bar{s}_k^p$. This process is repeated 200 times for each position in S, thus analyzing 100 millions of perturbations. The security border size is the highest distance measured for a point in the step (3). We found out that this distance is 0.98 m.

5.3 Experimental Results

We tested our methodology using several obstacles topologies. In this section we present the results relative to the map shown in Fig. 5 where the obstacles and the security borders are highlighted.

Table 1 shows the results of the first phase of our algorithm (see Sect. 4.2). We repeated the controller generation using two different approximations for the real state variables x and y, rounding them to 0.5 and 0.2 meters.

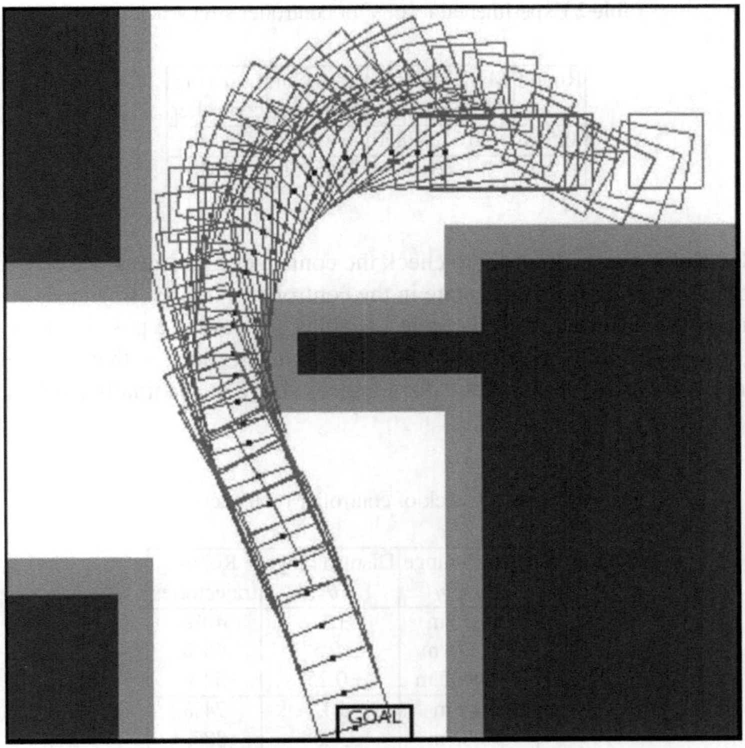

Fig. 5 Optimal trajectory generated by CGMurφ from initial position $x = 12$, $y = 16$, $\theta_s = 0$, $\theta_c = 0$.

Indeed, an higher precision extends the reachable state space and, consequently, the number of transitions in the controller. The results in Table 1 show that we are able to deal with system having millions of states.

In the second phase, we refined the controller by applying 36 disturbs to each state in the controller table and finding the appropriate rules to reconnect each new state to the optimal controller paths, as described in Sect. 4.3. As shown in Table 2, in this phase a significant number of transitions is added, due to the complexity of the truck-trailer dynamics.

Table 1 Experimental results for optimal raw controller synthesis

Round	Reachable states	Rules fired	Trans. in controller	Time Sec
0.5 m	2233997	64785913	382262	8160
0.2 m	12227989	354611681	1749586	32847

Table 2 Experimental results for controller strengthening

Round	MAX VARS	Trans. added	Trans. in controller	Size of controller
0.5 m	36	257850	640112	14 Mb
0.2 m	36	646364	2395950	50 Mb

Controller Robustness. In order to check the controller robustness, we considered all the trajectories starting from each state in the controller. For each trajectory state s, we applied a random disturbance on the state variables, generating a possibly new state s^p, and then we applied to s^p the rule associated to controller state s' that is nearest to s^p. A trajectory is *robust* if, applying the disturbances above, it eventually reaches the goal state.

Table 3 Check of controller robustness

Round	Disturb range for x,y	Disturb range for θ_s,θ_c	Robust trajectories
0.5 m	±0.25 m	±1°	40%
	±0.125 m	±0.5°	45%
	±0.0625 m	±0.25°	57%
0.2 m	±0.1 m	±1°	74%
	±0.05 m	±0.5°	88%

We checked all trajectories by applying different disturb ranges. As shown in Table 3, the fraction of robust trajectories increases with the controller precision (i.e., the real values approximation). Note that the percentages of robust trajectories in the second round are completely satisfying considering:

- the optimality of the trajectories;
- the extreme complexity of this parking problem;
- the unavailability of correction maneuvers.

6 Conclusions

The controller tables generated through our methodology contain millions of state-rule pairs. Thus, if we are working with small embedded systems, the table size could be a potential issue. This problem can be mitigated by applying various compression techniques on the table.

A completely different solution that we are also experimenting is the generation of *hybrid controllers*, that are optimal controllers working in parallel with e.g. a *fuzzy* controller. In this case, the optimal controller ensures the execution of the optimal plans (i.e., it is the optimal raw controller generated in Sect. 4.2), whereas the fuzzy controller is able to bring the system back to the optimal plans from any state outside the optimal raw controller. Thus, the fuzzy controller substitutes the extended knowledge generated by the algorithm in Sect. 4.3 with a set of inference rules. These rules may be in turn generated by an iterative learning process driven by an algorithm similar to the one of Sect. 4.3.

References

1. Kautz, H., Thomas, W., Vardi, M.Y.: 05241 executive summary – synthesis and planning. In Kautz, H., Thomas, W., Vardi, M.Y., eds.: Synthesis and Planning. Number 05241 in Dagstuhl Seminar Proceedings (2006)
2. Lygeros, J., Tomlin, C., Sastry, S.: Controllers for reachability specifications for hybrid systems Special Issue on Hybrid Systems, **35** (1999)
3. Åstrom, K.J., Hägglund, T.: PID controllers - Theory, Design, and Tuning. International Society for Measurement and Con; 2nd edn (2005)
4. Li, H., Gupta, M.: Fuzzy Logic and Intelligent Systems. Kluwer Academic Publishers (1995)
5. Jin, J.: Advanced Fuzzy Systems Design and Applications. Physica-Verlag (2003)
6. Bertsekas, D.P.: Dynamic Programming and Optimal Control. Athena Scientific (2005)
7. Sniedovich, M.: Dynamic Programming. Marcel Dekker (1992)
8. Burch, J.R., Clarke, E.M., McMillan, K.L., Dill, D.L., Hwang, L.J.: Symbolic model checking: 10^{20} states and beyond. Information and Computation **98** (1992) 142–170
9. Holzmann, G.J.: The SPIN Model Checker. Addison-Wesley (2003)
10. Dill, D.L., Drexler, A.J., Hu, A.J., Yang, C.H.: Protocol verification as a hardware design aid. In: Proceedings of the 1991 IEEE International Conference on Computer Design on VLSI in Computer & Processors, IEEE Computer Society (1992) 522–525
11. Hu, A.J., York, G., Dill, D.L.: New techniques for efficient verification with implicitly conjoined bdds. In: DAC '94: Proceedings of the 31st Annual Conference on Design Automation, New York, USA, ACM Press (1994) 276–282
12. http://www.dsi.uniroma1.it/~tronci/cached.murphi.html (2006)
13. Della Penna, G., Intrigila, B., Melatti, I., Tronci, E., Venturini Zilli, M.: Exploiting transition locality in automatic verification of finite state concurrent systems. STTT **6** (2004) 320–341
14. Murphi Web Page: http://sprout.stanford.edu/dill/murphi.html (2004)
15. Stern, U., Dill, D.: Using magnetic disk instead of main memory in the murφ verifier. In Hu, A.J., Vardi, M.Y., eds.: Computer Aided Verification, 10th International Conference, CAV '98, Vancouver, BC, Canada, June 28-July 2, Proceedings. Volume 1427 of Lecture Notes in Computer Science, Springer (1998) 172–183
16. Stern, U., Dill, D.L.: Improved probabilistic verification by hash compaction. In: CHARME '95: Proceedings of the IFIP WG 10.5 Advanced Research Working Conference on Correct Hardware Design and Verification Methods, London, UK, Springer-Verlag (1995) 206–224
17. Nguyen, D., Widrow, B.: The truck backer-upper: an example of self learning in neural networks. In: W.T. Miller, R.S. Sutton, and P.J. Werbos, eds.: Neural Networks For Control, Mit Press Series In Neural Network Modeling and Connectionism. MIT Press, Cambridge, MA (1990) 287–299

PART II

Robotics and Automation

Autonomous Gait Pattern for a Dynamic Biped Walking

Christophe Sabourin[1], Kurosh Madani[1] and Olivier Bruneau[2]

[1] Laboratoire Images, Signaux, et Systèmes Intelligents (LISSI EA / 3956)
Université Paris-XII, IUT de Sénart, Avenue Pierre Point, 77127 Lieusaint, France
sabourin@univ-paris12.fr, madani@univ-paris12.fr

[2] Laboratoire Vision et Robotique - Ecole Nationale Supérieure d'Ingénieurs de Bourges
10 Boulevard Lahitolle 18020 Bourges, France
olivier.bruneau@ensi-bourges.fr

Abstract. In this chapter, we propose an autonomous gait pattern for a dynamic biped walking based on a soft-computing approach. Our control strategy takes simultaneously advantage from a Fuzzy-CMAC based computation of robot's swing leg's desired trajectory and a high level control strategy allowing regulating the robot's average velocity. The main interest of this approach is to proffer to the walking robot autonomy and adaptability involving only one parameter: the average velocity. We present results about transition of velocities and we show that the presented control strategy allows to increase robustness of the walking robot according to perturbation forces.

Keywords. Autonomous gait pattern, biped dynamic walking, Fuzzy-CMAC neural networks.

1 Introduction

The design and the control of biped robots are one of the more challenging topics in the field of robotics and were treated by a large number of research works over past decades. The potential applications of this research area are very foremost in the middle and long term. Indeed this can lead firstly to a better comprehension of the human locomotion mechanisms, what can be very helpful for the design of more efficient orthosis. Secondly, the humanoid robots are intended to replace the human for interventions in hostile environments or to help him in the daily tasks. However, in addition to the problems related to autonomy and decision of such humanoid robots, their basic locomotion task is still today a big challenge. If it is true that a number of already constructed prototypes, among which the most remarkable are undoubtedly the robots Asimo [1] and HRP-2P [2], have proved the feasibility of such robots, it is also factual that the performances of these walking machines are still far from equalizing the human's dynamic locomotion process. The design of new control laws allowing real time control for real dynamic walking in unknown environments is thus today fundamental. Moreover, such robots must be able to adapt themselves automatically to indoor and outdoor human environments. Consequently, it is necessary to develop more autonomous biped robots with robust control strategies in order to allow them, on the one hand to adapt their gait to the real environment and on the other hand, to counteract external perturbations.

In the field of biped locomotion, the control strategies can be classified in two main categories. The first is based on a kinematics and dynamic modeling of the mechanical structure. This implies to identify perfectly the intrinsic parameters of biped robot's mechanical structure, requires a high precision measurement of the joints' angles, velocities and accelerations and needs a precise evaluation of interaction forces between feet and ground. Moreover, the control strategies based on a precise kinematics and dynamic modeling lead a large number of computational operations requiring heavy equipments. For all these reasons, the computing of the on-line trajectories generally are given by using a simplified modeling and the stability is ensured by the control of the Zero Moment Point (ZMP) [3] [4] [5] [6]. The second solution consists to use the soft-computing techniques (fuzzy logic, neural networks, genetic algorithm, etc.) and/or pragmatic rules resulting from the expertise of the walking human [7] [8] [9] [10] [11] [12]. Two main advantages distinguish this second class of approaches. Firstly, it is not necessary to know perfectly the characteristics of the mechanical structure. Secondly, this category of techniques takes advantage from learning (off-line and/or on-line learning) capabilities. This last point is very important because generally the learning ability allows increasing the autonomy of the biped robot.

In this chapter, we present a control strategy for an under-actuated robot: RABBIT (Fig. 1) [13] [14]. This robot constitutes the central point of a project, within the framework of CNRS ROBEA program [15], concerning the control of walking and running biped robots, involving several French laboratories.

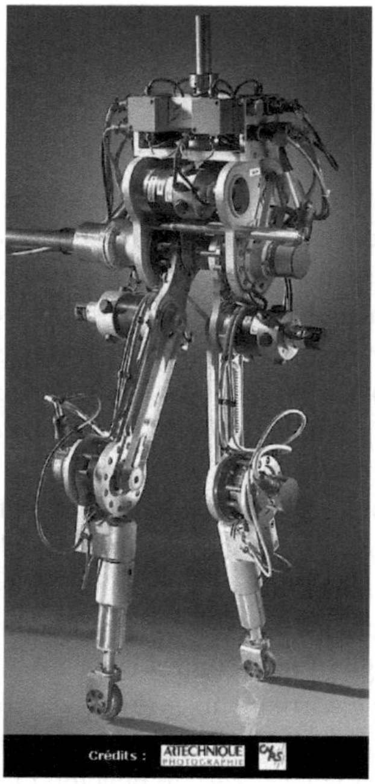

Fig. 1 Prototype RABBIT.

This robot is composed of two legs and a trunk and has no foot as shown on Fig. 1. If it is true, from design point of view, that RABBIT is simpler compared to a robot with feet, from the control theory point of view, the control of this robot is a more challenging task, particularly because, in phase of single support, the robot is under-actuated. In fact, this kind of robots allows studying real dynamical walking leading to the design of new control laws in order to improve biped robots' current performances. It is pertinent to note that the ZMP approach, generally used for humanoid robots, is not appropriated for the case of a biped without feet, because the contact surface between the foot and the ground is limited to a point.

This project has been the subject of many publications concerning the field of control strategies emerging on the one hand from rigorous mathematical modeling, and on the other hand issued from the use of CMAC neural networks. Developed approaches have been subject of experimental validations [16] [17] [18]. In this chapter, we present an extension of the control strategy using the CMAC neural network. In our previous work [16] [17], the CMAC was used to generate the joint trajectories of the swing leg but these trajectories were fixed. Consequently, the step length could not be changed during the walking. Today, our aim is to develop a control strategy able to generate a fully autonomous biped walking based on a soft-computing approach. In this chapter, we show how it is possible to change the walking gait by using the fusion of different trajectories learnt by several CMAC neural networks. In fact, our control strategy is based on two stages:

- The first one uses a set of pragmatic rules allowing to stabilize the pitch angle of the trunk and to generate the leg motions. This control strategy allows generating a stable dynamic walking with step length and velocity transitions [19]. During this first stage, the robot is supposed to move in an ideal environment (without disturbance). We also assume that frictions are negligible. However, in the case of our intuitive control, it is not possible to counteract external (pushed force) and internal (friction) disturbances. Consequently, we propose to use a neural network allowing to increase the robustness of our control strategy. In fact, in the first stage, the pragmatic rules are used as a reference control to learn, by a set of CMAC neural networks, a set of joint trajectories.
- In the second stage, we use these neural networks to generate and to modulate the trajectory of the swing leg. These trajectories are obtained by fusing outputs of several neural networks. In fact, the data contained in each CMAC represent a reference walking carried out during the first stage where each reference walking is characterized by a set of parameters as the step length and the average velocity. The fusion is realized by using fuzzy logic and allows to modulate, for example, step length according to average velocity. In fact, the fusion of several reference trajectories allows us to generate new gait and theoretically to carry out an infinity of trajectories only from a limited number of walking references.

This chapter is organized as follows. Section 2 presents the characteristics of our virtual under-actuated robot. In Sect. 3, we explain the method used to train CMAC neural network. Section 4 presents the control strategy using the Fuzzy-CMAC neural networks. In Sect. 5, we give the main results obtained in simulation. Conclusions and further developments are finally given.

2 Virtual Modeling of the Robot

The robot RABBIT has only four articulations: one for each knee, one for each hip. Motions are included in the sagittal plane by using a radial bar link fixed at a central column that allows to guide the direction of progression of the robot around a circle. Each articulation is actuated by one servo-motor RS420J. Four encoders make it possible to measure the relative angles between the trunk and the thigh for the hip, and between the thigh and the shin for the knee. Another encoder, installed on the bar link, allows to give the pitch angle of the trunk. Two binary contact sensors detect whether or not the leg is in contact with the ground. Based on the informations given by encoder, it is possible to calculate the step length L_{step} when the two legs are in contact with the ground. The duration of the step t_{step} is computed by using the contact sensor informations (duration from takeoff to landing of the same leg). Furthermore, it is possible to estimate the average velocity V_M by using Eq. (1).

$$V_M = \frac{L_{step}}{t_{step}} \tag{1}$$

The characteristics (masses and lengths of the limbs) are summarized in Table 1. Since the contact between the robot and the ground is just one point (passive DOF), the robot is under-actuated during the single support phase: there are only two actuators (at the knee and at the hip of the contacting leg) to control three parameters (vertical and horizontal position of the platform and pitch angle). In fact, this robot represents the minimal system able to generate a biped walking and running gaits.

Table 1 Masses and lengths of the limbs of the robot

Limb	Weight (Kg)	Length (m)
Trunk	12	0.2
Thigh	6.8	0.4
Shin	3.2	0.4

The numerical model of the robot previously described was designed with the software ADAMS[1] (Fig. 2). This software, from the mechanical system's modeling point of view (masses and geometry of the segments), is able to simulate the dynamic behavior of such system and namely to calculate the absolute motions of the platform as well as the limbs' relative motions when torques are applied on the joints by the virtual actuators. Figure 3 shows references for the angles and the torques required for the development of our control strategy. q_{i1} and q_{i2} are respectively the measured angles at the hip and the knee of the leg i. q_0 corresponds to the pitch angle. T^{sw}_{knee} and T^{sw}_{hip} are the torques applied respectively at the knee and at the hip during the swing phase, T^{st}_{knee} and T^{st}_{hip} are the torques applied during the stance phase.

[1] ADAMS is a product of MSC software.

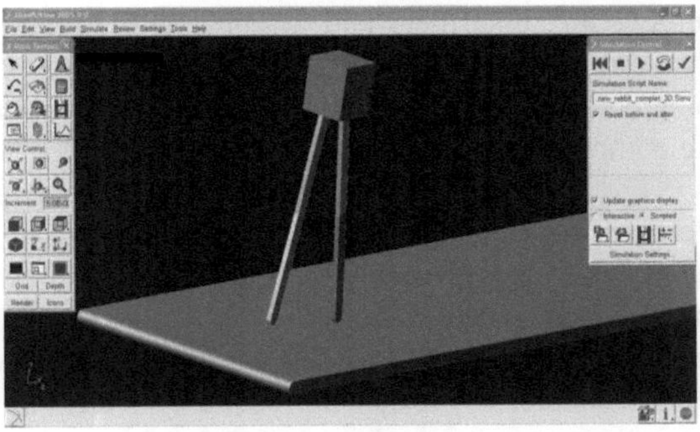

Fig. 2 Modeling of the robot with ADAMS.

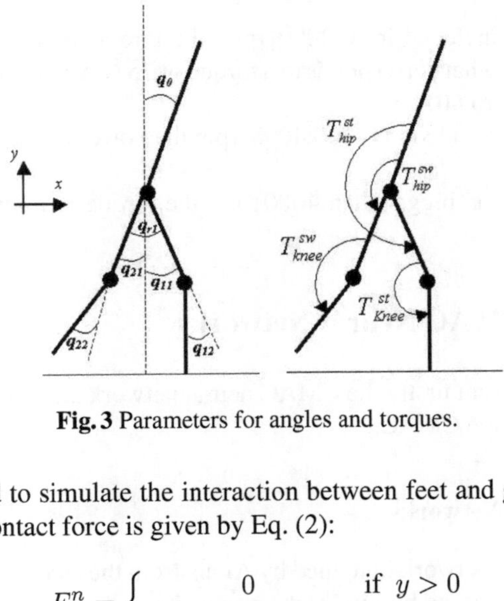

Fig. 3 Parameters for angles and torques.

The model used to simulate the interaction between feet and ground is exposed in [20]. The normal contact force is given by Eq. (2):

$$F_c^n = \begin{cases} 0 & \text{if } y > 0 \\ -\lambda_c^n \, |y| \, \dot{y} + k_c^n \, |y| & \text{if } y \leq 0 \end{cases} \tag{2}$$

where y and $\dot{y}$ are respectively the position and the velocity of the foot (limited to a point) with regard to the ground. k_c^n and λ_c^n are respectively the generalized stiffness and damping of the normal forces. They are chosen in order to avoid the rebound and to limit the penetration of the foot in the ground. The tangential contact forces are computed with the Eq. (3) in the case of a contact without sliding or with the Eq. (4) if sliding occurs.

$$F_c^t = \begin{cases} 0 & \text{if } y > 0 \\ -\lambda_c^t \dot{x} + k_c^t (x - x_c) & \text{if } y \leq 0 \end{cases} \tag{3}$$

$$F_c^t = \begin{cases} 0 & \text{if } y > 0 \\ -(sgn(\dot{x}))\lambda_g F_c^n - \mu_g \dot{x} & \text{if } y \leq 0 \end{cases} \tag{4}$$

Where x and $\dot{x}$ are respectively the position and the velocity of the foot with regard to the position of the contact point x_c at the instant of impact with ground. k_c^t and λ_c^t

are respectively the generalized stiffness and damping of the tangential forces. λ_g is the coefficient of dynamic friction depending on the nature of surfaces in contact and μ_g a viscous damping coefficient during sliding. After each iteration, the normal and tangential forces are computed from the Eqs. (2) and (3). But, if F_c^t is located outside the cone of friction ($\|F_c^t\| > \mu_s \|F_c^n\|$ with μ_s the static friction coefficient), then the tangential force of contact is computed with Eq. (4). The interest of this model is that it is possible to simulate walking with or without sliding phases allowing us to evaluate the robustness of the designed control strategy.

Within the framework of a real robot's control, the morphological description of this one is insufficient. It is thus necessary to take into account the technological limits of the actuators in order to implement the control laws used in simulation on the experimental prototype. From the characteristics of servo-motor RS420J used for RABBIT, we thus choose to apply the following limitations:

- when velocity is included in [0;2000] rpm, the torque applied to each actuator is limited to 1.5 Nm what corresponds to a torque of 75 Nm at the output of the reducer (ration gear equal to 50),
- when velocity is included in [2000;4000] rpm the power of each actuator is limited to 315 W,
- when the velocity is bigger than 4000 rpm, the torque is imposed to be equal to zero.

3 Training of CMAC Neural Networks

In this section, we present firstly the CMAC neural network and secondly, the principle which we use to train the CMAC.

3.1 CMAC Neural Networks

The CMAC is a neural network imagined by Albus from the studies on the human cerebellum [21] [22]. Despite its biological relevance, its main interest is the reduction of the training and computing times in comparison to other neural networks [23]. This is of course a considerable advantage for real time control. Because of these characteristics, the CMAC is thus a neural network relatively well adapted for the control of complex systems with a lot of inputs and outputs and has already been the subject of some researches in the field of the control of biped robots [8] [11].

The CMAC is an associative memory type neural network which is a set of N detectors regularly distributed on several C layers. The receptive fields of these detectors are distributed on the totality of the limited range of the input signal. On each layer, the receptive fields are shifted of a quantification step q. Consequently, the widths of the receptive field are not always equal. The number of detectors N depends on the one hand of the width of the receptive fields and on the other hand of the quantification step q. When the value of the input signal is included in the receptive fields of a detector, this one is activated. For each value of the input signal, the number of activated detector is equal to the number of layers C (parameter of generalization). Figure 4 shows a simplified organization of the receptive fields having 14 detectors distributed on 3 layers. Being given that there is an overlapping of the receptive fields, neighboring inputs will activate common detectors. Consequently, this neural network is able to carry out a generalization of the output calculation for inputs close to those presented during learning.

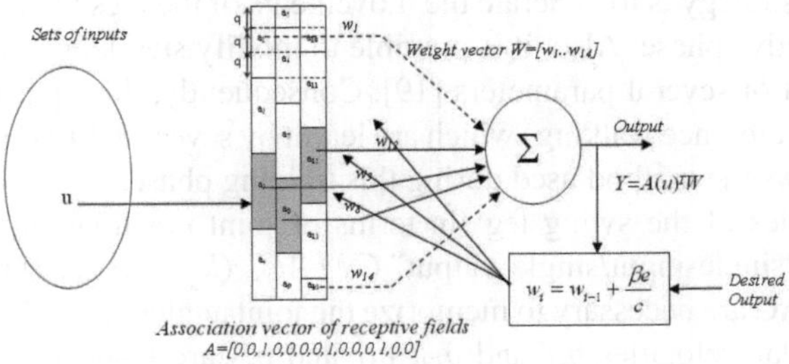

Fig. 4 Description of the simplified CMAC with 14 detectors distributed on 3 layers. For each value of the input signal, the number of activated detector is equal to 3. $A = [0,0,1,0,0,0,0,1,0,0,0,1,0,0]$, $Y = A(u)^T W = w_3 + w_8 + w_{12}$.

The output Y of the CMAC is computed by using two mappings:

- The first mapping projects an input space point u into a binary associative vector $A = [a_1 \dots a_N]$. Each element of A is associated with one detector and N is the number of detector. When one detector is activated, the corresponding element A of this detector is 1 otherwise it is equal to 0.
- The second mapping computes the output Y of the network as a scalar product of the association vector A and the weight vector $W = [w_1 \dots w_N]$ (Eq. 5).

$$Y = A(u)^T W \tag{5}$$

The weights W of CMAC are updated by using Eq. (6).

$$w_i = w_{i-1} + \frac{\beta e}{C} \tag{6}$$

w_{i-1} and w_i are respectively the weights before and after the training at each sample time t_i. C is the generalization number of each CMAC and β is a parameter which is included in $[0, 1]$. e is the error between the desired output Y^d of the CMAC and the computed output Y of the CMAC.

3.2 Learning Phase

During this learning phase, we use an intuitive control allowing us to perform dynamic walking of our virtual under-actuated robot without references trajectories. This intuitive control strategy is based on three points:

- the observation of the relations between joint motions and the evolution of the parameters describing the trajectory of the robot platform,
- an interpretation of the muscular behavior,
- the analysis of the intrinsic dynamics of a biped.

Based on these considerations, it is possible to determine a set of pragmatic rules. The objective of this strategy is to generate the movements of the legs by using a succession of passive and active phase. Also, it is possible to modify step length, average velocity by an adjustment of several parameters [19]. Consequently, this approach allow us to generate several reference walkings which are learnt by several CMAC neural networks.

Figure 5 shows the method used during this training phase.

The trajectories of the swing leg (in terms of joint positions and velocities) are learnt with four "single-input/single-output" $CMAC_k$ $(k = 1, \ldots, 4)$ neural networks. Indeed, two CMAC are necessary to memorize the joint angles q_{i1} and q_{i2} and two other CMAC for angular velocities $\dot{q}_{i1}$ and $\dot{q}_{i2}$. q_{i1} and q_{i2} are respectively the measured angles at the hip and the knee of the leg i; $\dot{q}_{i1}$ and $\dot{q}_{i2}$ are respectively the measured angular velocities at the hip and the knee of the leg i. When leg 1 is in support ($q_{12} = 0$), the angle q_{11} is applied to the input of each $CMAC_k$ ($u = q_{11}$) and when leg 2 is in support ($q_{22} = 0$), this is the angle q_{21} which is applied to the input of each $CMAC_k$ ($u = q_{21}$). Consequently, the trajectories learnt by the neural networks are function of the geometrical pattern of the robot. Moreover, we consider that the trajectories of each leg in swing phase are identical. This allow on the one hand to divide by two the number of CMAC and on the other hand to reduce the training time. The weights of each $CMAC_k$ are updated by using the error between the desired output Y_k^d of each $CMAC_k$ and the computed output Y_k of each $CMAC_k$.

4 Control Strategy using FUZZY-CMAC

Figure 6 shows the global strategy which is used to control the walking robot. It should be noted that the architecture of this control can be decomposed into three parts:

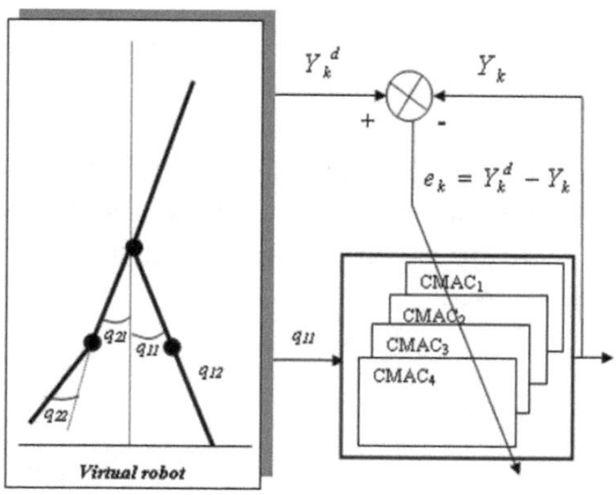

Fig. 5 Principle of the learning phase of CMAC neural networks ($u = q_{11}$).

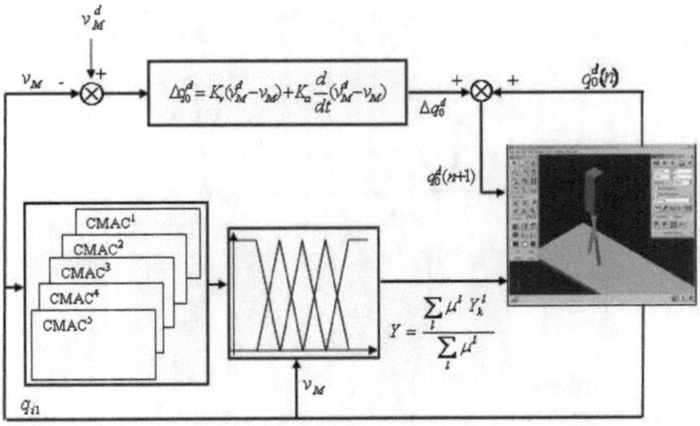

Fig. 6 Principle of the control strategy used Fuzzy-CMAC trajectories.

- The first is used to compute the trajectory of the swing leg from several output of the $CMAC_k$ neural networks and a Fuzzy Inference System.
- The second allows regulating the average velocity from a modification of the desired pitch angle.
- The third is composed by four PD control in order to ensure the tracking of the reference trajectories at the level of each articulation.

In Sect. 4.1, we give the characteristics of the reference walkings learnt by CMACs. Section 4.2 presents the principle of the Fuzzy-CMAC trajectories. Sections 4.3 and 4.4 describe respectively the high level control and PD control.

4.1 Reference Trajectories

During the training stage, five trajectories with an average velocity V_M included in [0.4...0.8] m/s have been learnt by five $CMAC^l$. Each $CMAC^l$ is composed of four single input/single output $CMAC_k$ (two $CMAC_k$ for the angular positions q_{i1} and q_{i2} and two other $CMAC_k$ for the angular velocities $\dot{q}_{i1}$ and $\dot{q}_{i2}$). Table 2 gives the main parameters which are used during the learning phase according to the desired average velocity V_M, where V_M is calculated by using Eq. (1). q_r^d and q_0^d are respectively the desired relative angle between the two thighs and the desired pitch of the trunk. q_{sw}^d corresponds to the desired angle of the knee at the end of the knee extension of the swing leg just before the double contact phase.

Each reference walking is characterized by a set of parameter (q_r^d, q_{sw}^d, q_0^d) allowing to generate different walking gaits (V_M, L_{step}). Figure 7 shows stick-diagrams repre-

Table 2 Parameters used during the learning stage

	V_M(m/s)	q_r^d(°)	q_{sw}^d(°)	q_0^d(°)
$CMAC^1$	0.4	20	−7	3.5
$CMAC^2$	0.5	25	−10	3
$CMAC^3$	0.6	30	−15	2.5
$CMAC^4$	0.7	35	−20	8
$CMAC^5$	0.8	40	−25	8

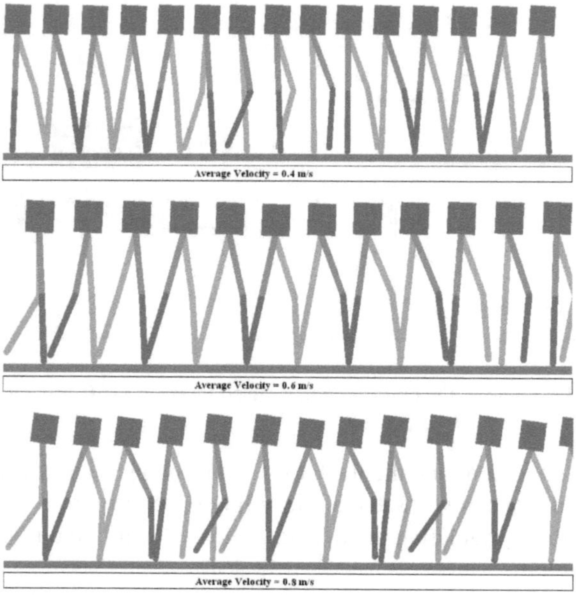

Fig. 7 Stick-diagram of the walking robot for three different velocities. V_M is equal to, from the top to the bottom, 0.4 m/s, 0.6 m/s and 0.8 m/s.

senting the biped robot for three average velocity. V_M is equal to, from the top to the bottom, 0.4 m/s, 0.6 m/s and 0.8 m/s.

It must be pointed out that the step length L_{step} increases when V_M increases. Table 3 gives L_{step} according to V_M.

Table 3 V_M and L_{step} for the five reference trajectories

	V_M(m/s)	L_{step}(m)
$CMAC^1$	0.4	0.23
$CMAC^2$	0.5	0.28
$CMAC^3$	0.6	0.31
$CMAC^4$	0.7	0.36
$CMAC^5$	0.8	0.4

4.2 Fuzzy-CMAC Trajectories

During the walking, the measured angle q_{11}, if leg 1 is in support, or q_{21}, if leg 2 is in support, is applied at each input of each $CMAC_k^l$ (see Fig. 6). The desired angles q_{i1}^d and q_{i2}^d, and the desired angular velocities $\dot{q}_{i1}^d$ and $\dot{q}_{i2}^d$ are carried out by using a fusion of the five learnt trajectories. This fusion is realized by using a Fuzzy Inference System (FIS). This FIS is composed of five rules:

- IF V_M IS $VerySmall$ THEN $Y = Y^1$
- IF V_M IS $Small$ THEN $Y = Y^2$

- IF V_M IS $Medium$ THEN $Y = Y^3$
- IF V_M IS Big THEN $Y = Y^4$
- IF V_M IS $VeryBig$ THEN $Y = Y^5$

where Y^l corresponds at the output of the $CMAC^l$. Figure 8 shows the membership functions. The average velocity is modeled by five fuzzy sets (VerySmall, Small, Medium, Big, VeryBig). Each desired trajectory Y_k is computed by using Eq. (7).

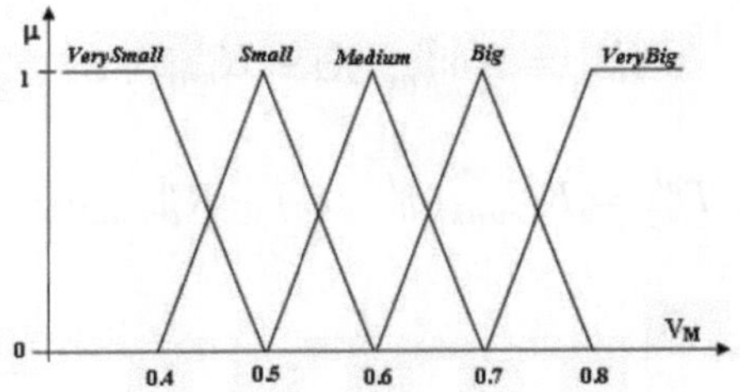

Fig. 8 Membership functions used to compute the Fuzzy-CMAC trajectories.

$$Y_k = \frac{\sum_{l=1}^{5} \mu^l Y_k^l}{\sum_{l=1}^{5} \mu^l} \tag{7}$$

Consequently, the trajectories depends on the one hand of the geometrical position of the stance leg (q_{i1}) and on the other hand of the measured average velocity (V_M).

4.3 High Level Control

The high level control allows us to regulate the average velocity by adjusting the pitch angle of the trunk at each step by using the error between the measured average velocity V_M and the desired average velocity V_M^d and of its derivative as described in Fig. 6.

At each step, Δq_0^d, which is computed by using the error between V_M and V_M^d and of its derivative (Eq. 8), is then added to the pitch angle of the previous step $q_0^d(n)$ in order to carry out the new desired pitch angle of the following step $q_0^d(n+1)$ as shown in Eq. (9).

$$\Delta q_0^d = K^p(V_M^d - V_M) + K^v \frac{d}{dt}(V_M^d - V_M) \tag{8}$$

$$q_0^d(n+1) = q_0^d(n) + \Delta q_0^d \tag{9}$$

4.4 PD Control

In order to ensure the tracking of the desired trajectories, the torques T_{knee} and T_{hip} (see Fig. 3) applied respectively at the knee and at the hip are computed by using PD control. During the swing stage, the torques are carried out by using Eqs. (10) and (11). q_{ij}^d and $\dot{q}_{ij}^d$ are respectively the reference trajectories (position and velocity) of the swing leg from the output of the Fuzzy-CMAC.

$$T^{sw}_{hip} = K^p_{hip}(q^d_{i1} - q_{i1}) + K^v_{hip}(\dot{q}^d_{i1} - \dot{q}_{i1}) \tag{10}$$

$$T^{sw}_{knee} = K^p_{knee}(q^d_{i2} - q_{i2}) + K^v_{knee}(\dot{q}^d_{i2} - \dot{q}_{i2}) \tag{11}$$

Secondly, the knee of the stance leg is locked, with $q^d_{i2} = 0$ and $\dot{q}^d_{i2} = 0$ (Eq. 12), and the torque applied to the hip allows to control the pitch angle of the trunk (Eq. (13)). q_0 and $\dot{q}_0$ are respectively the measured absolute angle and angular velocity of the trunk. q^d_0 is the desired pitch angle.

$$T^{st}_{knee} = -K^p_{knee} q_{i2} - K^v_{knee} \dot{q}_{i2} \tag{12}$$

$$T^{st}_{hip} = K^p_{trunk}(q^d_0 - q_0) - K^v_{trunk} \dot{q}_0 \tag{13}$$

5 Results

The control strategy presented in the previous section allows:

- To generate the joint trajectories of the swing leg from the geometrical configuration of the robot and the real average velocity.
- To regulate, at each step, the average velocity thanks to an adjustment of the pitch angle.

The main interest of our approach is that it allows walking robot autonomy and adaptability. A significant example of that is the robot's adaptability to adapt its step length using only one parameter: the average velocity. In this section, first of all we present results regarding transition of velocities (Sect. 5.1) and secondly we show that the presented control strategy allows increasing robustness of the walking robot with reference to perturbation forces (Sect. 5.2).

5.1 Autonomous Walking Gait

Figure 9 shows the stick-diagram of the biped robot representing the walking of the robot during the transition velocity and Fig. 10 shows the desired average velocity V^d_M, the measured velocity V_M (see Eq. (1)) and the step length L_{step} during the transition velocity.

Fig. 9 Stick-diagram of the biped robot during one transition velocity.

When the desired average velocity V_M^d is modified from 0.3 m/s to 0.9 m/s, V_M increases gradually and converges fowards the new value of V_M. The regulation at each step of the average velocity is obtained thanks to an adequate adjustment of the pitch angle. Moreover, L_{step} increases automatically from 0.25 m to 0.4 m.

Figures 11 and 12 show the angle q_{11} and q_{12}. It is interesting to note that, as it has been mentioned before, the trajectory depends on the one hand on the stance leg's geometrical position and on the other hand on the measured average velocity. It is pertinent to note that the control strategy allows to adapt automatically walking gait based on the five learning reference walking.

It must be pointed out that the walking gait can depends to the desired average velocity V_M^d. But in this case, the step length transition is abrupt. Furthermore, the fact that the walking gait depends of the measured average velocity allows increasing the robustness of the walking of the robot.

5.2 Evaluation of the Robustness According to a Pushed Force

During walking, a robot moving in real environment can be subjected to external forces involving an imbalance. Consequently, the control strategy must react quickly in order to compensate this perturbation and to avoid the fall of the robot. Generally, when human being is pushed forwards, he increases the step length. In the case of the proposed control strategy, the step length depends of the real average velocity V_M. If the robot is pushed forwards, the duration of the step t_{step} decreases and consequently, V_M increases. And if the robot is pushed backwards, the average velocity decreases and the step length too. In fact, at the next step after one perturbation force, L_{step} is adjusted according to the modification of the average velocity. In this manner, it is easier for the robot to compensate this kind of perturbation.

In order to illustrate this intrinsic robustness, we present in this section two results showing how the biped reacts when a forwards or backwards forces are applied on the trunk of the robot.

- Forwards pushed force: Figure 13 shows the stick-diagram representing the walking robot before and after the forwards perturbation force. At $t = 15$ s, we applied on the trunk of the robot an impulsive pushed force. The duration and the amplitude

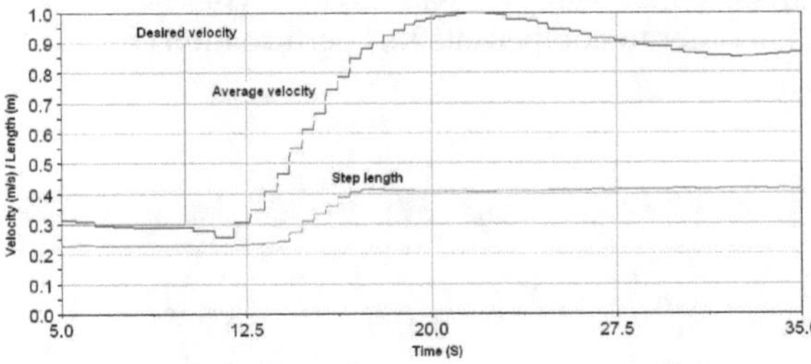

Fig. 10 Average velocity and step length during one transition velocity when the desired average velocity is modified from 0.3 m/s to 0.9 m/s.

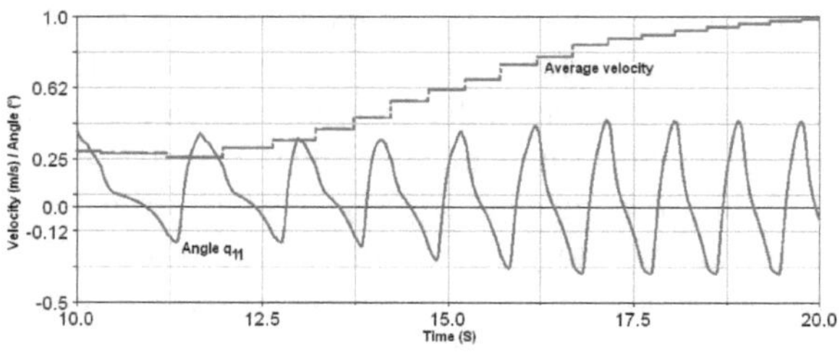

Fig. 11 Angle q_{11} when V_M increases gradually from 0.3 m/s to 1 m/s.

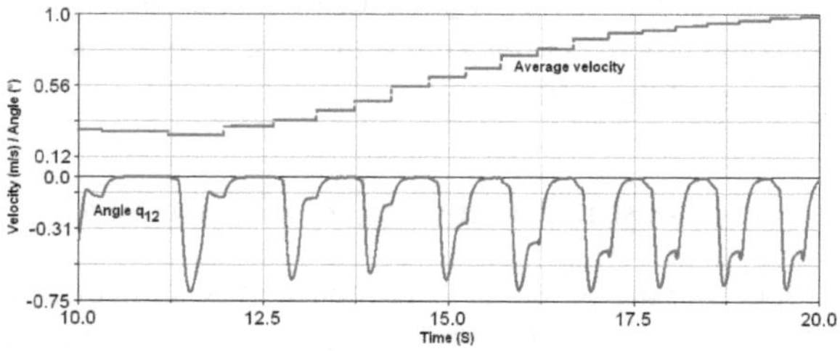

Fig. 12 Angle q_{11} when V_M increases from 0.3 m/s to 1 m/s.

of this force are respectively 0.2 s and 50 Nm. It must be pointed out that the trajectories of the biped robot are updated automatically in order to compensate this perturbation.
As the Fig. 14 shows it, before this perturbation, the robot walks with an average velocity to 0.7 m/s which corresponds to the desired average velocity.
After this perturbation, V_M increases considerably but the step length increases according to average velocity. Moreover, the pitch angle decreases thanks to the hight level control (Fig. 15). Consequently, our control strategy allow to compensate slowly this kind of perturbation.

- Backwards pushed force: Figure 16 shows the stick-diagram representing the walking robot before and after the backwards force perturbation. At $t = 15$ s, we applied on the trunk of the robot an impulsive pushed force. The duration and the amplitude of this force are respectively 0.2 s and -25 Nm.
In this case,V_M decreases and the step length decreases too (Fig. 17). The pitch angle increases in order to compensate this perturbation (Fig. 18).

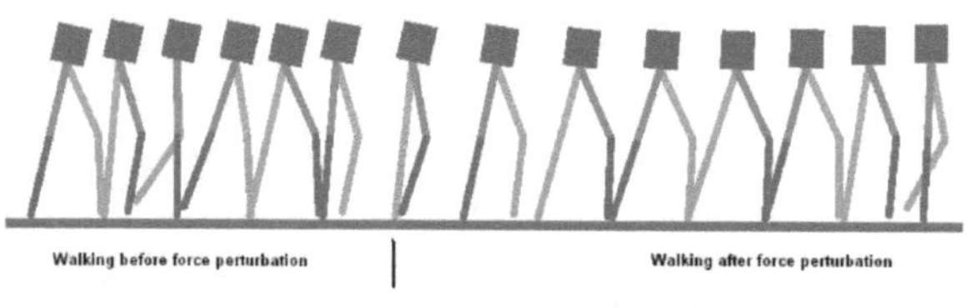

Fig. 13 Stick-diagram representing the walking robot before and after the forwards perturbation force. The duration and the amplitude of this force are respectively 0.2 s and 50 Nm.

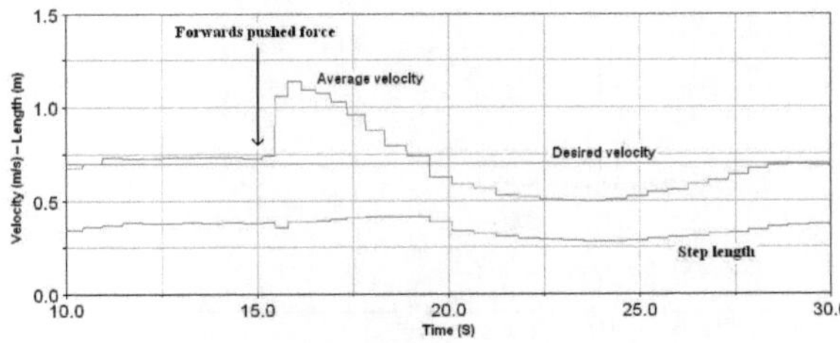

Fig. 14 Average velocity V_M when the robot is pushed forwards at $t = 15$ s.

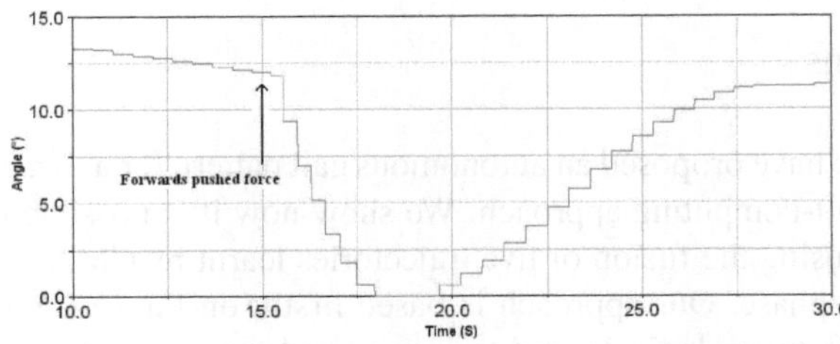

Fig. 15 Pitch angle of the trunk when the robot is pushed forwards at $t = 15$ s.

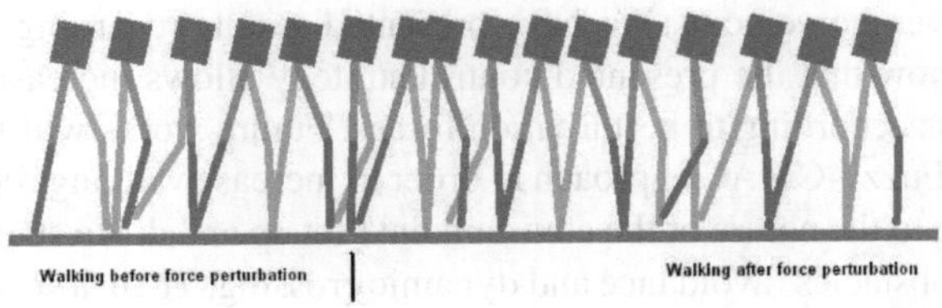

Fig. 16 Stick-diagram representing the walking robot before and after the backwards perturbation force. The duration and the amplitude of this force are respectively 0.2 s and −25 Nm.

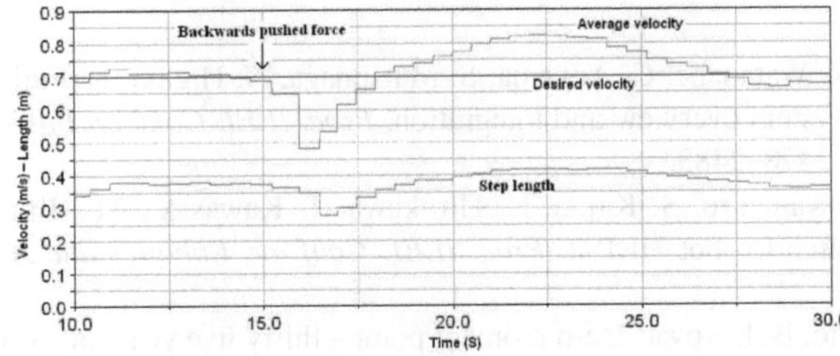

Fig. 17 Average velocity V_M when the robot is pushed backwards at $t = 15$ s.

In summary, our control strategy allows to compensate slowly pushed force because:

- L_{step} is adjusted automatically according the measured average velocity.
- The average velocity is updated thanks to the high level control.

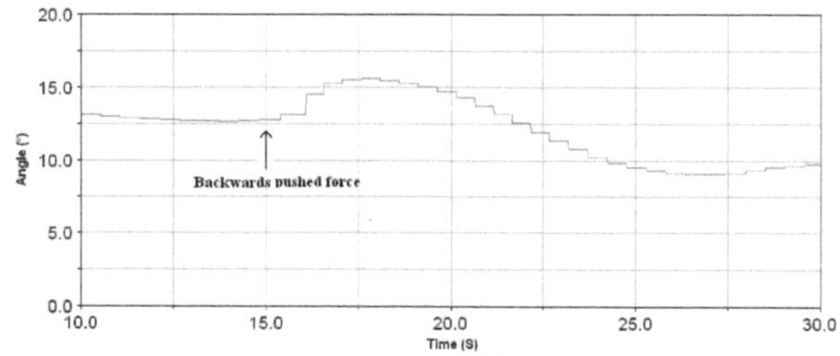

Fig. 18 Pitch angle of the trunk when the robot is pushed backwards at $t = 15$ s.

6 Conclusions

In this chapter, we have proposed an autonomous gait pattern for a dynamic biped walking based on a soft-computing approach. We show how it is possible to generate new walking gaits by using the fusion of five trajectories learnt by CMAC neural networks during a learning phase. Our approach is based firstly on Fuzzy-CMAC issued computation of robot's swing leg's desired trajectory and secondly on a high level control strategy allowing regulation of the robot's average velocity. The main interest of this approach is to proffer to the walking robot autonomy and adaptability involving only one parameter: the average velocity. We have presented results regarding velocity's transitions and we show that the presented control strategy allows increasing the walking robot's robustness according to perturbation forces. Future works will firstly focus the extension of our Fuzzy-CMAC approach in order to increase walking robot's autonomy as well according to the nature of the environment (get up and down stairs for instance), as regarding the obstacles' avoidance and dynamic crossing. Then, a second perspective will focus the experimental validation of our approach.

References

1. Y. Sakagami, R. Watanabe, C. Aoyama, S. Matsunaga, N. Higaki, K. Fujimura. The intelligent ASIMO: system overview and integration. *Proc. IEEE Conf. on Intelligent Robots and Systems*, 2002, 2478–2483.
2. K. Kaneko, F. Kanehiro, S. Kajita, H. Hirukawa, T. Kawasaki, M. Hirata, K. Akachi, T. Isozumi. Humanoid robot HRP-2. *Proc. IEEE Conf. on Robotics and Automation*, 2004, 1083–1090.
3. M. Vukobratovic, B. Borovac. Zero moment point – thirty five years of its live. *International Journal of Humanoid Robotics*, 2004, Vol. 1 No 1, 157–173.
4. S. Kajita, F. Kanehiro, K. Kaneko, K. Fujiwara, K. Harada, K. Yokoi, H. Hirukawa. Biped walking pattern generation by using preview control of Zero-Moment Point. *Proc. IEEE Conf. on Robotics and Automation*, 2003, 1620–1626.
5. Q. Huang, K. Yokoi, S. Kajita, K. Kaneko, H. Arai, N. Koyachi, K. Tanie. Planning walking patterns for a biped robot. *IEEE Transactions on Robotics and Automation*, 2001, Vol. 17, No 3, 280–289.
6. K. Hirai, M. Hirose, Y. Haikawa, T. Takenaka. The development of honda humanoid robot. *Proc. IEEE Conf. on Robotics and Automation*, 1998, 1321–1326.

7. M. Vukobratovic, O. Timcenko. Stability analysis of certain class of bipedal walking robots with hybridization of classical and fuzzy control. *Proc. International Conference on Advanced Robotics and Intelligent Automation*, 1995, 290–295.
8. A. Brenbrahim, J. Franklin. Biped dynamic walking using reinforcement learning. *Robotics and Autonomous Systems*, 1997, Vol. 22, 283–302.
9. C. Zhou, Q. Meng. Reinforcement learning with fuzzy evaluative feedback for a biped robot. *Proc. IEEE Conf. on Robotics and Automation*, 2000, 3829–3834.
10. F. Yamasaki, K. Endo, H. Kitano, M. Asada. Acquistion of humanoid walking motion using genetic algorithm – Considering characteristics of servo modules. *Proc. IEEE Conf. on Robotics and Automation*, 2002, 3123–3128.
11. A. L. Kun, T. Miller. The design process of the unified walking controller for the UNH biped. *Proc. IEEE Conf. on Humanoid Robots*, 2000.
12. J. Nakanishi, J. Morimoto, G. Endoa, G. Chenga, S. Schaala, and M. Kawatoa, Learning from demonstration and adaptation of biped locomotion. *Robotics and Autonomous Systems*, 2004, Vol. 47, No 2–3, 79–91.
13. C. Chevallereau, G. Abba, Y. Aoustin, F. Plestan, E. R. Westervelt, C. Canudas-de-Wit, J. W. Grizzle. RABBIT: A testbed for advanced control theory. *IEEE Control Systems Magazine*, 2003, Vol. 23, No 5, 57–79.
14. http://robot-rabbit.lag.ensieg.inpg.fr/
15. http://www.laas.fr/robea/
16. C. Sabourin, O. Bruneau, G. Buche. Control strategy for the robust dynamic walk of a biped robot. *The International Journal of Robotics Research (IJRR)*, 2006, Vol. 25, No 9, 843–860.
17. C. Sabourin, O. Bruneau. Robustness of the dynamic walk of a biped robot subjected to disturbing external forces by using CMAC neural networks. *Robotics and Autonomous Systems*, 2005, Vol. 23, 81–99.
18. E. R. Westervelt, G. Buche, J. W. Grizzle. Experimental validation of a framework for the design of controllers that induce stable walking in planar bipeds. *The International Journal of Robotics Research*, 2004, Vol. 23 No 6, 559–582.
19. C. Sabourin, O. Bruneau, J.-G. Fontaine. Start, stop and transition of velocities on an underactuated bipedal robot without reference trajectories. *Internationnal Journal of Humanoid Robotics*, 2004, Vol. 1, No 2, 349–374.
20. O. Bruneau, F.B. Ouezdou. Distributed ground/walking robot interactions. *Robotica, Cambridge University Press*, 1999, Vol. 17, No 3, 313–323.
21. J. S. Albus. A new approach to manipulator control: the cerebellar model articulation controller (CMAC). *Journal of Dynamic Systems, Measurement and Control*, 1975, Vol. 97, 220–227.
22. J. S. Albus. Data storage in the cerebellar model articulation controller (CMAC). *Journal of Dynamic Systems, Measurement and Control*, 1975, Vol. 97, 228–233.
23. W. T. Miller, F. H. Glanz, L. G. Kraft. CMAC: An associative neural network alternative to backpropagation. *Proceedings of the IEEE, Special Issue on Neural Networks*, 1990, Vol. 78, No 10, 1561–1567.

Particle-filter Approach for Cooperative Localization in Unstructured Scenarios

Fernando Gomez Bravo[1], Alberto Vale[2] and Maria Isabel Ribeiro[2]

[1] Departamento de Ingeniería Electrónica, Sistemas Informáticos y Automática
Univ. de Huelva, Campus de la Rábida
Crta. Huelva-Palos de la Frontera s/n, 21819 Huelva, Spain
fernando.gomez@diesia.uhu.es

[2] Institute for Systems and Robotics, Instituto Superior Técnico
Av. Rovisco Pais 1, 1049-001 Lisboa, Portugal
{vale,mir}@isr.ist.utl.pt

Abstract. This chapter presents a Particle Filter approach to solve the metric localization of a team of three robots. The method considers the existence of a fixed active beacon which has sensorial capabilities. The localization strategy is based on the distance and orientation measurements among the robots and the robots and the fixed active beacon. By means of this technique the team of robots are capable of navigating in isolated and unstructured scenarios. In spite of common differential vehicles, this approach is intended to be applied in car-like vehicles, still a large scope of mobile robots in real environments.

Keywords. Particle-filter, cooperative localization, mobile robots.

1 Introduction

One of the cutting-edges of technology is the presence of robots in unexplored scenarios to accomplish the most different type of missions. Since most of the tasks in these scenarios can not be accomplished by humans, mobile robots development are a present real necessity. The localization of a mobile robot is in the sequel of the implementation to accomplish different types of predetermined tasks coupled with high level of precision. Moreover, a mobile robot is not an isolated mobile platform, since other mobile robots can be working for a similar goal. Consequently, and if visible between them, the different mobile robots may work together, to improve their localization [1] [2].

Cooperative localization schemes for teams of robots explore the decentralized perception that the team supports to enhance the localization of each robot. These techniques have been studied in the recent past [3] [4] [5] [6].

Cooperative localization is a key component of cooperative navigation of a team of robots. The use of multi-robot systems, when compared to a single robot, has evident advantages in many applications, in particular there where the spatial or temporal coverage of a large area is required. An example on the scientific agenda is the planet surface exploration, in particular Mars exploration. At the actual technology stage, the difficulty in having human missions to Mars justifies the development, launching and operation of teams of robots that may carry out autonomously or semi-autonomously a set of exploration tasks. The mission outcomes in the case of surface exploration may largely benefit from having a team of cooperative robots rather than having a single robot or a set of isolated robots. In either case, and depending on the particular allocated mission, the robots may have to be localized relative to a fixed beacon or landmark, most probably the launcher vehicle that carried them from earth.

This chapter proposes a methodology for cooperative localization of a team of robots based on a Particle-Filter (PF) approach, relying on distance and orientation measurements among the robots and among these and a fixed beacon. It is considered that the fixed beacon has sensorial capabilities of range and orientation measurement, but has a maximum detectable range. Each robot has limited range detection along a limited field of view. For localization purposes, the proposed approach propagates a PF for each robot. The particle weight update is based on the measurements (distance and orientation) that each robot acquires relative to the other robots and/or the fixed beacon. A motion strategy is implemented in such a way that each robot is able to detect, at a single time instant, at least one teammate or the fixed beacon. Therefore, either the fixed beacon (if detected) or the other robots (if detected) play the role of an external landmark for the localization of each robot.

The developed strategy is targeted to an exploration mission and it is considered that all the robots start the mission close to the fixed beacon, evolve in the environment (the team of robots could be carried to the target scenario by a single platform which could be used as a reference beacon) and eventually may loose detection of the fixed beacon. Finally, the robots have to return to the starting location. Even more, due the sensorial limitations of the robots they also could loose detection among them.

In the approach presented in this chapter, besides cooperative localization, the team navigation involves obstacle avoidance for each robot and a motion strategy where one of the robots, the master, follows a pre-specified path and the other teammates, the slaves, have a constrained motion aimed at having the master in a visible detectable range from, at least, one of the slaves. The master stops whenever it is to loose visible contact with the slaves and these move to approach the master and to provide it with visible landmarks.

The localization problem has to deal with the uncertainty in the motion and in the observations. Common techniques of mobile robot localization are based on probabilistic approaches, that are robust relative to sensor limitations, sensor noise and environment dynamics. Using a probabilistic approach, the localization turns into an estimation problem with the evaluation and propagation of a probability density function (pdf). This problem increases when operating with more than one robot, with complex sensor's model or during long periods of time. A possible solution for the localization problem is the PF approach [7] [8], that tracks the variables of interest. Multiple copies (particles) of the variable of interest (the localization of each robot) are used, each one associated with a weight that represents the quality of that specific particle.

The major contribution of this chapter is the use of a PF approach to solve the cooperative localization problem allowing the robots to follow a reference path in scenarios where no map and no global positioning systems are available and human intervention is not possible. The simulation is implemented with a fleet of three car-like vehicles.

This chapter is organized as follows. Section 1 presents the chapter motivation and an overview of related work on mobile robot navigation using PF. Section 2 introduces the notation and the principles of Particle-Filter. Section 3 explains the main contributions of the chapter, namely the cooperative localization and motion strategy using PF in a team of robots. Simulation results obtained with some robots in an environment with obstacles are presented in Sect. 4. Section 5 concludes the chapter and presents directions for further work.

2 Single Robot Navigation

This section describe the robots used in the work and presents the basis for the PF localization approach for each single robot in the particular application described in Sect. 1.

2.1 Robot Characterization

It is assumed that each robot is equipped with a 2D laser scanner, with a maximum range capability, ρ_{max}, over a limited rear and front angular field of view of width $2\varphi_{max}$, as represented in Fig. 1-right. Consequently, the robots are able to measure the distance, ρ, and the direction, φ, to obstacles in their close vicinity and are prepared to avoid obstacles. This sensor supports the robot perception to evaluated the distance and the orientation under which the beacons (the fixed beacon and/or the other robots in the team) are detected. It is also assumed that each robot is able to recognize the other robots and the fixed beacon based on the same laser scanner or using, for instance, a vision system.

Maneuvering car/cart-like vehicles is a difficult task, when compared with other type of vehicles, for instance mobile robots with differential kinematic system. Due to this fact, car/cart-like kinematic system has been considered to emphasize the capabilities of the purposed localization technique when applied on these type of vehicles. The kinematic model of a car/cart-like vehicle, presented in Fig. 1, is expressed by

$$\begin{bmatrix} \dot{x} \\ \dot{y} \\ \dot{\theta} \end{bmatrix} = \begin{bmatrix} \cos(\theta) & 0 \\ \sin(\theta) & 0 \\ 0 & 1 \end{bmatrix} \cdot \begin{bmatrix} v(t) \\ v(t)\frac{\tan\phi(t)}{l} \end{bmatrix} \tag{1}$$

where (x, y) is the position, θ is the vehicle's heading, both relative to a global referential, $v(t)$ is the linear velocity and $\phi(t)$ is the steering angle that defines the curvature of the path and l is the distance between the rear and front wheels (see Fig. 1-left).

The proposed navigation strategy is optimized for car/cart like vehicles (as described in Sect. 3), but the approach can be implemented in robots with similar capabilities of motion and perception.

2.2 Localization based on Particle-filter

The pose of a single robot is estimated based on a PF [7]. Each particle i in the filter represents a possible pose of the robot, i.e.,

$$^{i}p = \left(^{i}x, {}^{i}y, {}^{i}\theta\right). \tag{2}$$

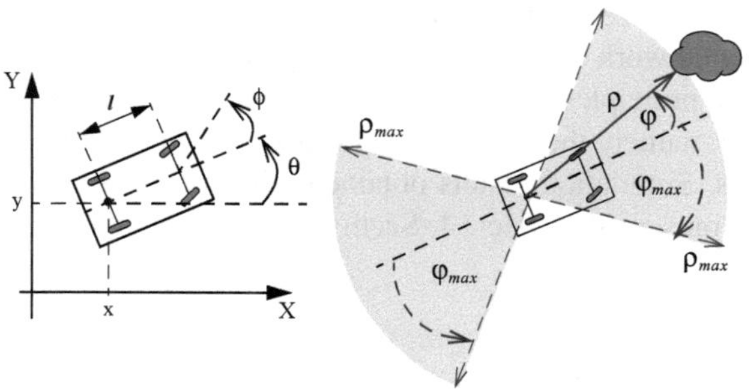

Fig. 1 Robot's kinematic model (left) and sensorial capabilities (right).

In each time interval, a set of possible poses (the cloud of particles) is obtained. Then, the particles are classified according to the measurement obtained by the robot at its real pose. Simulations of the measurements are performed, considering each particle as the actual pose of the robot and the probabilistic model of the sensor. The believe of each particle being the real pose of the robot is evaluated taking into account these measurements. The real pose of the robot is estimated by the average of the cloud of particles based on their associated believes. An important step of the method is the re-sampling procedure, where the less probable particles are removed from the cloud, bounding the uncertainty of the robot pose [8].

Differently from other approaches, where a set of fixed beacons are distributed for localization purposes [9], this chapter considers the existence of a single fixed beacon L at (x_L, y_L), as represented in Fig. 2a, that provides a fixed reference each time the robot observes it. The fixed beacon has sensorial capabilities similar to those of the robots, with limited range detection but over a 360° field of view (see Fig. 2a).

The particles associated to the robot are weighted taking into account the observations made by the robot and by the fixed beacon. Let ρ_{LB} and φ_{LB} be the distance and the relative orientation of the robot B obtained from the fixed beacon measure-

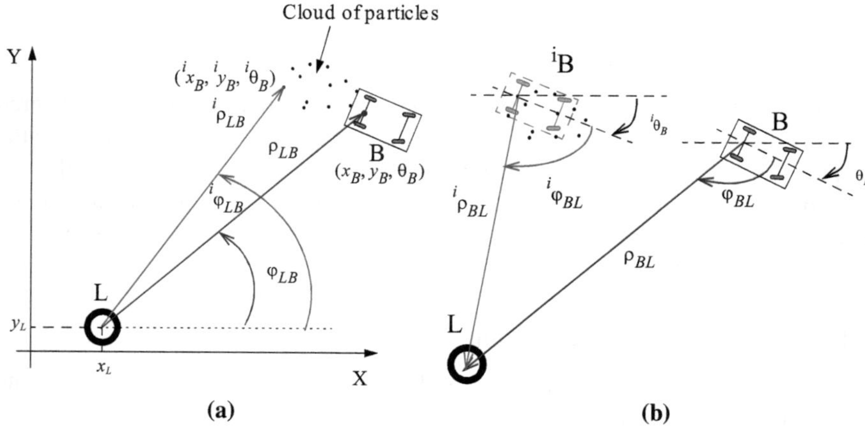

Fig. 2 **(a)** Robot and a fixed beacon observation and **(b)** each particle and robot observations.

ment (Fig. 2a). Similarly, ρ_{BL} and φ_{BL} are the distance and the relative orientation of the beacon obtained from the robot measurement (Fig. 2b). Additionally, each particle associated to the robot B defines ${}^{i}\rho_{LB}$ and ${}^{i}\varphi_{LB}$ (which represents the fixed beacon measurement if the robot was placed over the particle i):

$$ {}^{i}\rho_{LB} = \sqrt{(x_L - {}^{i}x_B)^2 + (y_L - {}^{i}y_B)^2} + \xi_\rho \tag{3} $$

$$ {}^{i}\varphi_{LB} = \arctan\left(\frac{{}^{i}\rho_{LB_y}}{{}^{i}\rho_{LB_x}}\right) + \xi_\varphi \tag{4} $$

where ξ_ρ and ξ_φ represent the range and angular uncertainties of the laser sensor, considered as Gaussian, zero mean, random variables. The simulated measurements for each particle define two weights, ${}^{i}P_\rho$ and ${}^{i}P_\varphi$, related to the acquired sensorial data as,

$$ {}^{i}P_\rho = \kappa_\rho \cdot \left|\rho_{LB} - {}^{i}\rho_{LB}\right|^{-1} \tag{5} $$

$$ {}^{i}P_\varphi = \kappa_\varphi \cdot \left|\varphi_{LB} - {}^{i}\varphi_{LB}\right|^{-1} \tag{6} $$

where κ_ρ and κ_φ are coefficients that allow to compare both distances. Thus, a weight that determines the quality of each particle according to the beacon measurement is given by

$$ {}^{i}w_{LB} = \begin{cases} \eta_{LB} \cdot {}^{i}P_\rho \cdot {}^{i}P_\varphi & if \;\; V_{LB} = 1 \\ 1 & if \;\; V_{LB} = 0 \end{cases} \tag{7} $$

where η_{LB} is a normalization factor and V_{LB} is a logic variable whose value is 1 if L observes B, and 0 otherwise. On the other hand, each particle also defines ${}^{i}\rho_{BL}$, ${}^{i}\varphi_{BL}$ (which represents the robot measurement of the distance and angle with which the fixed beacon will be detected if the robot were placed over the particle i, i.e., in ${}^{i}B$, see Fig. 2b) and V_{BL} (a logic variable whose value is 1 if B observes L and 0 otherwise). The weight that determines the quality of each particle according to the measurement to the beacon is similar to (7), by switching the indices "B" and "L". Therefore, the total weight of the particle i associated with the robot B, ${}^{i}w_B$, is given by

$$ {}^{i}w_B = {}^{i}\lambda_B \cdot {}^{i}w_{LB} \cdot {}^{i}w_{BL} \tag{8} $$

where ${}^{i}\lambda_B$ is a normalization factor. Given the formulation of the weights, if the vehicle is not able to observe the fixed beacon and the fixed beacon is not able to detect the robot, or equivalently $V_{LB} = 0$ and $V_{BL} = 0$, all the particles have the same importance. As will be shown in Sect. 4, when no observations are available, re-sampling is not possible and therefore localization uncertainty increases. Once the mobile robot observes again the fixed beacon or vice-versa, re-sampling is possible and uncertainty decreases.

2.3 Path Following and Obstacle Avoidance

Due to the limited range of perception featured by the sensors, the robot may navigate in areas where the perception of the fixed beacon is not available. To cope with this constraint the navigation strategy should assure that the vehicle will return to the initial configuration, i.e., the initial position where the vehicle and the beacon are able to detect each other. Different strategies can be applied to drive the robot to the initial configuration.

One strategy is based on a set of way points along a closed path. However, the environment is still unknown and some way points could lay over an obstacle. When this occurs the navigation algorithm may endows to an unreachable point leading to circular paths around obstacles. To overcome this problem another strategy is adopted. The idea is to build a continuous path that the robot has to follow with an obstacle avoidance capability based on reactive navigation techniques.

Different approaches have been proposed for path following, where the robot position is estimated by Extended Kalman Filter (EKF) using Global Position System (GPS) and odometry [10]. In the present work no GPS is available, and rather than using EKF, a PF approach is used.

To accomplish an accurate navigation, each robot follows a path previously defined and the "Pure-pursuit" algorithm [11] is applied for path-tracking. This algorithm chooses the value of the steering angle as a function of the estimated pose at each time instant. Without a previous map, different obstacles may lay over the path and, consequently, a reactive control algorithm will be applied if the vehicle is near an obstacle [12] [13]. The path-tracking and the obstacle avoidance behaviours are combined by means of a hybrid control structure in such a way that the robot is able to avoid the obstacle and continue the path-tracking when the vehicle is near the path [14] [15].

3 Cooperative Localization and Navigation

This section illustrates the cooperation among robots to improve the robots localization when the detection of the fixed beacon is not available. The method takes advantage of the measurement of the different teammates in order to estimate the pose of each robot. Moreover, a cooperative motion strategy is presented that allows the team of robots to perform a more efficient localization. Robots are identified by numbers 1 to 3 and the fixed beacon is numbered with 4 as represented in Fig. 3.

3.1 Cooperative Robots Localization based on Particle-filter

The pose of each robot is estimated by the PF technique, considering all the robots involved, i.e., the measurements acquired by all the robots. When a robot moves, only the weights of the particles of that robot are updated. However, the estimation of the position using PF is performed taking into account the measurements acquired by the sensors of that robot and the other robots.

Similarly for the case involving the fixed beacon, if B is a robot moving and A is a static robot, the observation of robot B performed by the robot A defines the values ρ_{AB} and ϕ_{AB}. In the same way, the observation of robot A performed by robot B defines ρ_{BA} and ϕ_{BA}, which is similar to Fig. 2, replacing the beacon L by the robot A. Thus, each particle associated with the robot B defines the values ${}^{i}\rho_{AB}$, ${}^{i}\rho_{BA}$, ${}^{i}\phi_{AB}$ and ${}^{i}\phi_{BA}$. These values can be obtained applying expressions similar to (3) and (4).

Nevertheless, in this case, instead of using the position of the fixes beacon, the estimated position of the robot A $(\hat{x}_A, \hat{y}_A, \hat{\theta}_A)$ is considered. Therefore, from the measurement of robot A and B, the weights ${}^i w_{AB}$ and ${}^i w_{BA}$, which determine the quality of each particle according to both measurements, can be obtained. The weights are calculated by expressions similar to (5), (6) and (7).

Therefore, all the measurements acquired by a robot depend on the estimated position of the other robots. The evaluation for the poses estimation includes cumulative errors and, consequently, the weight of the particles are influenced by the measurement provided by the fixed beacon (when it is able to observe the robots or the robots are able to observe the fixed beacon). Then, the weight of the particle i of the robot j is given by

$$ {}^i w_j = \begin{cases} \prod_{n=1, n \neq j}^{3} {}^i w_{nj} \cdot {}^i w_{jn} & if V_{4j} + V_{j4} = 0 \\ {}^i w_{4j} \cdot {}^i w_{j4} & if V_{4j} + V_{j4} = 1 \end{cases}, $$

whose value depends on the logic variables V_{j4} (equal to 1 if the robot j can observe the fixed beacon) and V_{4j} (equal to 1 if the fixed beacon is able to observe the robot j).

3.2 Motion Strategy

In this approach, the objective of the team navigation is the exploration of an unknown environment. A closed path is defined to be followed by one of the robots of the team. For a correct path following, localization is carried out using the PF approach described in Sects. 2.1 and 2.2. As the path may be such that the fixed beacon is not detected, to achieve this goal some robots operate as beacons to their teammates, with this role interchanged among them according to a given motion strategy.

There are other approaches exploring the same idea. In [16] and [17] a team of Milibots is organized in such a way that part of the team remains stationary providing known reference points for the moving robots. This approach, named as "leap-frogging" behaviour, uses trilateration to determine the robots position. To apply trilateration, three beacons are required. During the main part of the navigation problem considered in this chapter, only two robots (acting as beacons) are detected from the robot whose localization is under evaluation, as the fixed beacon is not detected. There are situations where only a single robot is visible and trilateration is not useful.

In [8] a collaborative exploration strategy is applied to a team of two robots where one is stationary, and acts as a beacon, while the other is moving, switching the roles between them. In that approach, PF technique is applied for cooperative localization and mapping. Nevertheless, it is supposed that the heading of the observed robot can be measured. Since this estimation is not trivial, this chapter presents a different approach for PF.

In the proposed methodology the team is divided in two categories of robots. One of the robots is considered the "master" having the responsibility of performing an accurate path following of a previously planned trajectory. The other two robots, the "slaves", play the role of mobile beacons. The master robot is identified by the number 1, and the slave robots with 2 and 3 (see Fig. 3). Initially, the master follows the planned path until it is not able to detect, at least, one slave. Once this occurs, the master stops and the slave robots start to navigate sequentially. They try to reach different poses where they will act as a beacon for the master. These poses can be determined previously, taking into account the path and a criterion of good coverage, or can be established during the navigation, considering the position where the master stopped. In both cases, the objective is to "illuminate" the master navigation by the slave mobile beacons in such a way that it can be located accurately. At this stage, once the master stops and one of the slaves is moving, the master and the other slave play the role of beacon for the slave which is moving.

According to the previous statement and considering the type of sensors and the angular field of view, a strategy for motion of the slaves is considered. The strategy is based on the idea of building a triangular formation after the master stops, what happens when the master is not able to detect, at least, one slave. Using this triangular configuration the vehicles are allowed to estimate their position by means of the measurement of their relative positions with the minimum possible error. Furthermore, slave robots give the master a large position estimation coverage. This configuration is fundamental if the view angle of the sensors is constrained, otherwise, the robots can see each other in any configuration.

Once the master has stopped, two points P_2 and P_3 are generated (see Fig. 3) and the slaves will have to reach these points with the same orientation of the master so that they can see each other. P_2 and P_3 are calculated by taking into account the angular field of view of the robots. Hence, P_2 is located along the master's longitude axis and P_3 is such that the master, P_2 and P_3 define an isosceles triangle.

Therefore, when the master robot stops, a path is generated in such a way that it connects each current slave position with the goal point (P_2 or P_3). This path has also to accomplish the curvature constraint and allow the slave vehicles to reach the goal point with a correct orientation. For this purpose, β-Splines curves [18], have been applied as represented in Fig. 3.

Master and slave perform the path following and the collision avoidance in a sequential way. The slave 2 starts moving when the master has stopped. The slave 3 begins to move once the slave 2 reaches its goal. Finally, the master starts moving when the slave 3 has reached its target configuration.

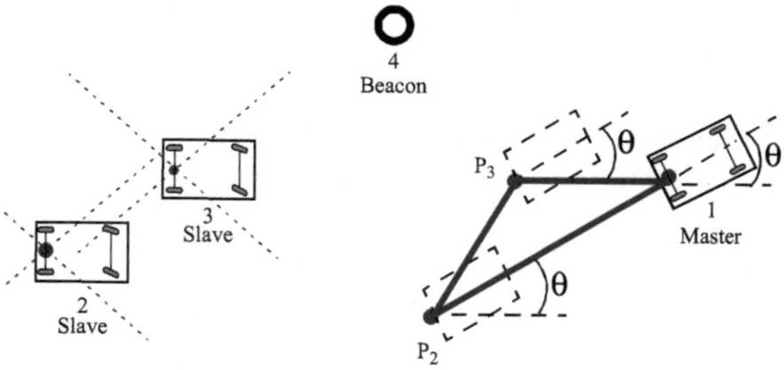

Fig. 3 β-spline generation.

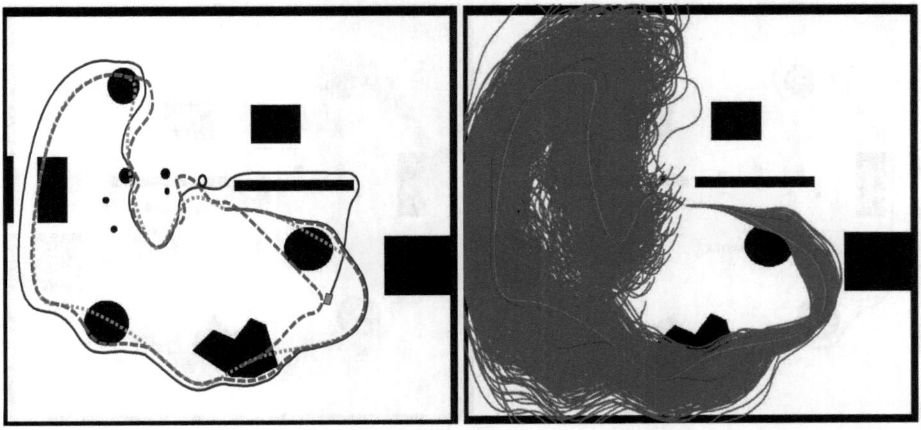

(a) (b)

Fig. 4 (a) Single robot navigation and **(b)** evolution of the cloud of particles.

4 Simulation Results

Different experiments have been implemented to test the proposed approach, performed with different numbers of teammate: one, two and three robots. Several trajectories have been generated to evaluate the influence of the length and the shape of the path on the localization performance.

The algorithm was tested by considering that the robots perform both clockwise or anti-clockwise loops. Figure 4a presents the desired path (dotted-line), the real path (continuous line) and the estimated path (dashed-line) obtained by the navigation of a single robot, i.e., no cooperative localization is considered. The fixed beacon is represented by the non-filled circle. Figure 4b presents the evolution of the cloud of particles along the navigation process, which conveys the associated uncertainty. The cloud

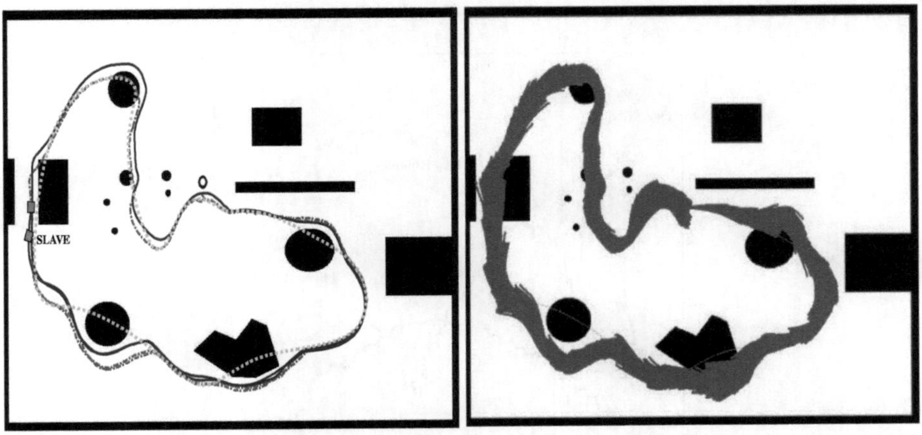

(a) (b)

Fig. 5 (a) Two robots in a big loop and **(b)** evolution of the cloud of the master robot.

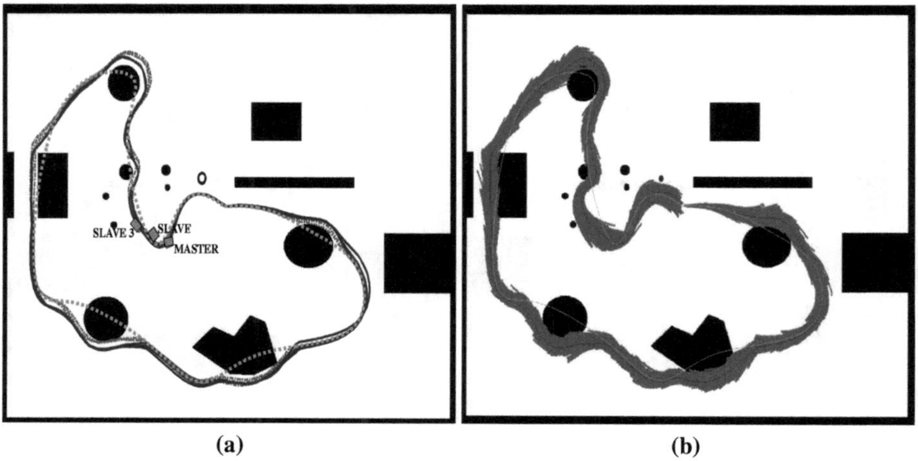

Fig. 6 (a) Three robots in a big loop and **(b)** evolution of the cloud of the master robot.

shrinks when the robot observes the fixed beacon and enlarges when no observation is acquired.

Figures 5a and 6a illustrate the evolution of the real and the estimated path when the cooperative navigation is applied on a team of two and three robots, respectively, along the same desired path. Figures 5b and 6b present the evolution of the cloud of particles for the master robot. The estimation is improved, mainly when using three robots and, consequently, the navigation of the master robot is closer to the desired path. In each experiment the robots closed the loop three times. The improvement in the pose estimation when applying the cooperative localization technique is shown in these figures as the real and the estimated path are closer than in the previous experiment, with a single robot. Figure 7 presents the error of the position estimation in the previous three experiments along one loop, with different number of teammate. It is remarkable

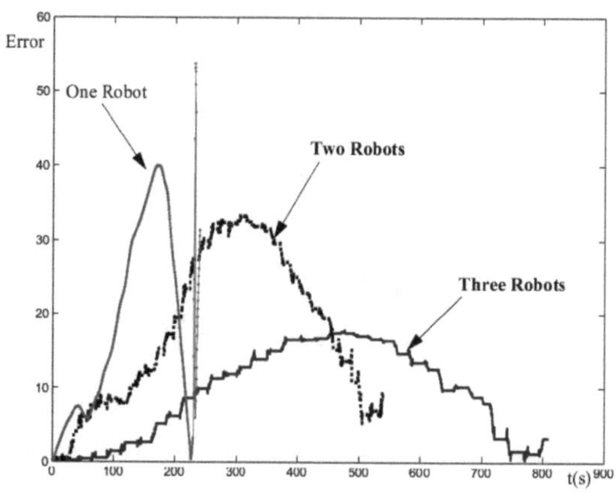

Fig. 7 Position estimation error along the big loop for teams with 1, 2 and 3 robots.

that the error decreases when the number of the robots in a team increases. However, the time the robots take to close the loop increases with the number of teammate, since the motion strategy for cooperative localization requires that part of the team remain stationary while one of the robot is navigating.

Figure 8 illustrates two experiments with different paths in which a team of three robots closed the loop several times. Both paths are shorter than the path of Fig. 5a. In the path of Fig. 8b (the one whose shape looks like a Daisy) the robots navigate close to the beacon for three times.

Figure 9 presents the error of the position estimation along one loop of the experiment of Figs 8a,b and 5a. It illustrates that shorter paths provide lower error and therefore better performance. In addition, this figure also illustrates that the pose estimation can be improved by modifying the shape of the path as is the case of the experiment of the "Daisy path" (Fig. 8b). In this experiment, the error remains bounded presenting lower values than the other experiments. This improvement is achieved due to the effect of navigating near the beacon and applying its perception for updating the weight of the robots particles. Obviously, the pose estimation improves when the time periods where the robot navigates near the beacon increases. A conclusion is, if the exploration of wide and large spaces is needed, trajectories similar to the daisy path are preferred, i.e., trajectories where the team navigates near the beacon several times.

5 Conclusions and Open Issues

This chapter presented a Particle-Filter approach to solve the cooperative localization problem for a small fleet of car-like vehicles in scenarios without map or GPS and no human intervention. A team of three robots and a fixed beacon have been considered in such a way that the robots take advantage of cooperative techniques for both localization and navigation. Each robot serves as an active beacon to the others, working as

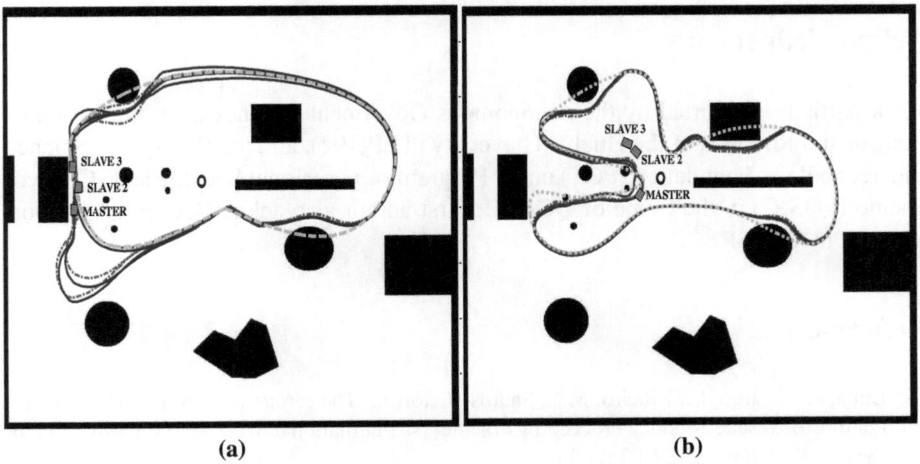

Fig. 8 Cooperative navigation along two different paths: **(a)** short loop and **(b)** Daisy loop.

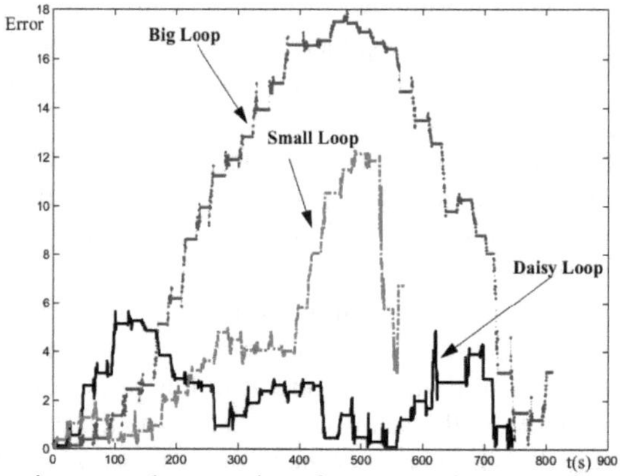

Fig. 9 Position estimation error along one loop for a team of three robots along different paths.

a fixed reference and, at the same time, providing observations to the robot which is moving. With this approach, robots are able to follow a previous calculated path and avoid collision with unexpected obstacles.

Moreover, the proposed approach has been validated by different simulated experiments. Different path and number of teammates have been considered. The experiments illustrate that the number of teammates decreases the estimation error. Likewise, the length of the planned path affects to the quality of the estimation process. However, for paths with similar length, the estimation procedure can be improved by changing its shape.

The following step is the implementation of this approach in a team of real car-like robots. There are still open issues requiring further research, in particular the number of particles and other tuning parameters, the matching between beacons and other mobile robots and the implementation of active beacons.

Acknowledgements

Work partially supported by the Autonomous Goverment of Andalucia, the Group of Control and Robotics of the Huelva University (TEP-192), and the Portuguese Science and Technology Foundation (FCT) under Programa Operacional Sociedade do Conhecimento (POS_C) in the frame of QCA III. First author also acknowledges the support received by the ISR.

References

1. Lima, P., Custdio, L., Ribeiro, M.I., Santos-Victor, J.: The rescue project: cooperative navigation for rescue robots. Proceedings of the 1st International Workshop on Advances in Service Robotics, ASER'03 (2003)
2. Madhavan, R., Fregene, K., Parker, L.E.: Distibuted cooperative outdoor multirobot localization and mapping. Autonomous Robots **2** (2004) 23–39

3. Gustavi, T., Hu, X., Karasalo, M.: Multi-robot formation control and terrain servoing with limited sensor information. Preprints of the 16th IFAC World Congress (2005)
4. Tang, K.W., Jarvis, R.A.: An evolutionary computing approach to generating useful and robust robot team behaviours. Proc. of 2004 IEEE/RSJ IROS **2** (2004) 2081–2086
5. Ge, S.S., Fua, C.H.: Complete multi-robot coverage of unknown environments with minimum repeated coverage. Proc. of the 2005 IEEE ICRA (2005) 727–732
6. Martinelli, A., Pont, F., Siegwart, R.: Multi-robot localization using relative observations. Proc. of the 2005 IEEE Int. Conf. on Robotics and Automation (2005) 2808–2813
7. Thrun, S., Fox, D., Burgard, W., Murphy, F.D.: Robust monte carlo localization for mobile robot. Artificial Intelligence Magazine **128** (2001) 99–141
8. Rekleitis, I.: A particle filter tutorial for mobile robot localization Technical Report TR-CIM-04-02, Centre for Intelligent Machines, McGill University, Montreal, Quebec, Canada (2004)
9. Betke, M., Gurvits, L.: Mobile robot localization using landmarks. IEEE Transaction on Robotics and Automation **13** (1997) 251–263
10. Grewal, M.S., Andrews, A.P.: Kalman Filtering Theory and Practice. Prentice-Hall, Englewood Cliffs and New Jersey (1993)
11. Ollero, A.: Robotica: Manipuladores y Robots moviles. Marcombo (2001)
12. Cuesta, F., Ollero, A., Arrue, B., Braunstingl, R.: Intelligent control of nonholonomic mobile robots with fuzzy perception. Fuzzy Set and Systems **134** (2003) 47–64
13. Cuesta, F., Ollero, A.: Intelligent Mobile Robot Navigation. Springer, The Netherlands (2005)
14. Cuesta, F., Gomez-Bravo, F., Ollero, A.: A combined planned/reactive system for motion control of vehicles manoeuvres. Proc. of the IFAC W. on Motion Control (1998) 303–308
15. Gomez-Bravo, F., Cuesta, F., Ollero, A.: A new sequential hybrid control structure for car and tractor-trailer parallel parking. Proc. of the 7th IFAC Symposium on Robot Control (SYROCO) (2003) 605–610
16. Grabowski, R., Khosla, P.: Localization techniques for a team of small robots. Proc. of the IEEE/RSJ Int. Conf. on Intelligent Robots and Systems **2** (2001) 1067–1072
17. Navarro-Serment, L.E., Grabowski, R., Paredis, C., Khosla, P.: Localization techniques for a team of small robots. IEEE Rob. and Aut. Magazine **9** (2002) 31–40
18. Barsky, B.: Computer graphics and geometric modelling using β-Splines. Springer-Verlag, New York (1987)

Interaction Control Experiments for a Robot with one Flexible Link

[enidh]L. F. Baptista, [ist]N. F. S. Bóia, [ist]J. M. M. Martins and [ist]J. M. G. Sá da Costa

[ist]Technical University of Lisbon, Instituto Superior Técnico
Department of Mechanical Engineering
{martins, sadacosta}@dem.ist.utl.pt
GCAR/IDMEC, Avenida Rovisco Pais, 1049-001 Lisboa Codex, Portugal
[enidh] Escola Náutica Infante D. Henrique, Department of Marine Engineering
Av. Eng. Bonneville Franco, 2770-058 Paço de Arcos, Portugal
luisbaptista@enautica.pt

Abstract. One of the major drawbacks of flexible-link robot applications is its low tip precision, which is an essential characteristic for applications with interaction control with a contact surface. In this work, interaction control strategies considering rigid and flexible contact surfaces are applied on a two degrees of mobility flexible-link manipulator. The interaction strategies are based on the closed-loop inverse kinematics algorithm (CLIK) to obtain the angular references to the joint position controller. The control schemes were previously tested by simulation and further implemented on the flexible-link robot. The obtained experimental results exhibit a good force tracking performance, especially for a rigid surface, and reveal the successful implementation of these control architectures for a robot with one flexible link.

Keywords. Flexible-link manipulator, closed-loop inverse kinematics, interaction control, real-time control.

1 Introduction

The evolution of industrial manufacturing, lead to the necessity of optimize the production, where the main goal is to achieve best quality products at lower prices. The manipulator robot is a crucial automation equipment that fulfill these requirements, due to its high productivity and easy adaptation to a large number of complex and repetitive tasks. The manipulator robots have also the ability to work in adverse environments to the human workers. Due to these characteristics, the study of robot manipulator control has received a growing attention by a lot of researchers during the last decades, in order to design robots with high performance [1].

In general, industrial robots have rigid mechanical elements which leads to a high power consumption. To overcome this disadvantage, lightweight and flexible links have been considered in the construction of new robots. These new links allow the same mobility capacity as the rigid robots with a lower power consumption. Also, due to the lighter weight of the links, the interaction with the environment, especially in the case of collision, cause less damage.

When a manipulator robot executes an interaction task, the tip or end-effector enters in contact with the environment and a certain force is exerted on the surface. Since it's necessary to achieve an high precision tip position to obtain a good interaction force control, advanced control algorithms have been developed to obtain a high force tracking performance [2]. However, flexible-link manipulators exhibit an important drawback in comparison with rigid robots, due to the difficulty in control its tip or end-point position. The flexibility rises the dynamic coupling, the non-linearities, and gives to the robot infinite degrees of freedom derived from the vibration modes of the flexible elements. Due to these vibrations, the system becomes a non-minimum phase system [3]. The zeros in the right semi-plan, due to the non minimum phase lead to an unstable system, when the tip position is directly controlled through feedback.

To avoid these drawbacks, several techniques to efficiently control flexible-link robots have been studied. The control of a flexible manipulator at the joint level has been established by a lot of authors like Khorrami and Jain [4] for the tracking problem and Vandegrift et al. [5] for the regulation problem, among others. One of the proposed strategies to solve the inverse kinematics problem for flexible arms, was derived from the closed loop inverse kinematics algorithm (CLIK) developed for rigid manipulators [6]. The inverse kinematics formulation with feedback of joint coordinates and deflection variables for constrained flexible manipulators was developed by Siciliano [7] and Siciliano and Villani [8]. Finally, more complex algorithms for solving the inverse kinematics problem at high speed velocities with flexible manipulators have been proposed by Cheong et al. [9].

The purpose of this work is to obtain experimental results with interaction control algorithms for a planar robot with two revolute joints and two links, where the second link is flexible. The control strategies were implemented considering the CLIK algorithm to obtain the desired angular references to the joint position controller. The interaction control algorithm was applied considering rigid and flexible contact surfaces, respectively.

The outline of the chapter is as follows. Section 2 describes the flexible link forward and inverse kinematics formalism and the closed loop inverse kinematics algorithm (CLIK). In Sect. 3 a brief overview of the CLIK-based interaction controllers for rigid and flexible surfaces are described. Section 4 describes the planar flexible-robot setup, the hardware and software control architecture considered for the real-time experiments. In Sect. 5 the obtained interaction control results in real-time are presented. Finally, in Sect. 6 some conclusions are drawn.

2 Flexible Link Kinematics

Let us consider a planar robot with two degree of mobility, where the first link is rigid and the second link is flexible, as depicted in Fig. 1.

The robot's flexible link can be modeled as an Euler-Bernoulli cantilever, with length L. The flexible link is attached to a rigid joint. When a torque τ is applied to the rigid joint, the flexible link is rotated by an angle θ between the body frame $\{X,Y,Z\}$ and the base reference frame $\{X_0,Y_0,Z_0\}$, as illustrated in Fig. 2.

In this figure, $v(x,t)$, represents the lateral displacement point x along the flexible link, relative to X axis. Thus, the projection of x on X axis will be given by $L - u(x,t)$ coordinate. Two models are considered to obtain the length reduction coordinate

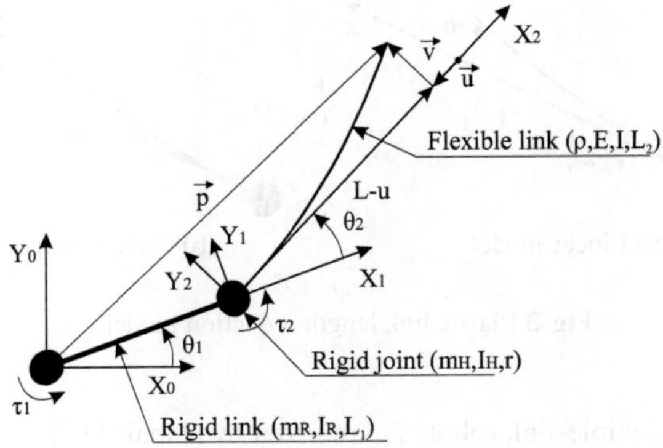

Fig. 1 Planar flexible robot schematics.

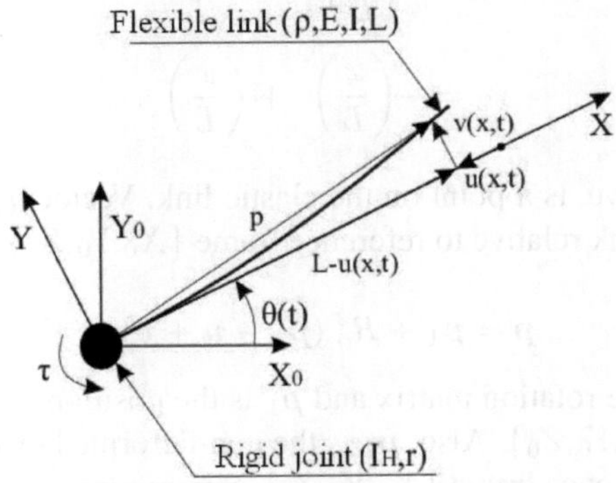

Fig. 2 Flexible link scheme deflection caused by an applied torque.

$u(x,t)$: linear and quadratic models. The linear model approach considers that length reduction is null, i.e. $u(x,t) = 0$. In the quadratic model approach, length reduction is calculated by the following expression

$$u(x,t) = -\frac{1}{2}\int_r^x \left(\frac{dv}{d\xi}\right)^2 d\xi \tag{1}$$

In Fig. 3 both length reduction approach due to an elastic link deflection are represented.

In the following, the Assumed Modes discretizing method [10] will be considered for the lateral displacement v calculation. This method consider

$$v(x,t) = \chi(x)\delta(t) \tag{2}$$

where χ are the normalized mode shapes and δ are the generalized elastic coordinates. Considering only the two first vibration modes, v is given by

$$v(x,t) = \sum_{k=1}^{2} \chi_i(x)\,\delta_i(t) \tag{3}$$

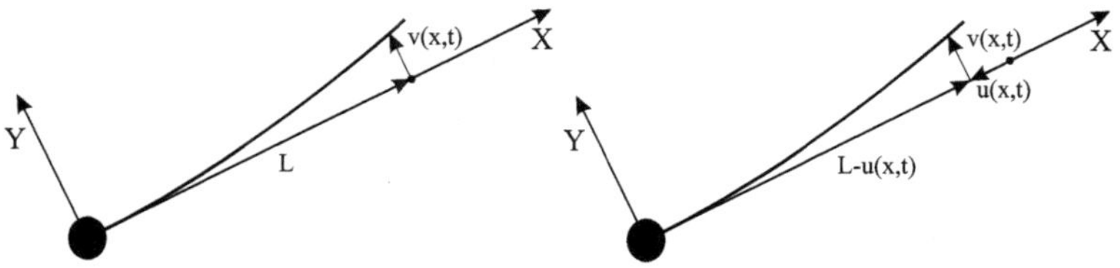

(**a**) Linear model (**b**) Quadratic model

Fig. 3 Elastic link length reduction models.

For this particular flexible-link robot, χ_i are given by Nabais [11]:

$$\chi_1 = 3\left(\frac{x}{L}\right)^2 - 2\left(\frac{x}{L}\right)^3$$
$$\chi_2 = -\left(\frac{x}{L}\right)^2 + \left(\frac{x}{L}\right)^3 \tag{4}$$

where x as refered above, is a point on the elastic link. Vector $\boldsymbol{p}$ represented in Fig. 1 is the position on the link relative to reference frame $\{X_0,Y_0,Z_0\}$. This vector is represented by

$$\boldsymbol{p} = \boldsymbol{p_1} + R_0^2\,(\boldsymbol{p_2} - \boldsymbol{u} + \boldsymbol{v}) \tag{5}$$

where R_0^2 represents the rotation matrix and $\boldsymbol{p_1}$ is the position on the first link relative to reference frame $\{X_0,Y_0,Z_0\}$. Also, $\boldsymbol{p_2}$ is the non-deformed second link end-effector position relative to reference frame $\{X_2,Y_2,Z_2\}$. Summing $\boldsymbol{p_2}$, $\boldsymbol{u}$ and $\boldsymbol{v}$, leads to:

$$\boldsymbol{p} = \boldsymbol{p_2} - \boldsymbol{u} + \boldsymbol{v} \tag{6}$$

The rotation matrix R_0^2 that describes the position p relative to reference frame $\{X_0,Y_0,Z_0\}$ is represented by:

$$R_0^2 = \begin{bmatrix} \cos(\theta_1+\theta_2) & -\sin(\theta_1+\theta_2) \\ \sin(\theta_1+\theta_2) & \cos(\theta_1+\theta_2) \end{bmatrix} \tag{7}$$

Considering that p is the end-effector or tip position, $\boldsymbol{p}$ represents the forward kinematics of the flexible-link robot. Thus, the forward kinematic equations are given by:

$$\begin{bmatrix} p_x \\ p_y \end{bmatrix} = \begin{bmatrix} L_1\cos(\theta_1) \\ L_1\sin(\theta_1) \end{bmatrix} + R_0^2 \begin{bmatrix} L_2 - u \\ v \end{bmatrix} \tag{8}$$

which leads to the following equations:

$$\begin{aligned} p_x &= L_1\cos(\theta_1) + (L_2-u)\cos(\theta_1+\theta_2) \\ &\quad - v\sin(\theta_1+\theta_2) \\ p_y &= L_1\sin(\theta_1) + (L_2-u)\sin(\theta_1+\theta_2) \\ &\quad + v\cos(\theta_1+\theta_2) \end{aligned} \tag{9}$$

where u and v are given by Eqs. (1) and (3), respectively.

The inverse kinematics equations relate the cartesian position coordinates, given by Eq. (9), and the joint θ and deflection δ coordinates. Replacing the Eqs. (1)–(3) into Eq. (9), two equations and four unknown variables $(\theta_1, \theta_2, \delta_1, \delta_2)$ are obtained. Thus, the system is undetermined and other methods should be exploited to overcome this problem.

2.1 CLIK Algorithm

To solve the problem presented above, the Closed Loop Inverse Kinematics algorithm (CLIK) developed for rigid robots was adopted in this work, according to Siciliano [7]. This algorithm feeds back the joint angles θ calculated by the CLIK algorithm in a closed loop dynamic system in order to obtain the reference values to the joint position controller. This algorithm is given by Siciliano and Villani [8]:

$$\dot{\theta}_d = J_p^T(\theta) K_P (p_d - p) \tag{10}$$

where

- $\dot{\theta}_d$ are the desired joint velocities,
- $J_p = J_\theta$ i.e., the rigid part of the Jacobian matrix,
- p_d is the desired tip position,
- p is the cartesian position determined by forward kinematics with coordinates θ,
- K_P is a proportional gain matrix.

To obtain the joint references θ_d for real-time control, it is necessary to integrate the joint velocities given by Eq. (10). Thus, the discrete version of θ_d, is given by:

$$\theta_d(t_{k+1}) = \theta_d(t_k) + T_s J_p^T(\theta_d(t_k)) K_P (p_d(t_k) - p(t_k)) \tag{11}$$

where

- θ_d are the reference joint angles for the position controller,
- T_s is the sampling period.

3 Interaction Control

3.1 Rigid Surface

Let us consider that the robot is in contact with a rigid surface. The restriction imposed by the surface, is described by

$$\phi(p) = \phi(k(\theta, \delta)) = 0 \tag{12}$$

Assuming that the robot is in a static condition, the deflections satisfy the following equation:

$$K\delta = -J_\delta^T(\theta, \delta)\lambda j_\phi \tag{13}$$

where K is the robot stiffness matrix, j_ϕ is the equation gradient (12), defined by

$$j_\phi = \left(\frac{\partial \phi}{\partial p}\right)^T \tag{14}$$

and λ is the Lagrange multiplier associated with the restriction. From Eq. (13), δ is given by

$$\delta = -K^{-1} f \tag{15}$$

where,

$$f = J_\delta^T(\theta, \delta)\lambda j_\phi \tag{16}$$

Differentiating Eq. (15) leads to

$$\dot{\delta} = -K^{-1} J_f(\theta, \delta)\dot{\theta} \tag{17}$$

where

$$J_f = \frac{\partial f}{\partial \theta} = \lambda j_\phi \frac{\partial J_\delta^T(\theta, \delta)}{\partial \theta} \tag{18}$$

and,

$$\dot{\theta} = J_p^T(\theta, \delta) K_P (p_d - p) \tag{19}$$

The Jacobian J_p is defined by

$$J_p = J_\theta(\theta, \delta) - J_\delta(\theta, \delta) K^{-1} J_f(\theta, \delta) \tag{20}$$

where e_p is the difference between the desired tip position and the cartesian position determined by the robot forward kinematics. The joint angles and the deflection coordinates are given by the CLIK algorithm. Discretizing these coordinates, leads to [7]

$$\theta_d(t_{k+1}) = \theta_d(t_k) + T_s J_p^T(\theta_d(t_k), \delta_d(t_k)) K_P e_p(t_k) \tag{21}$$

and

$$\delta_d(t_{k+1}) = -K^{-1}(J_\delta^T(\theta_d(t_k), \delta_d(t_k))\lambda(t_k) j_\phi \tag{22}$$

Notice that the Jacobian J_p not only depends on θ coordinates, but also depends on δ coordinates. This happens because δ depends on f_d, i.e. the force that should be applied by the robot on the surface.

The overall interaction controller applies a joint position PD controller plus the desired force f_d. The control law is described by

$$\tau = K_p(\theta_d - \theta) - K_d\dot{\theta} + J_\theta^T(\theta, \delta) f_d n \tag{23}$$

where n is the normal to the surface and f_d is the desired force. Notice that the PD controller doesn't control directly the interaction force between the tip of the link and the contact surface. In fact, only the joint angles are controlled. In Fig. 4 the simplified block diagram of the CLIK-based interaction controller considering a rigid surface is represented.

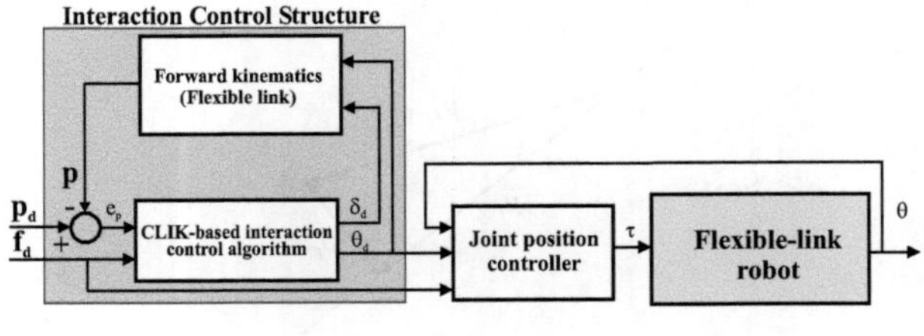

Fig. 4 Simplified block diagram of the CLIK-based interaction controller considering a rigid surface.

3.2 Flexible Surface

In this case, the interaction control algorithm is similar to the previous presented algorithm for the rigid surface. The main difference concerns with the calculation method of the reference force, which is based on the estimated stiffness surface coefficient k_e. Notice that, for simplicity, the environment is modeled as a linear spring. Due to this assumption, the deflection coordinates δ, the Jacobian J_p and the trajectory planning algorithm will be slightly different [8].

In the interaction control algorithm considering a rigid surface described above, the contact force is represented by the Lagrange multiplier λ. In this case, the force is represented by

$$f_d = k_e \boldsymbol{p}_f \tag{24}$$

In the interaction controller with a flexible surface, the desired trajectory has two components, one tangent to the contact surface $\boldsymbol{p}_s$ and another component normal to the surface $\boldsymbol{p}_f$. With these two components, the desired reference trajectory is represented by the following equation

$$\boldsymbol{p}_d = \boldsymbol{p}_s + \boldsymbol{p}_f \tag{25}$$

where $\boldsymbol{p}_d$ is the desired tip cartesian position, and $\boldsymbol{p}_f$ is determined by

$$\boldsymbol{p}_f = k_e^{-1} f_d \boldsymbol{n} \tag{26}$$

In Fig. 5 the geometric representation of the desired tip position, considering the desired force f_d on the surface is presented.

For the flexible contact surface, the Jacobian J_p is then given by

$$J_p = J_\theta(\theta,\delta) - k_e J_\delta(\theta,\delta) K^{-1} J_f(\theta,\delta) \tag{27}$$

where,

$$J_f = \frac{\partial J_\delta^T n}{\partial \theta}(n^T p - n^T p_e) + J_\delta^T n \frac{\partial n^T p}{\partial \theta} \tag{28}$$

The discrete version of the deflection coordinates algorithm, is given by

$$\begin{aligned}\delta_d(t_{k+1}) = &-K^{-1}(k_e J_\delta^T(\theta_d(t_k), \delta_d(t_k)) \\ &\times n(n^T p(t_k) - n^T p_e))\end{aligned} \tag{29}$$

where

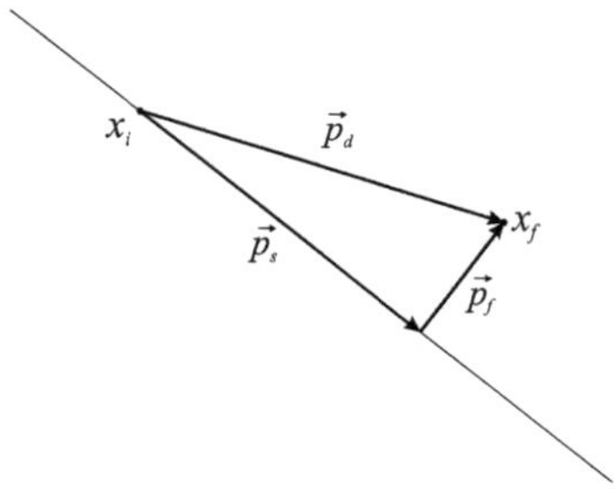

Fig. 5 Geometric representation of the desired tip position.

- p_e is the non-deformed coordinate of the surface,
- J_θ is the rigid part of the robot's Jacobian,
- J_δ is the flexible part of the robot's Jacobian.

In Fig. 6 the simplified block diagram of the CLIK-based interaction controller considering a flexible surface is represented.

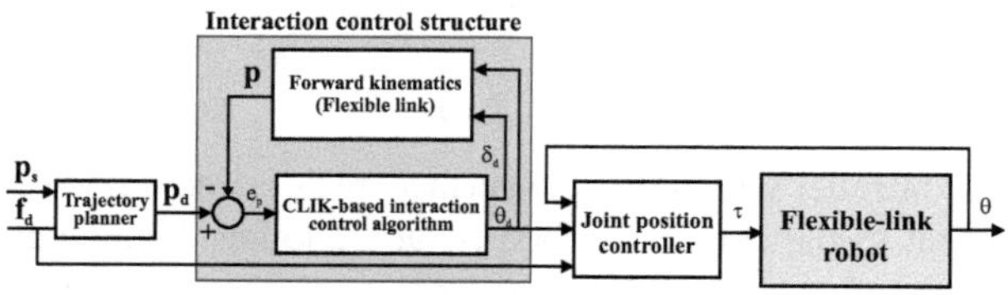

Fig. 6 Simplified block diagram of the CLIK-based interaction controller considering a flexible surface.

In [7, 8], the Jacobian J_f is considered as only dependent on θ. This assumption is valid when small link deflections are considered, i.e. they can be neglected. In this work, the robot's link is extremely flexible, and this assumption is not valid. For this reason, it is assumed that Jacobian J_f is fully dependent on θ and δ.

When the robot is in contact with the environment, the interaction controller described above, apply the desired force f_d on the surface through the joint position PD algorithm described by Eq. (23).

4 Robot Experimental Setup

For the purpose of analyze the interaction control algorithm performance, an experimental setup was built at the Robotics Laboratory. In Fig. 7, a picture of the planar flexible-link robot used for the experiments, is presented.

Table 1 exhibits the most important physical parameters of the joints and links of this robot.

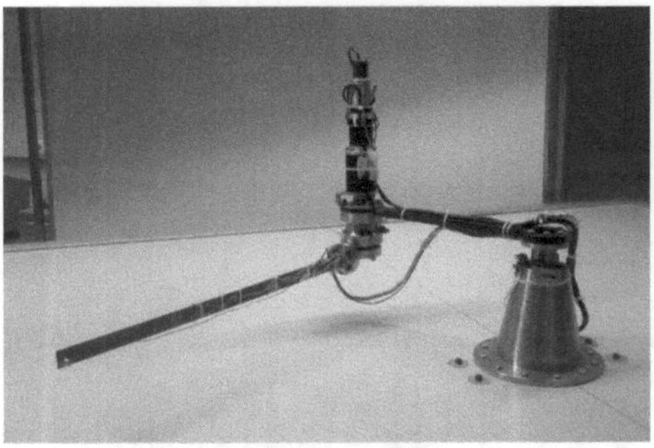

Fig. 7 Picture of the experimental robotic setup.

Table 1 Physical parameters of the robot

Joint 1 and rigid link 1	
L_R – link length	0.32 m
I_{R0} – Inertia of the link	0.25 kgm^2
I_{m1} – Inertia of the actuator	0.093 kgm^2
Joint 2	
r – Radius of the joint	0.075 m
I_H – Rotating inertia	13.22×10^{-4} m^4
M_H – Mass of the joint	0.47 kg
I_{m2} – Inertia of the actuator	0.024 kgm^2
Flexible link 2	
L – Link length	0.5 m
e – Link thickness	0.001 m
h – Link width	0.02 m
I – Cross section inertia	1.67×10^{-12} m^4
I_b – Link inertia	99×10^{-4} kgm^2
m_b - Link mass	0.0785 kg

The control hardware used to drive the flexible robot consists of a host PC Pentium IV 3 GHz computer that runs the Matlab/Simulink software and a target PC Pentium 200 MHz computer, where the real-time target software runs under the Matlab/xPC environment. The signals are processed through a low cost ISA-bus servo I/O board from SERVO TO GO, INC., and the electric d.c. joint motors are driven by linear power amplifiers configured to operate as current amplifiers. In this functioning mode, the input control signal is a voltage in the range of ± 10 V with current ratings in the interval $[-3,\ 3]$ A. The deflection of the elastic link is measured by three full bridge strain gage sensors located along the link and processed by HOTTINGER BM instrumentation amplifiers. The contact forces are measured by a JR^3 6-axis force/torque sensor mounted on the contact surface (see Fig. 10 for details). The force sensor hardware provide decoupled and digitally filtered data at a frequency rate of 8 KHz for each channel. Figure 8 represent the overall hardware and software control architecture for the flexible-link robot.

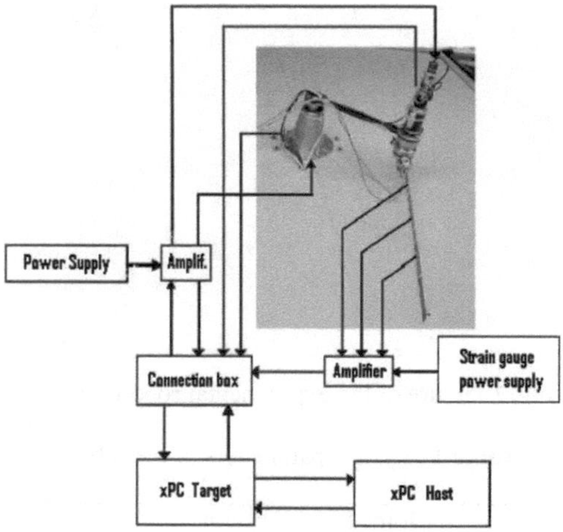

Fig. 8 Hardware and software control architecture for the flexible-link robot.

5 Experimental Results

For the purpose of analyzing the interaction control performance, the control methodologies presented in Sects. 3.1 and 3.2 are applied through experimentation to the planar robot represented in Fig. 7. Notice that all the interaction tasks described on this section were previously tested by simulation in Matlab/Simulink in order to obtain the best performance. For this purpose, Virtual Reality Toolbox from Matlab was used to build the tridimensional (3D) robot model, as depicted in Fig. 9.

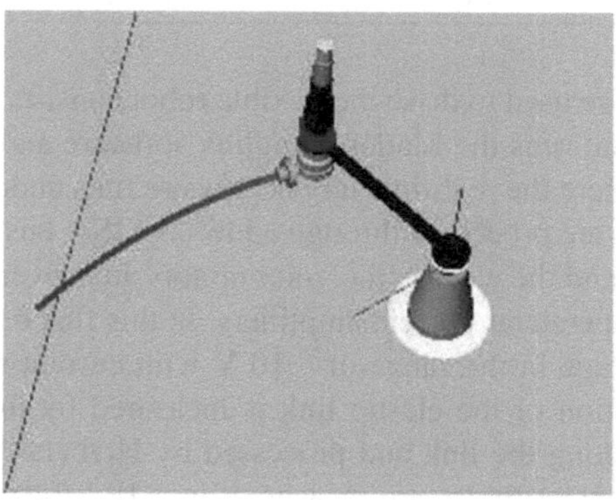

Fig. 9 Picture of the 3D planar robot model built in Matlab/Virtual Reality Toolbox. Notice that the dot line represents the undeformed flexible contact surface.

5.1 Rigid Surface

The first task consists of applying a force profile on the rigid surface while maintaining the robot's position. In the second task, the robot should move along the rigid surface with simultaneously application of the desired force profile on the surface, as represented in Fig. 10.

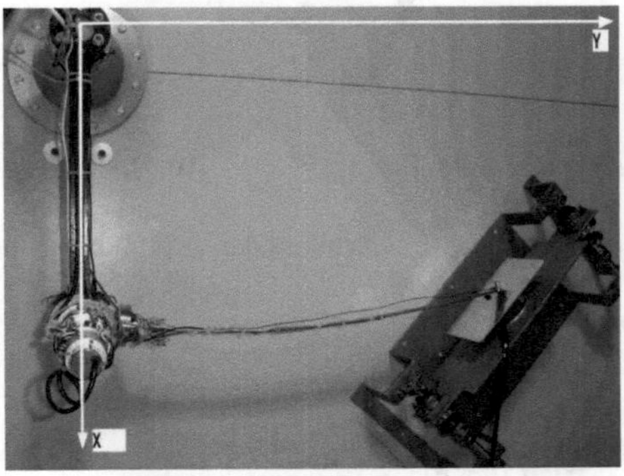

Fig. 10 Top view of the flexible robot executing an interaction task.

The sampling frequency used is 1 kHz and the following controller gains were used in all the experimental tasks: $K_P = [2000;\ 2000]$ for the CLIK algorithm and $K_p = [3000;\ 600]$, $K_d = [20;\ 10]$ for the PD controller. Notice that all the experiments were executed considering that tip is already in contact with the surface before the execution of the task.

Due to the maximum allowed values for the deflections adjusted in the robot's supervision and control software and the high degree of flexibility of the link, the maximum force that is possible to apply by the flexible link on the surface, is 1 N. In Fig. 11 the results for a task where only a desired force trajectory is applied on the surface are presented. The force is applied at the initial contact point, P$= [0.32;\ 0.575]$ m. The contact surface has 45 of inclination with the reference base frame x-axis. The reference force profile has a maximum value of 0.9 N, a growing time of 15 seconds and a full evolution time of 50 seconds.

In Fig. 12 the results for a task with force and position reference trajectories are presented. The reference force has the same profile of the first task and the position reference trajectory executes a straight line movement with a cycloidal profile of 5 cm in 5 seconds along the rigid surface.

From the analysis of the plots, is possible to observe a good force tracking performance in static conditions (Fig. 11). Also, when the robot executes a movement along the rigid surface, while executing the desired force profile, an acceptable force tracking performance with low force errors is observed along the trajectory (Fig. 12). In all these experiments an overshoot is observed when the applied force begins to decrease, due to tip/surface contact friction effects. Notice that the desired force is applied on the surface without force feedback, but the force errors are kept small. These results validate the interaction control strategy described in Sect. 3.1.

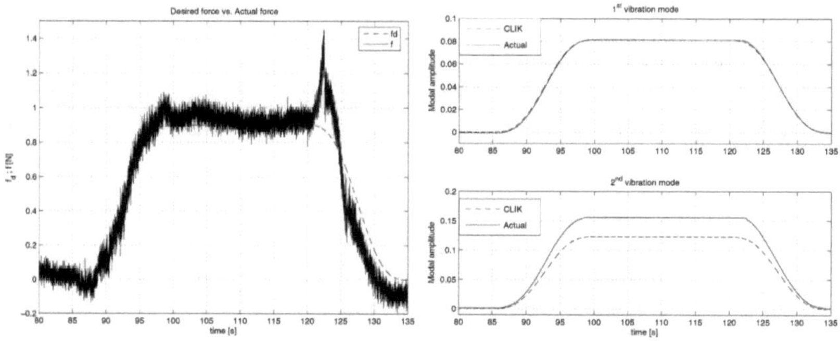

(a) Desired force vs. actual force (b) Modal link amplitude

Fig. 11 Rigid surface: contact force and modal amplitudes of the flexible link.

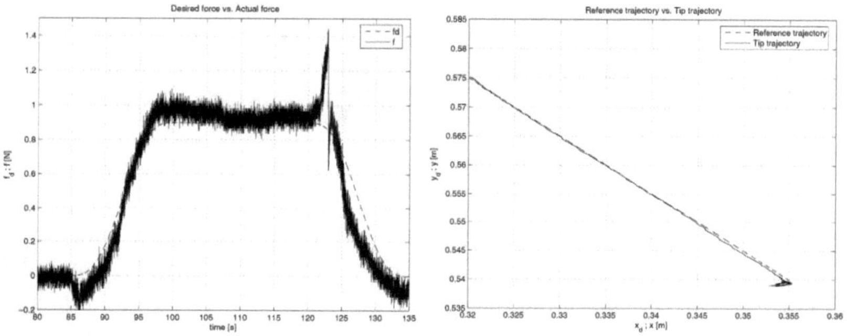

(a) Desired force vs. actual force (b) $x - y$ trajectory evolution

Fig. 12 Rigid surface: contact force and cartesian trajectory along the surface.

5.2 Flexible Surface

In order to obtain preliminary experimental results for a flexible environment, a soft foam was fixed on the plate attached to the force sensor. The overall estimated stiffness coefficient of the device is $k_e \approx 110$ N/m. Since the interaction controller for the flexible surface revealed to be extremely sensitive for k_e values larger than 15 N/m, a desired path profile f_d with a maximum value of 1 N was planned considering the estimated stiffness environment but setting k_e=15 N/m in Eq. (29) in order to observe the correspondent applied force and deformation of the environment.

From Fig. 13 it is possible to observe an acceptable force tracking behavior in static conditions. In this case, a force overshoot is observed at the end of the growing path due to the flexible environment characteristics. From the force trajectory plot, is possible to observe that applied force reach the desired value of 1 N. However, since there are a significant gap between the estimated stiffness coefficient and the k_e value used in Eq. (29), the modal amplitudes of the flexible link will not match the desired ones calculated by the CLIK algorithm (Fig. 13b). Also, due to the k_e mismatch described above, the cartesian trajectory evolution along $x-$axis will exhibit a poor tracking per-

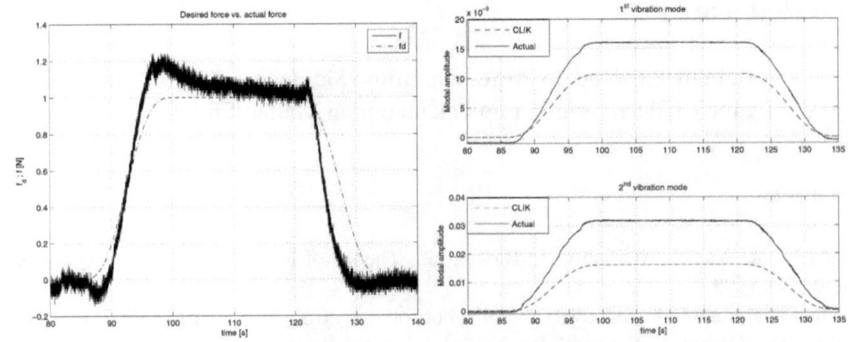

(a) Desired force vs. actual force **(b)** Modal link amplitude

Fig. 13 Flexible surface: contact force and modal amplitudes of the flexible link.

formance, as depicted on Fig. 14a. Finally, is possible to observe that joint position controller reveal an excellent tracking performance (see Fig. 14b for details).

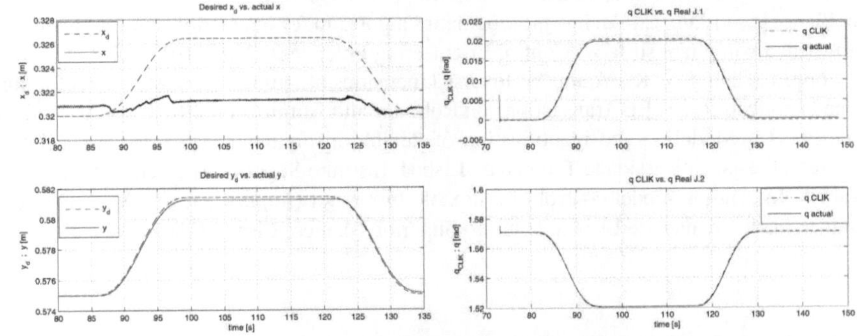

(a) Desired trajectory vs. actual trajectory **(b)** CLIK reference angles vs. actual angles

Fig. 14 Flexible surface: cartesian and joint trajectories.

6 Conclusions

In this chapter interaction control strategies for a manipulator robot with a two degrees of mobility and a flexible link were analyzed by simulation and experimentation. The interaction control results reveal the successful implementation of the control algorithms in real-time for a robot with one flexible link. The interaction control results were obtained considering rigid and flexible contact surfaces.

Future research will concentrate on the improvement of the real-time software functionality, the study of more complex inverse kinematic algorithms for flexible arms and the improvement of the interaction controller robustness for the flexible contact surface.

Acknowledgements

The authors would like to thank to Professors Bruno Siciliano and Luigi Villani for their support during the CLIK algorithm implementation in Matlab/Simulink.

References

1. Canudas de Wit, C., Siciliano, B., Bastin, G.: Theory of Robot Control. Springer Verlag, New York (1997)
2. Zeng, G., Hemami, A.: An overview of robot force control. Robotica **15** (1997) 473–482
3. Talebi, H., Khorasani, K., Patel, R.: Neural network based control schemes for flexible-link manipulators: simulations and experiments. Neural Networks **11** (1998) 1357–1377
4. Khorrami, F., Jain, S.: Nonlinear control with end-point acceleration feedback for a two-link flexible manipulator: experimental results. Journal of Robotic Systems **11** (1994) 591–603
5. Vandegrift, M., Lewis, F., Zhu, S.: Flexible-link robot arm control by feedback linearization/singular perturbation approach. Journal of Robotic Systems **11** (1994) 591–603
6. Siciliano, B.: A closed-loop inverse kinematic scheme for on-line joint based robot control. Robotica **8** (1990) 231–243
7. Siciliano, B.: Closed-loop inverse kinematics algorithm for constrained flexible manipulators under gravity. Journal of Robotics Systems **16** (1999) 353–362
8. Siciliano, B., Villani, L.: An inverse kinematics algorithm for interaction control of a flexible arm with a compliant surface. Control Engineering Practice **9** (2001) 191–198
9. Cheong, J., Chung, W.K., Youm, Y.: Inverse kinematics of multilink flexible robots for high-speed applications. IEEE Transactions on Robotics and Automation **20** (2004) 269–282
10. Martins, J.: Modeling and identification of flexible manipulators towards robust control. Master's thesis, Universidade Técnico de Lisboa, Instituto Superior Técnico, Lisbon (2000)
11. Nabais, J.: Sliding model control of a flexible link robot (*in portuguese*). Master's thesis, Universidade Técnico de Lisboa, Instituto Superior Técnico, Lisbon (2002)

Smooth Trajectory Planning for Fully Automated Passengers Vehicles: Spline and Clothoid Based Methods and its Simulation

Larissa Labakhua[1], Urbano Nunes[2], Rui Rodrigues[2] and Fátima S. Leite[2]

[1] University of Algarve, Escola Superior de Tecnologia/ADEE, Faro, Portugal
llabak@ualg.pt

[2] Institute of Systems and Robotics, University of Coimbra, Coimbra, Portugal
urbano@isr.uc.pt, ruicr@isec.pt, fleite@mat.uc.pt

Abstract. A new approach for mobility, providing an alternative to the private passenger car, by offering the same flexibility but with much less nuisances, is emerging, based on fully automated electric vehicles. A fleet of such vehicles might be an important element in a novel individual, door-to-door, transportation system to the city of tomorrow. For fully automated operation, trajectory planning methods that produce smooth trajectories, with low associated accelerations and jerk, for providing passenger's comfort, are required. This chapter addresses this problem proposing an approach that consists of introducing a velocity planning stage to generate adequate time sequences for usage in the interpolating curve planners. Moreover, the generated speed profile can be merged into the trajectory for usage in trajectory-tracking tasks like it is described in this chapter, or it can be used separately (from the generated 2D curve) for usage in path-following tasks. Three trajectory planning methods, aided by the speed profile planning, are analysed from the point of view of passengers' comfort, implementation easiness, and trajectory tracking.

Keywords. Trajectory planning, splines, clothoids, trajectory tracking, fully-automated vehicles, passenger comfort.

1 Introduction

Negative side effects of car use in build-up areas jeopardise the quality of life. Technology driven inventions like cybernetic transport systems may contribute to sustainable urban mobility. In this context, a new approach for mobility providing an alternative to the private passenger car, by offering the same flexibility but with much less nuisances, is emerging, based on fully automated electric vehicles, named cybercars [2] [7]. A fleet of such vehicles might be an important element in a novel individual, door-to-door, transportation system to the city of tomorrow. These vehicles must be user-friendly, easy to handle and safe, not only for passengers but also for the other road users. These vehicles are already in operation in specific environments featuring short trips at low speed [1] [7].

For fully automated operation, trajectory planning methods that produce smooth trajectories, with low associated accelerations and jerk, are required. Although motion planning of mobile robots has been thoroughly studied in the last decades, the requisite of producing trajectories with minimum accelerations and jerk (integrating both lateral and longitudinal accelerations) has not been traceable in the technical literature. A global minimum-jerk trajectory planning approach is proposed in [8] but in the context of joint space trajectories of robot manipulators.

This chapter addresses the problem of generating smooth trajectories with low associated accelerations, proposing an approach that consists of introducing a velocity planning stage to generate adequate time sequences to be fed into the interpolating curve planners. The generated speed profile can be merged into the trajectory for usage in trajectory-tracking tasks like it is described in this chapter, or it can be used separately (from the generated 2D curve) for usage in path-following tasks [9]. Three trajectory planning methods used to generate smooth trajectories from a set of waypoints, embedding a given speed profile, are analysed from the point of view of passengers' comfort, easiness of implementation, and trajectory tracking performance. The trajectory-planning methods studied are the following ones: cubic spline interpolation, trigonometric spline interpolation, and a combination of clothoids, circles and straight lines. For its evaluation, the well-known Kanayama trajectory-tracking controller was used [4]. The kinematics model of a Robucar vehicle (platform used in our autonomous navigation experiments) is used to evaluate through simulations the studied trajectory planning methods.

2 Acceleration Effects on the Human Body

For a vehicle following a trajectory at speed v, accelerations are induced on the passengers, which can be expressed as

$$\boldsymbol{a} = \frac{d\boldsymbol{v}}{dt} = \frac{dv}{dt}\boldsymbol{e}_T + v\frac{d\theta}{dt}\boldsymbol{e}_N \tag{1}$$

where $\boldsymbol{v}$ denotes the longitudinal velocity (tangent to the trajectory), θ is the vehicle orientation, and $\boldsymbol{e}_T$ and $\boldsymbol{e}_N$ are unit vectors in the tangent and normal trajectory directions, respectively. Moreover

$$\frac{d\theta}{dt} = \frac{1}{\rho}v \tag{2}$$

where ρ is the curvature radius. From (1) and (2) one gets the longitudinal acceleration (tangential component), induced by variations in speed,

$$a_T = \frac{dv}{dt} \tag{3}$$

and lateral accelerations (normal component), originated by changes in vehicle's orientation, whose values are also affected by the vehicle speed:

$$a_L = \frac{d\theta}{dt}v = \frac{1}{\rho}v^2 \tag{4}$$

The lateral acceleration is function of the trajectory curvature and speed. Assuming constant speed, the smaller is the curvature the smaller is the induced lateral acceleration, and therefore less harmful effects on the passengers. The ISO 2631-1 standard (Table 1) relates comfort with the overall r.m.s. acceleration, acting on the human body, defined as

$$a_w = \sqrt{k_x^2 a_{wx}^2 + k_y^2 a_{wy}^2 + k_z^2 a_{wz}^2} \tag{5}$$

where a_{wx}, a_{wy}, a_{wz}, are the components of the r.m.s. acceleration w.r.t. x, y, z axes and k_x, k_y, k_z, are multiplying factors. For a seated person $k_x = k_y = 1.4$, $k_z = 1$. For motion on the xy-plane, $a_{wz} = 0$. The local coordinate system is chosen so that its x-axis is aligned with the longitudinal axis of the vehicle, and it's y-axis defines the trajectory lateral direction.

2.1 Speed Profile

Trajectory planning for passenger's transport vehicles must generate smooth trajectories with low associated accelerations and jerk. As expressed by (3) and (4), lateral and longitudinal accelerations depend on the vehicle's speed. Thus, the trajectory planner should not only generate a smooth curve (spatial dimension) but also its associated speed profile (temporal dimension).

Table 1 ISO 2631-1 Standard

Overall acceleration	**Consequence**
$a_w < 0.315$ m/s^2	Not uncomfortable
$0.315 < a_w < 0.63$ m/s^2	A little uncomfortable
$0.5 < a_w < 1$ m/s^2	Fairy uncomfortable
$0.8 < a_w < 1.6$ m/s^2	Uncomfortable
$1.25 < a_w < 2.5$ m/s^2	Very uncomfortable
$a_w > 2.5$ m/s^2	Extremely uncomfortable

Using Table 1 and (5), for "not uncomfortable" accelerations, the longitudinal and lateral r.m.s. accelerations must be less than 0.21 m/s^2. Speed profiles can be calculated under this constraint, and consequently appropriate time-interval values sequences obtained to be used by the curve planners. Assuming a constant speed and a perfect arc cornering with a radius r, the reference speed in corners (segment between waypoints i and j) is

$$v_{ij} \leq \sqrt{a_T \cdot r},\ a_T \leq 0.21 \text{ m/s}^2 \tag{6}$$

It makes sense to consider a straight course segment just before each corner for reducing speed, and others after corners for increasing speed. So, the reference speed on the straight segments begin (end), designated by the waypoint k, can be calculated as

$$v_k = v_i \pm a_L \Delta t, \quad a_L \leq 0.21 \text{ m/s}^2 \tag{7}$$

$$\Delta t = \sqrt{2l / a_L} \tag{8}$$

where the waypoint i designates the corner begin (end), and is the straight segment length.

Figure 1 shows an urban road way with very close corners, a roundabout, and a set of waypoints defined by stars.

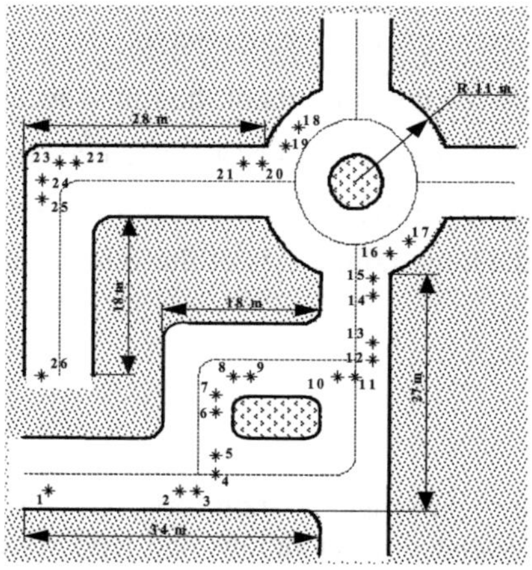

Fig. 1 Urban road way with very close corners, a roundabout, and waypoints defined by stars.

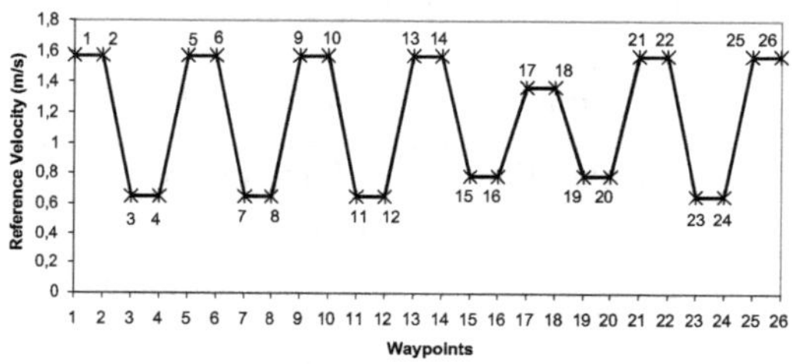

Fig. 2 Speed profile defined by speed values at the waypoints specified in Fig. 1: $v_{ri}, i = 1,2,\ldots,26$.

For the purpose of comparison of the three trajectory planning algorithms, applied to the scenario depicted in Fig. 1, and somehow to observe the above acceleration constraints, it was empirically defined the speed profile shown in Fig. 2. A simple algorithm to generate a C^1 speed profile curve using a second order polynomial is presented in [9] which is a step in an iterative trajectory planning method that generates smooth curves with bounded associated accelerations.

3 Kinematics Model

Cybercars are expected to be used in urban areas, airport terminals, pedestrian zones, etc., i.e. in places where the vehicle will move at relatively low speed. Therefore, kinematics-based trajectory control can be considered.

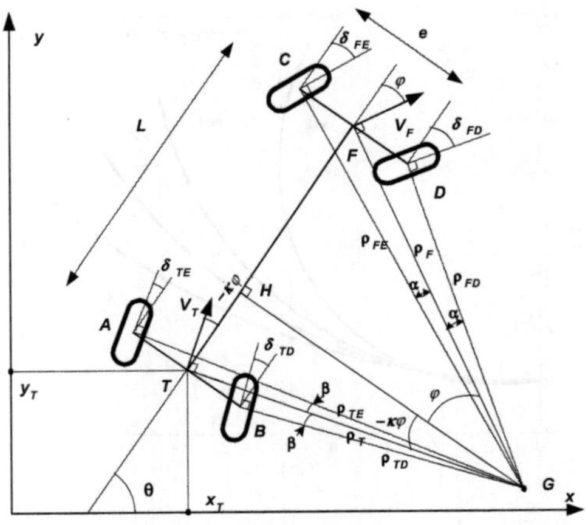

Fig. 3 Kinematics model of a 4-wheel car-like vehicle with front and rear steering capability.

These vehicles are under-actuated systems with two controls, speed and steering angle, but evolving in a $3D$ configuration space $\{x, y, \theta\}$, the first 2 coordinates for the $2D$ position and the third for the vehicle orientation. A representation of the kinematics model of Robucar (bi-steerable, 4-wheels actuated vehicle manufactured by Robosoft) is shown in Fig. 3. The model shows the possibility to steer both the rear and front pairs of wheels. The rear steering angle is proportional by a factor $-k$ to the front steering angle. If the angle φ represents the front wheels' steering command, the back wheels will be deflected from the central axis of the vehicle by an angle $-k\varphi$. Assuming that the wheels roll without slipping, the rear and front steering angles give the directions of the velocities at points F and T, respectively. Hence, the position of the instantaneous turning centre of the solid, point G in Fig. 3 can be deduced. Using the geometrical model of Fig. 3, the kinematics model of the vehicle, with the possibility of steering both the rear and front wheels, can be derived [11]:

$$\dot{q} = \begin{bmatrix} \dot{x}_T \\ \dot{y}_T \\ \dot{\theta} \\ \dot{\varphi}_F \\ \dot{\varphi}_T \end{bmatrix} = \begin{bmatrix} \cos(\theta - k\varphi_F) \\ \sin(\theta - k\varphi_F) \\ \dfrac{\sin(\varphi_F + k\varphi_F)}{L \cdot \cos(\varphi_F)} \\ 0 \\ 0 \end{bmatrix} \cdot v_T + \begin{bmatrix} 0 \\ 0 \\ 0 \\ 1 \\ -k \end{bmatrix} \cdot v_2 \tag{9}$$

where L is the vehicle length and v_2 defines the front wheels steering angular speed. The rear wheels steering angular speed is kv_2.

The results shown in Fig. 4 were calculated using model (9). For a given front steering angle φ, the effect of the rear steering angle is shown.

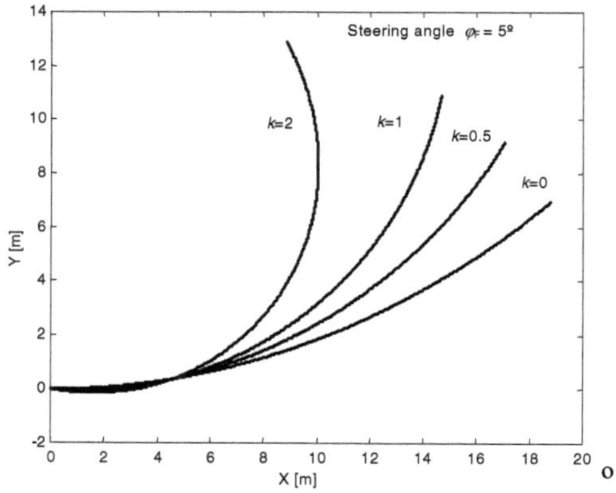

Fig. 4 Car-like vehicle trajectories, using the same front wheel steering angle $\varphi_F = 5^{\circ}$ and different values of the rear steering angle, given by the coefficient *k*.

Autonomous vehicles are expected to be used in places such as city centres with narrow areas and wherever it is needed to share the space with pedestrians. So, it is also important to know the position of each wheel, in order to avoid any kind of casualties, sidewalks, etc. Solving the kinematics model (9), and knowing the vehicle length *L* and its width *e*, it is possible to derive an output equation for the wheels' positions.

4 Trajectory Planning Methods

4.1 Cubic Splines

We assume that $t_0 < t_1 < \cdots < t_m$ is a chosen partition of the time interval $[t_0, t_m]$, and that $p_0, p_1, \ldots, p_m$ are given distinct points in $\Re^2$. We are interested in the construction of a smooth curve in $\Re^2$ which goes through the point p_k at time t_k, for all $k = 0, 1, \ldots, m$, with prescribed initial and final velocities (v_0 and v_m respectively). The instants of time are chosen in order that the trajectory satisfies a reasonable criterion of performance. Typically, this interpolation problem can be solved by a cubic spline, which is roughly a smooth concatenation of simple cubic polynomial curves. More precisely, a curve $S(t)$, $t \in [t_0, t_m]$, is a cubic spline in $\Re^2$ if it fulfils simultaneously the following:

1. $S(t)$ is defined in each subinterval $[t_k, t_{k+1}]$ by:

$$S_k(t) = \begin{pmatrix} a_{1k} + b_{1k}t + c_{1k}t^2 + d_{1k}t^3 \\ a_{2k} + b_{2k}t + c_{2k}t^2 + d_{2k}t^3 \end{pmatrix} \tag{10}$$

2. $S(t)$ is C^2-smooth in $[t_0, t_m]$, i.e., $S(t)$, $S'(t)$, $S''(t)$ are continuous functions in $[t_0, t_m]$;

3. $S(t_k) = p_k$, $k = 0, \ldots, m$ (interpolation conditions);

4. $S'(t_0) = v_0$ and $S'(t_m) = v_m$ (boundary conditions).

All the coefficients in (10) are uniquely determined by solving a set of linear algebraic equations arising from conditions 2–4. The cubic spline $S(t)$ is a smooth concatenation of each spline segment $S_k(t)$ and thus uniquely computed [3].

4.2 Trigonometric Splines

An alternative to build a C^2-smooth trajectory in a two-dimensional environment satisfying all the requirements at the beginning of this section is based on the construction of a trigonometric interpolating curve, described in [6] and [10]. This curve is again obtained by putting together smaller pieces (spline segments). However, one particular but important feature of this construction is that each piece can be computed separately. As a consequence, one may reduce the computations of each spline segment to the time interval $[0,1]$, thus simplifying notations.

The piece connecting point p_k (at $t = 0$) to point p_{k+1} (at $t = 1$) is denoted by $S_k(t)$ and given by the following convex combination of two other curves, $L_k(t)$ and $R_k(t)$:

$$S_k(t) = \cos^2(\pi t/2)\, L_k(t) + \sin^2(\pi t/2)\, R_k(t).$$

The curves L_k and R_k are called respectively the left component and the right component of the spline segment and will be computed from the local data as follows. The name "trigonometric spline" is suggested by the expression which defines the spline components.

- Computation of L_k ($k \neq 0$):

If the points p_k, p_{k+1} and p_{k-1} define a straight line, then $L_k(t)$ is the line segment connecting p_k (at $t = 0$) to p_{k+1} (at $t = 1$). Otherwise, consider the circle defined by the 3 points and let $L_k(t)$ be the circular arc joining p_k (at $t = 0$) and p_{k+1} (at $t = 1$) that does not contain p_{k-1}.

- Computation of R_k ($k \neq m$):

The previous algorithm (for the left component) is also implemented to compute the right component, but uses instead the points p_k, p_{k+1} and p_{k+2}.

The computation of the left component L_0 of the spline segment S_0 and the right component R_m of the spline segment S_m is slightly different. The computation of L_0 requires the use of the prescribed initial direction (at time t_0) in addition to the points p_0 and p_1. For R_m it is required to use the prescribed final direction (at time t_m) besides the points p_{m-1} and p_m. More details can be found in [10]. Properties of the trigonometric spline:

a. The final curve is guaranteed to be C^2-smooth;

b. The procedure used to compute L_0 and R_m shows how to compute a trigonometric spline when directions are prescribed at each instant of time t_k. This is an important issue in trajectory planning in a real environment. However, in this case S will no longer be C^2-smooth;

c. Another important property is due to the fact that only four data points are used to compute each spline segment. This is of particular importance in real trajectory planning. Indeed, under the presence of an unpredictable change of a data point (resulting, for instance, from the appearance of a sudden obstacle), at most the two previous and the two following segments of the spline, have to be recalculated. This contrasts with the classical cubic spline, mentioned previously, which would have to be entirely recalculated.

4.3 Clothoids

Using clothoid curves it is also possible to produce smooth trajectories with smooth changes in curvature (see Fig. 5). Clothoids allow smooth transitions from a straight line to a circle arc or *vice versa*. The clothoid curvature can be defined as in [5] by:

$$k(s)=\sigma s+k_0\,, \tag{11}$$

where σ is the curvature derivative, k_0 the initial curvature, s the position variable $s\in[0,l]$, and l the curve length. The orientation angle at any clothoid point is obtained integrating (11):

$$\theta(s)=\int_0^s k(u)du=\frac{\sigma}{2}s^2+k_0 s+\theta_0 \tag{12}$$

where θ_0 is the initial orientation angle. The parametric equations of a clothoid in the xy-plane are given by:

$$\begin{bmatrix} x \\ y \end{bmatrix}(s)=r_l\sqrt{2\pi\theta_l}\,R(\theta_0-\theta_0')* \\ *\left(\begin{bmatrix} CF\left(\sqrt{\frac{2\theta'(s)}{\pi}}\right) \\ SF\left(\sqrt{\frac{2\theta'(s)}{\pi}}\right) \end{bmatrix}-\begin{bmatrix} CF\left(\sqrt{\frac{2\theta_0'}{\pi}}\right) \\ SF\left(\sqrt{\frac{2\theta_0'}{\pi}}\right) \end{bmatrix}\right)+\begin{bmatrix} x_0 \\ y_0 \end{bmatrix}, \tag{13}$$

where θ_l and r_l are respectively the orientation angle and the radius of the clothoid at the point $s=l$, R is a $2D$ rotation matrix, x_0 and y_0 are the co-ordinates of the clothoid at $s=0$, CF and SF denote respectively the cosine and sine Fresnel integrals

$$CF(x)=\int_0^x \cos\left(\frac{\pi}{2}u^2\right)du\,,\quad SF(x)=\int_0^x sen\left(\frac{\pi}{2}u^2\right)du$$

and

$$\theta'(s)=\frac{\sigma}{2}s^2+k_0 s+\theta_0'\,,\ \text{and}\ \theta_0'=\frac{k_0^2}{2\sigma} \tag{14}$$

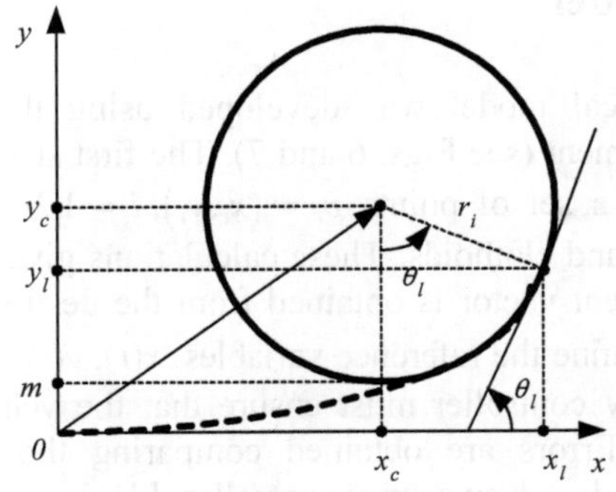

Fig. 5 Transition from a straight line to a circumference arc using a clothoid curve. *m* represents the distance between the straight line and the circumference.

The smooth transition from a straight line to a circumference is shown in Fig. 5, where the x-axis represents a straight line tangent to the clothoid trajectory. The dashed line clothoid curve should handle the smooth transition between the straight line and the circumference arc with centre at (x_c, y_c), radius r_l and curvature $k=1/r_l$. Solving the equations, one can find the clothoid parameters σ and l :

$$\sigma = \frac{1}{2r_l^2\theta_l} \text{ and } l = 2r_l\theta_l \tag{15}$$

The trajectory planning using clothoids is not an interpolation method. The trajectory results from the concatenation of straight line segments, clothoid curves, and circumference arcs.

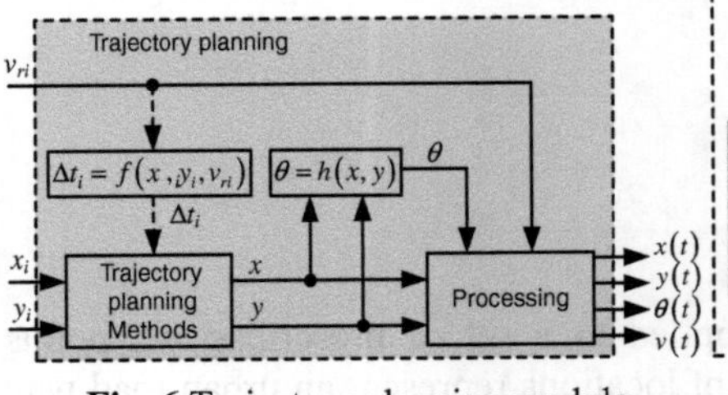

Fig. 6 Trajectory planning module.

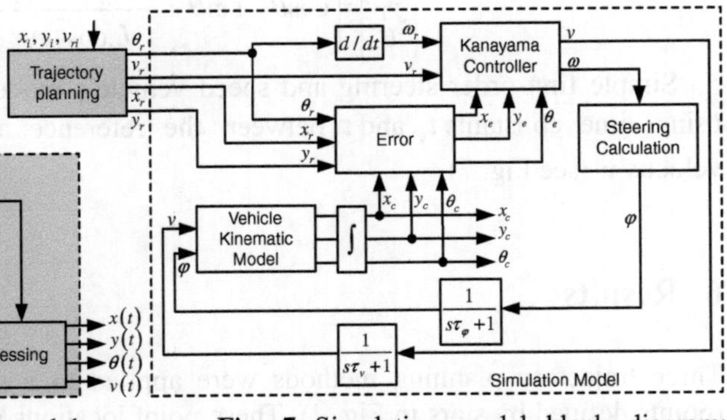

Fig. 7 Simulation model block diagram: trajectory planning and trajectory-tracking modules.

Thus, the trajectory is obtained by means of a geometric construction, and it is not possible to use the prescribed points in the same way as in interpolation methods. A previous processing is needed for assigning new points, circumference arcs radius, and the distance between the straight line segments and circumference arcs.

5 Simulation Model

A simulation numerical model was developed using the MATLAB/SIMULINK programming environment (see Figs. 6 and 7). The first step consisted on calculating the trajectories, from a set of points $p_i = (x_i, y_i)$, $i = 1,2,\ldots,n$, using cubic splines, trigonometric splines and clothoids. These calculations give the reference positions x and y. A time-dependent vector is obtained from the desired trajectory speed values v_{ri} which is used to define the reference variables $x(t), y(t), \theta(t)$ and $v(t)$, as shown in Fig. 6. A trajectory controller must ensure that the vehicle follows the planned reference trajectory. Errors are obtained comparing the reference position with vehicle's position, and a Kanayama controller [4] is used to calculate velocity commands v and ω. The angle φ is calculated in order to model the steering input of a car-like vehicle. For a front wheels only steering, $k = 0$,

$$\varphi = \arctan(\omega \cdot L/v) \tag{16}$$

while for both front and rear wheels steering, and for equal front and rear angles, k=1, results:

$$\varphi = \arcsin(\omega \cdot L/2v) \tag{17}$$

For other values of k it is more complicated to find the value of angle φ. One possible way is to expand the sine and cosine in Taylor series and solve the resulting equation. The kinematics model (9) is related to the velocity v_T this velocity is in the direction of the rear wheels, as shown in Fig. 3. On the other hand, the target velocity is in the direction of the vehicle axis. Hence,

$$v_T = v/\cos(k\varphi) \tag{18}$$

and the kinematics model becomes

$$\begin{bmatrix} \dot{x}_T \\ \dot{y}_T \\ \dot{\theta} \end{bmatrix} = \begin{bmatrix} \cos\theta & \sin\theta & \dfrac{\sin(\varphi_F + k\varphi_F)}{L.\cos^2 \varphi_F} \end{bmatrix}^T v \tag{19}$$

Simple first order steering and speed vehicle's model were used in simulations, using time constants τ_φ and τ_v between the reference and the targets angle φ and velocity v (see Fig. 7).

6 Results

Three trajectory planning methods were applied to a set of prescribed waypoints (points defined by stars in Fig. 4). These point locations represent an urban road way with very close corners and a roundabout. As an example, a planned trajectory using trigonometric splines is depicted in Fig. 8. Figures 9 to 14 show results of the trajectory planning and vehicle's path-following for the three trajectories obtained using the planning methods described in Sect. 4.

Figures 9, 11 and 13 show the orientation angle, curvature, and longitudinal and lateral accelerations behaviour. The curvature is a non time-depending parameter, which shows the smoothness of the planned curve. The acceleration results allow an evaluation of the trajectory comfort. However, the accelerations also depend on linear speed variation. So, using a different speed profile other results would be obtained. Subsequently, the planned trajectories were applied to the simulation model for trajectory tracking, using a Kanayama controller. The tracking errors obtained from the simulation are shown in Figs. 10, 12 and 14. The angle, longitudinal and lateral errors are shown for cubic splines, trigonometric splines and clothoid curves planned trajectories tracking. Table 2 summarises results of the applied trajectory planning methods.

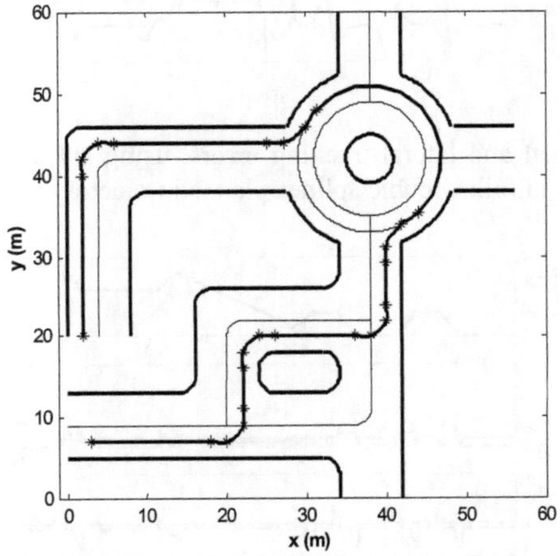

Fig. 8 Generated trajectory using trigonometric splines.

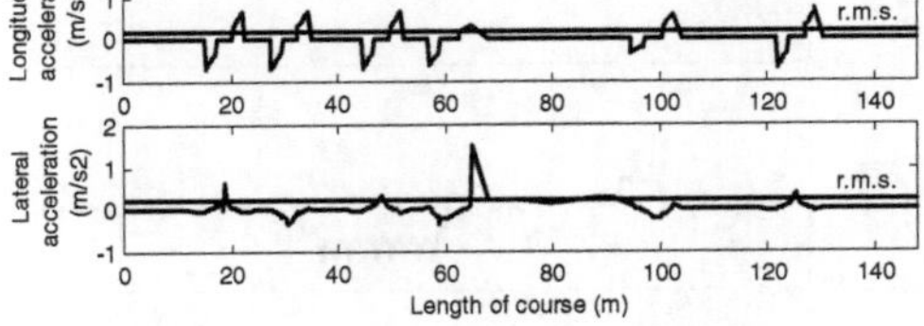

Fig. 9 Orientation angle θ, curvature, longitudinal and lateral acceleration behaviour along the course for the given reference velocity vector, using cubic splines trajectory planning.

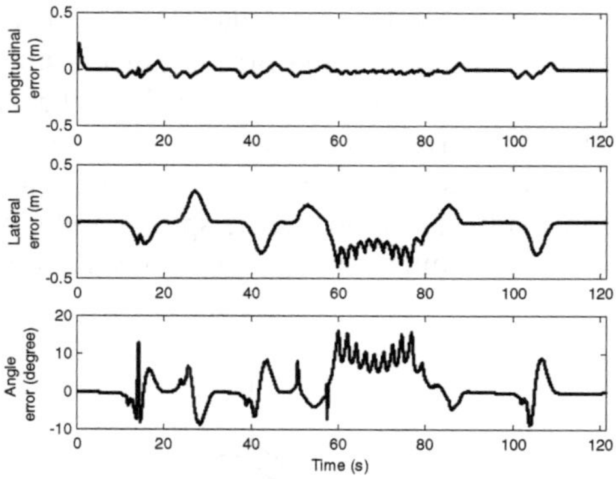

Fig. 10 Angle, longitudinal and lateral tracking errors, using a Kanayama controller and the vehicle kinematics model to follow cubic splines planned trajectory.

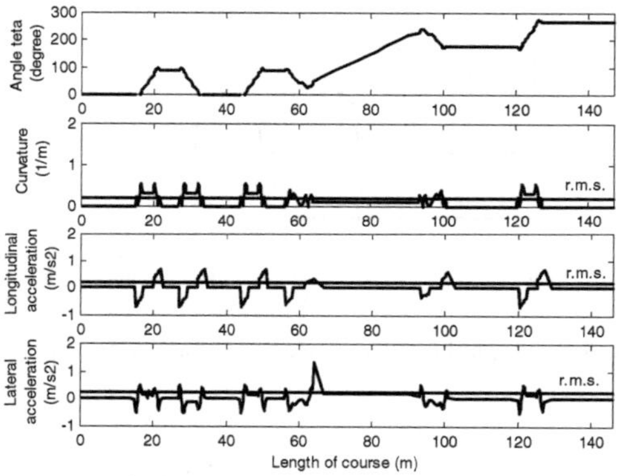

Fig. 11 Orientation angle θ, curvature, longitudinal and lateral acceleration behaviour along the course for the given reference velocity vector, using trigonometric splines trajectory planning.

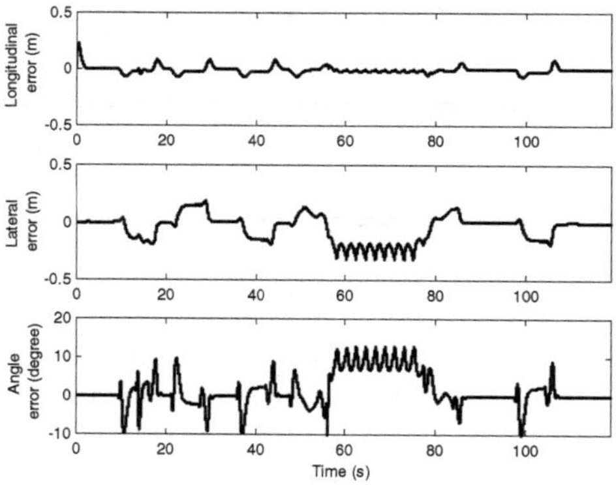

Fig. 12 Angle, longitudinal and lateral tracking errors, using a Kanayama controller and the vehicle kinematics model to follow cubic trigonometric planned trajectory.

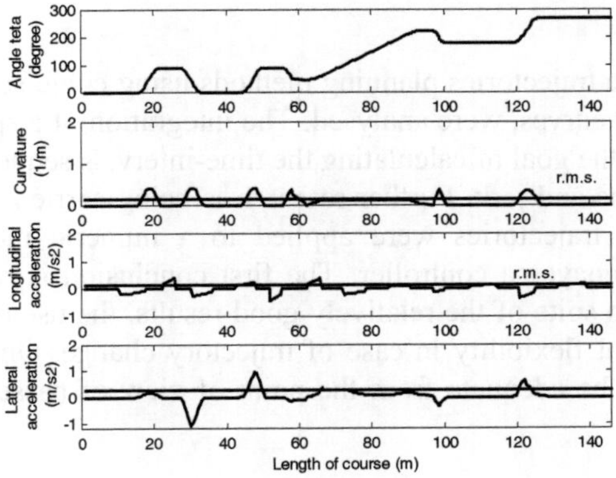

Fig. 13 Orientation angle θ, curvature, longitudinal and lateral acceleration behaviour along the course for the given reference speed profile, using clothoid curves trajectory planning.

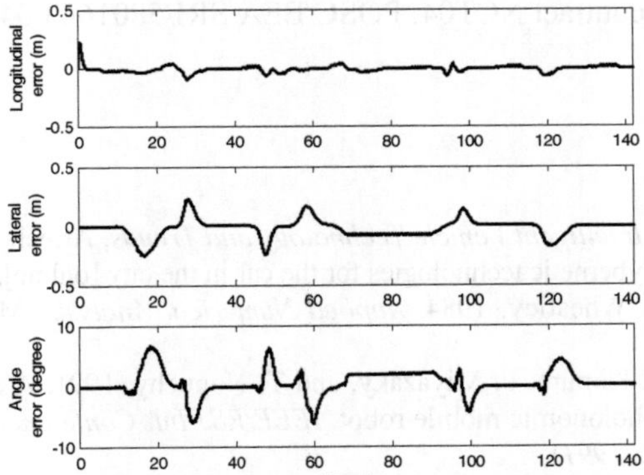

Fig. 14 Angle, longitudinal and lateral tracking errors, using a Kanayama controller and the vehicle kinematics model to follow clothoid curves planned trajectory.

Table 2 Planning methods results

Quantity	Cub.	Trig.	Clothoid
Max. Curvature (1/m)	0.87	0.56	0.41
r.m.s. Curvature (1/m)	0.21	0.20	0.16
Max. Long. Accel. (m/s^2)	0.69	0.69	0.42
r.m.s. Long. Accel. (m/s^2)	0.21	0.21	0.15
Max. Lateral Accel. (m/s^2)	1.50	1.32	0.95
r.m.s. Lateral Accel. (m/s^2)	0.24	0.25	0.25
Overall Acceleration (m/s^2)	0.43	0.46	0.40

7 Conclusions

In this chapter, three trajectories planning methods using cubic splines, trigonometric splines and clothoid curves, were analysed. The integration of a speed profile planner was proposed, with the goal of calculating the time-intervals sequence that lead to low level of accelerations and jerk. Further research is being carried out in this direction [9]. The generated trajectories were applied to a numeric model for trajectory-tracking, using a Kanayama controller. The first conclusion is related to the use of methods easiness. In spite of the relatively good results, the use of clothoid curves is complex and without flexibility in case of trajectory change. On the other hand, all methods showed to be adequate from the point of view of passengers' comfort and tracking.

Acknowledgements

This work was supported in part by ISR-UC and FCT (Fundação para a Ciência e Tecnologia), under contract NCT04: POSC/EEA/SRI/58016/2004.

References

1. Bishop, R., 2005. *Intelligent Vehicle Technology and Trends*, Artech House, London, UK.
2. Cybercars 2001. Cybernetic technologies for the car in the city [online], www.cybercars.org.
3. Gerald, C. and P. Wheatley, 1984. *Applied Numerical Analysis*, Menlo Park, California, Addison-Wesley.
4. Kanayama, Y., Y. Kimura, F. Miyazaky, and T. Noguchy, 1991. A stable tracking control method for a non-holonomic mobile robot. *IEEE/RSJ Int. Conference on Intelligent Robots and Systems (IROS 1991).*
5. Leao, D., T. Pereira, P. Lima, and L. Custódio, 2002. Trajectory planning using continuous curvature paths. *Journal DETUA*, Vol. 3, no. 6 (in Portuguese), 557–564.
6. Nagy, M. and T. Vendel, 2000. Generating curves and swept surfaces by blended circles. *Computer Aided Geometric Design*, Vol. 17, 197–206.
7. Parent, M., G. Gallais, A. Alessandrini, and T.Chanard, 2003. CyberCars: review of first projects. *Int. Conference on People Movers APM 03*, Singapore.
8. Piazzi, A. and A. Visioli, 2000. Global minimum-jerk trajectory of robot manipulators. *IEEE Transactions on Industrial Electronics*, Vol. 47, no. 1, 140–149.
9. Solea, R. and U. Nunes, 2006. Trajectory planning with velocity planner for fully-automated passenger vehicles. *IEEE 9th Intelligent Transportation Systems Conference*, Toronto, Canada.
10. Rodrigues, R., F. Leite, and S. Rosa, 2003. On the generation of a trigonometric interpolating curve in $\Re^3$. *11th Int. Conference on Advanced Robotics*, Coimbra, Portugal.
11. Sekhavat, S. and J. Hermosillo, 2000. The Cycab robot: a differentially flat system. *IEEE/RSJ Int. Conf. on Intelligent Robots and System (IROS 2000).*

Finding the Best Classifier for Evaluating Cork Quality in an Industrial Environment

Beatriz Paniagua-Paniagua, Miguel A. Vega-Rodríguez, Juan A. Gómez-Pulido and Juan M. Sánchez-Pérez

Dept. Informática, Univ. Extremadura, Escuela Politécnica
Campus Universitario s/n, 10071, Cáceres, Spain
{bpaniagua, mavega, jangomez, sanperez}@unex.es
http://arco.unex.es

Abstract. A classification problem existing in the cork industry is the cork stopper/disk classification according to their quality using a visual inspection system. Cork is a natural and heterogeneous material, therefore, its automatic classification (usually, seven different quality classes exist) is very difficult. In this work, we study the use of different classifiers to solve this problem. The classifiers, which we present here, work with several quality discriminators (features), that we think could influence cork quality. These discriminators (features) have been checked and evaluated before being used by the different classifiers that will be exposed here. In this chapter we attempt to evaluate the performance of a total of 4 different cork quality-based classifiers in order to conclude which of them is the most appropriate for this industry, and therefore, obtains the best cork classification results. In conclusion, our experiments show that the Euclidean classifier is the one which obtains the best results in this application field.

Keywords. Classifiers, cork quality, image processing, industrial application, automated visual inspection system.

1 Introduction

The production of stoppers and disks for sealing champagnes, wines and liquors is the most important industrial application of cork. In fact, according to the experts, cork is the most effective product, natural or artificial, for the sealed [2]. In the cork industry, stoppers and disks are classified in different quality classes based on a complex combination of their defects and particular features. Due to this, the classification process has been carried out, traditionally, by human experts manually.

At present, there are several models of electronic machines for the classification of cork stoppers and disks in the market. The performance of these machines is acceptable for high quality stoppers/disks, but for intermediate or low quality, the number of samples classified erroneously is large. In conclusion, the stoppers/disks should be re-evaluated by human experts later. This slows down and increases in price the process enormously. Think that, on average, a human expert needs a minimum training period of 6 months to attain a minimum agility, although the learning process lasts years (compare it with other experts: wine tasters, cured ham tasters, etcetera). Another negative aspect is the subjectivity degree added to the classification process due to the necessary human re-evaluation.

We have to add to these antecedents the fact that Spain is the 2nd world producer of cork [1], only surpassed by Portugal, and that in Extremadura (a south-western region of Spain), for its geographical situation, the cork industry is one of its more important industries: it produces 10% of the world cork [4].

All these motivations have lead us to the development of this research, whose main objective is the construction of a computer vision system for cork classification based on advanced methods of image processing and feature extraction in order to avoid the human evaluation in the quality discrimination process.

For this purpose we have performed a study of the features that could better inform us about the cork quality. We have focused this study on an analysis of thresholding techniques (segmenting the different cork defects) and textural features, in addition to other features (like holes and different-area defects). From this study we conclude that the features that better define the cork quality are: the total cork area occupied by defects, the cork texture contrast, the cork texture entropy, and the biggest size defect in the cork stopper/disk.

Later, and with these results, an analysis of different possible classifiers has been made. The studied classifiers have been a Back-Propagation neural network, the K-means classification algorithm, a K-nearest neighbours classifier and the minimum Euclidean distances classification algorithm. In this work, we evaluate all these classification algorithms with the purpose of knowing which of them is the most appropriate for our application environment.

The rest of the chapter is organized as follows: Section 2 describes briefly the tools and the data used for the development of our experiments. In Sect. 3, we present the features used by the classifiers. Then, Sect. 4 shows the theoretical bases for the analysis we have made and other important details. Finally, Sect. 5 presents the final results statistical evaluation for each classifier, while Sect. 6 exposes the conclusions and future work.

2 Data and Tools

At the moment, the computer vision system we use to acquire the cork stopper/disk images is formed by the elements shown in Fig. 1: the host (a Pentium processor), a colour Sony camera (SSC-DC338P model), the illumination source (fluorescent-light ring of high frequency – 25 KHz – of StockerYale), and a METEOR 2/4 frame-grabber of Matrox, with the software required for the image acquisition (MIL-Lite libraries of Matrox).

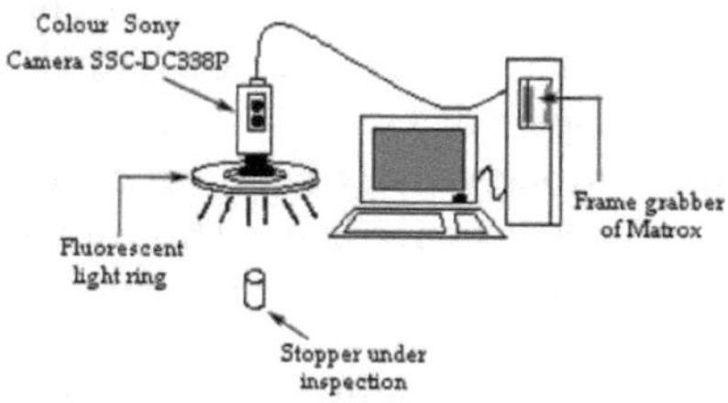

Fig. 1 Computer vision system.

On the other hand, the database used in our experiments consists in 700 images taken from 350 cork disks (we have taken two images of each disk, for both heads). There are seven different quality classes, 50 disks in each class. The initial classification, in which this study is based on, has been made by a human expert from ASECOR (in Spanish: "Agrupación Sanvicenteña de Empresarios del CORcho", in English: "Cork Company Group from San Vicente-Extremadura"). We suppose this classification is optimal/perfect and we want to know which classifier obtains the most similar classification results.

3 Features

To develop our classifiers study, different feature extraction methods were analysed: thresholding techniques, statistical texture analysis, etcetera.

As for automatic thresholding, we carried out a study of global and local thresholding techniques [5] [8]. The objective was to extract the defect area from the cork area, thus being able to extract the percentage of the cork area occupied by defects (an important feature in cork quality discrimination). Eleven global thresholding methods were studied: *static thresholding, min-max method, maximum average method, Otsu method, slope method, histogram concavity analysis method, first Pun method, second Pun method, Kapur-Sahoo-Wong method, Johannsen-Bille method and moment-preserving method*. In general, global thresholding methods are very limited in our problem. For a good global thresholding we need bimodal histograms, and the results obtained with unimodal histograms have been quite bad. These methods are suitable for the cork stopper/disk area extraction from the image background. In this situation we can find that all conditions for a good operation are fulfilled, but they are not suitable for the defect area extraction from the cork area. As for local thresholding, two methods have been studied: *statistical thresholding method and Chow-Kaneko method*. The local thresholding methods have been more suitable than the global methods for the solution of our problem. This has been due to they are able to find better thresholds in unimodal histograms. Nevertheless, the increase of the computational cost can make them unsuitable for our problem. Taking into account all these considerations, the best of all these methods applied to our problem was static thresholding method with a heuristically fixed threshold in the gray level 69.

Regarding texture analysis [3] [6], two main methods have been studied, both based on statistical texture analysis. The first was a method based on simple co-occurrence matrices and another was a method based on rotation-robust normalized co-occurrence matrices. Furthermore, we have studied nine quality discriminators (textural features) for each method: *energy, contrast, homogeneity, entropy, inverse difference moment, correlation, cluster shade, cluster prominence and maximum probability*. The best obtained results were with the contrast and the entropy, both calculated by using rotation-robust normalized co-occurrence matrices.

In addition to the total area occupied by defects (obtained after doing an image thresholding with the previous methods) and the texture analysis of the cork area, other features were analysed too. Concretely, additional studies were made on: *holes (perforations) in the cork area and size of the biggest defect in the cork*. In the case of cork holes, a quantitative comparison is done between the theoretical area of cork (computed using the cork stopper/disk perimeter) and the real area of cork. If the real area is smaller than the theoretical one (surpassing certain threshold) we consider that the cork has holes. In order to calculate the biggest defect in the cork stopper/disk, the followed methodology is to perform successive morphological erosions on the thresholded image (defects area). In each iteration, we control the remaining defect percentage. In this way, we can quickly observe the size that could have the biggest defect of the cork, analysing the number of required iterations for eliminating almost all the defect pixels (or required iterations for reaching certain threshold of defect pixels). The best results obtained in this case (the evaluation of these two additional features) were for the size of the biggest defect in the cork.

In conclusion, after an exhaustive feature study, the features chosen to be used in our classifier study were: the total cork area occupied by defects (thresholding with heuristic fixed value 69), the textural contrast, the textural entropy and the size of the biggest defect in the cork.

4 Used Methods

In this work, in order to classify a cork disk in a specific class, we will use the corresponding classification algorithm base on the four features selected: defects area, contrast, entropy, biggest defect size. The four classifiers chosen for this study are the following [7] [8]: a Back-Propagation neural network, a K-means classifier, the K-nearest neighbours classification algorithm, and a minimum Euclidean distance classifier.

4.1 Neural Classifier

Concretely, we have developed a Back-Propagation neural network. An artificial neural network represents a learning and automatic processing paradigm inspired in the form in which the nervous system of the animals works. It consists in a simulation of the properties observed in the biological neural systems through mathematical models developed with artificial mechanisms (like a computer). In the case of this problem, a Back-Propagation network architecture has been chosen, very suitable for pattern recognition and class detection. The network designed for this study has the following architecture:

- One input layer that is the one that receives external signals, which will be the four features selected during the course of this work. Therefore, the input layer has *4 neurons.*
- One hidden layer, whose number of neurons is based on the proportion given by the following equation:

$$N^o\ hidden\ neurons = \frac{N^o\ input\ neurons \times N^o\ output\ neurons}{2}$$

  Therefore, and knowing that the output layer has 3 neurons, the number of hidden neurons should be 6. But, at the end, we decided to increase the number of neurons in the hidden layer and increase the complexity of the weight matrix. In this way, we make easier the learning for the network. Due to this fact, our hidden layer has *7 neurons*.

- One output layer that gives back the results obtained by the neural network in binary format. As the classes to classify are seven, only *3 neurons* will be necessary to codify the results correctly.

The weights associated to the network interconnections are initialized randomly and are adjusted during the learning. The type of learning used by this neural network is *supervised.* That is, we present to the network pairs of patterns (an entrance and its corresponding wished exit). While we are showing patterns to the network, the weights are adjusted so that the error between the real results and the desired ones is diminished. This process is repeated until the network is stable. After this phase, we can run the neural network.

4.2 K-Means Classifier

As always, we have studied this classifier for the four selected features. We have decided to study the reliability of this classifier because of its consecrated fame in specialized literature. This classification algorithm makes reference to the existence of a number of K classes or patterns, and therefore, it is necessary to know the number of classes. We know, a priori, that we have 7 classes, reason why the algorithm is suitable for our necessities. K-means classification algorithm is a simple algorithm, but very efficient, and due to this fact it has been so used.

Beginning from a set of p objects to classify X_1, X_2, …, X_p, the K-means classification algorithm makes the following steps:

Step 1
Knowing previously the number of classes, we say K, K samples are randomly chosen and clustered into arrays (see the following equation), and these arrays will be the centroids (due to the fact of being the only elements) for each class.

$$\alpha_1 : Z_1(1); \alpha_2 : Z_2(1); \cdots \alpha_K : Z_K(1)$$

Step 2
Being this algorithm a recursive process with a counter n, we can say that in the generic iteration n we allocate all the samples $\{X\}_{1 \leq j \leq p}$ among the K-classes, as we can observe in the following equation:

$$X \in \alpha_j(n) \, sii \, \|X - Z_j(n)\| < \|X - Z_i(n)\| \, \forall i = 1,2,\ldots,K / i \neq j$$

In the previous equation we have indexed the classes (that are dynamic classes) and their centroids.

Step 3
In the moment we have allocated all the samples among the different classes, it is necessary to update the class centroids. With this calculation, we are looking for to minimize the profit index that is shown in the following equation:

$$J_i = \sum_{X \in \alpha_i(n)} \|X - Z_i(n)\|^2$$

This index can be minimized using the sample average of $\alpha_i(n)$ (see the following equation):

$$Z_i(n+1)=\frac{1}{N_i(n)}\sum_{X\in\alpha_i(n)} X$$

Being $N_i(n)$ the number of samples in class α_i, after the iteration n.

Step 4

We check if the classification algorithm has reached the stability, as it is shown in the following equation:

$$Z_i(n+1)=Z_i(n)\,\forall i=1,2,\ldots,K$$

If it does, the algorithm finishes. If not, we return to step 2 for repeating all the process again.

Finally, we have to say that, for the centroids allocation, the distance shown in the following equation was used. This is the Euclidean distance scaled with the standard deviation instead of with the variance, which gave better results in a previous study.

$$\textit{Modified Scaled Euclidean Distance}=\sqrt{\frac{(x_{i1}-\mu_1)^2}{\sigma_1}+\frac{(x_{i2}-\mu_2)^2}{\sigma_2}+\cdots+\frac{(x_{iN}-\mu_N)^2}{\sigma_N}}$$

4.3 K-Nearest Neighbours Classifier

Regarding the classification algorithm based on the K-nearest neighbours, we can say that also works with the four best features obtained in the study about cork quality, above-mentioned. The distance selected for this experimentation was the Euclidean distance scaled with the standard deviation (showed before). We have decided this according to the results obtained for the Euclidean classifier.

This algorithm is part of the methods group known as *correlations analysis methods*. It consists in classifying an unknown feature vector, depending on the sample or K samples of the training set that is/are more similar to it, or what is the same, which is/are nearer to this vector in terms of minimum distance. The used distance more suitable for this method is the Euclidean distance. This is what we know as *rule of the nearest neighbours*. The classification algorithm of the K-nearest neighbours even can be very efficient when the classes have overlapping, and this is very interesting for our problem (cork quality classes).

A first brute-force approach for this algorithm computes the distance between the unknown feature vector and all the samples in the database (training set), it stores all these distances, and then it classifies the unknown vector in the class whose samples gave more minimum distances (in this case, many distances have to be examined). One of the advantages of this approach is that new samples can be added to the database at any time, but it also has a higher calculation time.

A better approach is to examine only the K nearest neighbours (samples) to the unknown vector, and to classify it based on those K-neighbours. The class of the unknown feature vector will be the one that have most of the K-neighbours. This has been the approach implemented in our classification algorithm.

4.4 Euclidean Classifier

This classifier is one of the simplest and most efficient classifiers. This classifier has also been used to observe the tendency (goodness) of all the features previously studied, analysing which of all the studied features were more suitable for cork quality discrimination.

The classification algorithm supposes several classes with their respective prototypes (centroids). Given an unknown feature vector to classify, the Euclidean classifier will associate this vector to the class whose prototype is closest to it, that is, the prototype whose Euclidean distance is smallest.

Our study have been made for four versions of the Euclidean distance: simple Euclidean distance (see equation below), Euclidean distance with prefiltrate (certain corks were classified directly, without passing the Euclidean classifier, to low-quality classes if a hole in them was detected, that is, we used a set of decision rules in addition to the Euclidean classifier), scaled Euclidean distance (see equation below) and modified scaled Euclidean distance, according to the standard deviation (see equation in Sect. 4.2).

$$\textit{Euclidean Distance} = \sqrt{(x_{i1}-\mu_1)^2+(x_{i2}-\mu_2)^2+\cdots+(x_{iN}-\mu_N)^2}$$

$$\textbf{\textit{Scaled Euclidean Distance}} = \sqrt{\left(\frac{x_{i1}-\mu_1}{\sigma_1}\right)^2+\left(\frac{x_{i2}-\mu_2}{\sigma_2}\right)^2+\cdots+\left(\frac{x_{iN}-\mu_N}{\sigma_N}\right)^2}$$

The best results were obtained by the two last distances, but the modified scaled Euclidean distance was chosen for being more balanced in the results.

5 Obtained Results

The results of this section have been obtained using the 4 classification algorithms previously explained. We present these results by means of confusion matrices [7], due to their capability to show the conflicts among the different quality categories. Therefore, not only the definition of each class will be displayed, but also the main confusions among them.

5.1 Neural Classifier

The experimental results that are shown in this section correspond to a simplified version of the neural network. This decision was taken due to the non convergence of the network, when it was tried to learn the seven cork quality classes. Although we normalized the input data in a range from 0 to 24, and made a preselection of the cork disks that were more adapted to be training patterns, the convergence was impossible.

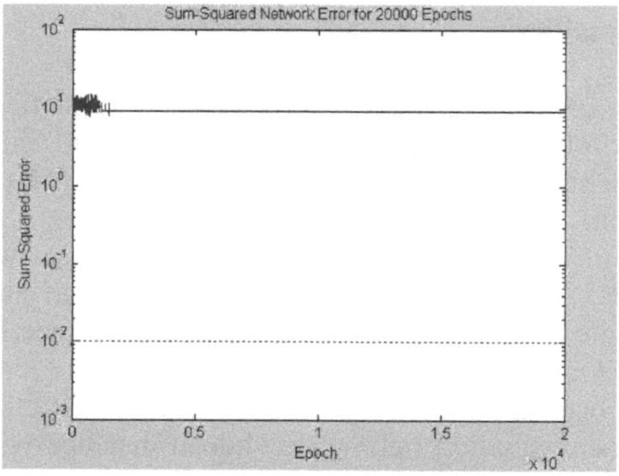

Fig. 2 Graph showing the non convergence.

As it can be observed in Fig. 2, with the first version of the neural network, it was impossible to reach the convergence of the network, and the error introduced in the classification was too high. The dotted line shows the level of ideal error established (0.01), and the solid line shows the real error in the classification (around 10 points). The shown result was obtained after 20000 iterations of the network.

After multiple tests with the neural network, it was verified that, probably due to the overlapping between contiguous classes, the network was only able to learn two classes, for example, class 0 and class 3.

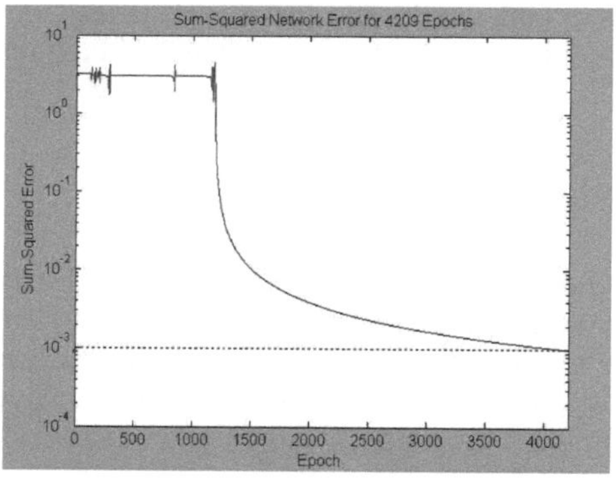

Fig. 3 Graph showing the convergence of the network.

As it is shown in Fig. 3, with this second version (and even after having lowered the maximum level of error to 0.001), the neural network reaches the convergence quickly, in only 4209 iterations.

After this explanation, we can present the results of the confusion matrix. Table 1 shows the confusion matrix for the neural classifier. As it was expected, we have obtained quite bad results due to the class overlapping. Since the neural network only recognizes two classes, all the corks are classified in class 0 or class 3. Anyway, the results are coherent, it can be observed as classes 0 and 3 are classified mainly in themselves. Classes 4, 5 and 6, which are more distant from class 0, are classified mainly in class 3. And classes 1 and 2 are those that present more confusion between class 0 and 3 for being between them.

Table 1 Confusion matrix for the neural classifier

	C0	C1	C2	C3	C4	C5	C6
C0	36	0	0	14	0	0	0
C1	28	0	0	22	0	0	0
C2	15	0	0	35	0	0	0
C3	12	0	0	38	0	0	0
C4	8	0	0	42	0	0	0
C5	3	0	0	47	0	0	0
C6	2	0	0	48	0	0	0

In conclusion, Table 2 presents the final results, with a wrong classification percentage of 78.85.

Table 2 Total results for the neural classifier

	C0	C1	C2	C3	C4	C5	C6	**TOT.**
Wrong	14	50	50	12	50	50	50	**276**
Right	36	0	0	38	0	0	0	**74**

5.2 K-Means Classifier

Table 3 displays the confusion matrix for the K-means classifier.

Table 3 Confusion matrix for the K-means classifier

	C0	C1	C2	C3	C4	C5	C6
C0	39	0	8	3	0	0	0
C1	25	0	15	9	0	0	1
C2	5	0	20	23	0	1	1
C3	1	0	7	24	4	3	11
C4	0	0	3	9	12	19	7
C5	0	0	0	7	17	24	2
C6	0	0	1	6	17	22	4

The confusion matrix we have obtained offers good results, although we can observe that there is a class that almost disappears, class 1. Nevertheless, the other classes have many right classifications, except class 6. In this sense, a great absorption power of class 5 over classes 4 and 6 is observed. The matrix presents only a little dispersion, which is very positive for the classification.

In conclusion, the final wrong classification percentage is 64.85 (Table 4).

Table 4 Total results for the K-means classifier

	C0	C1	C2	C3	C4	C5	C6	**TOT.**
Wrong	11	50	30	26	38	26	46	**227**
Right	39	0	20	24	12	24	4	**123**

5.3 K-Nearest Neighbours Classifier

For the calculation of the best size of K, three possible values have been checked. The chosen values were the following:

- A *little* value, K=10.
- A *big* value, K=49, the number of cork disks in a class (50) minus the disk under study.
- A *medium* value, K=20.

After a preliminary test, we finally concluded that the best size of K is 20. Once we have chosen the value of K, we have done our experiments using the Euclidean distance that has generated the best results, the scaled Euclidean distance according to the standard deviation (see equation in Sect. 4.2).

Table 5 presents the confusion matrix for the K-nearest neighbours classifier. As we can observe in the matrix, we have obtained good results. The matrix has a strong classification tendency around the main diagonal for all the classes, although it would be possible to say that still there are many erroneous classifications in some classes.

Table 5 Confusion matrix for the K-nearest neighbours classifier

	C0	C1	C2	C3	C4	C5	C6
C0	38	9	3	0	0	0	0
C1	24	15	11	0	0	0	0
C2	8	12	20	9	1	0	0
C3	1	8	10	16	10	2	3
C4	1	0	4	13	15	7	10
C5	0	0	2	7	12	10	19
C6	0	0	4	4	11	16	15

In conclusion, the final error rate (Table 6) is 63.14%.

Table 6 Total results for K-nearest neighbours classifier

	C0	C1	C2	C3	C4	C5	C6	TOT.
Wrong	12	35	30	34	35	40	35	**221**
Right	38	15	20	16	15	10	15	**129**

5.4 Euclidean Classifier

The obtained confusion matrix (Table 7) presents quite positive results. Using a classifier based on scaled Euclidean distances with the standard deviation, we can observe that class 6 acquires a great power of absorption, that even affects class 4. On the other hand, we can see a strong discrimination of classes 0, 6 and 3, with a great number of corks classified rightly in these classes.

Table 7 Confusion matrix for the Euclidean classifier

	C0	C1	C2	C3	C4	C5	C6
C0	33	12	4	1	0	0	0
C1	19	14	13	3	1	0	0
C2	6	9	15	18	2	0	0
C3	1	4	7	23	11	0	4
C4	2	0	1	10	13	3	21
C5	0	0	1	12	7	6	24
C6	1	0	1	7	7	3	31

The total results are shown in Table 8, with a final wrong classification percentage of 61.42.

Table 8 Total results for the Euclidean classifier

	C0	C1	C2	C3	C4	C5	C6	**TOT.**
Wrong	17	36	35	27	37	44	19	**215**
Right	33	14	15	23	13	6	31	**135**

6 Conclusions

The automatic visual inspection of cork is a problem of great complexity, in what refers to its quality-based classification, because cork is a natural material, and therefore, highly heterogeneous. This heterogeneity causes that cork quality depends on many combined factors, and among them, cork texture, defect area, size of the biggest defect, etc.

In this chapter we have performed a deep survey about several classifiers that includes each of these features (the best features we have found in a previous research). Concretely, we have focused on four important classifiers in the image processing field.

According to the experimental results we can say that, in case of cork, there are more suitable classifiers than others, although some of the studied classifiers have been very near in their final results. As final conclusion, we can say that the Euclidean classifier has been the more reliable in our application field. Figure 4 presents the wrong classification percentage obtained by the different classifiers.

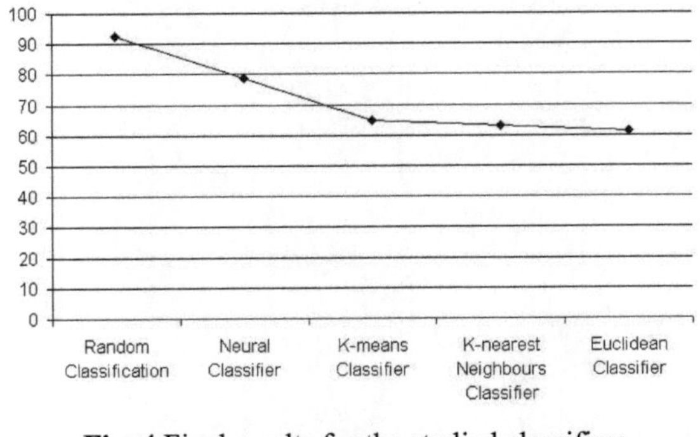

Fig. 4 Final results for the studied classifiers.

This graph also includes the wrong classification percentage that a random classification would have obtained if it was used.

As we can observe in the previous graph, the Euclidean classifier has produced the best results, but it is worthy to say that all the studied classification algorithms improve the results obtained by a random classification, although the goodness of the obtained results widely varies between some classifiers and others.

Furthermore, we think the results and conclusions obtained in this study can be useful to other visual inspection researches focused on other natural materials (wood, slate, etcetera), because they have common characteristics with the cork (heterogeneity, defects, changing texture according to their quality, etc.).

As future work we have planned to study other classifiers like, for example, fuzzy-neural networks. Also, we do not discard the inclusion and analysis of other features that could improve the classification results.

Acknowledgements

This work has been supported by the Spanish Government under Grant TIN2005-08818-C04-03.

References

1. CorkQC, The Natural Cork Quality Council, 2006. *Industry Statistics*. At http://www.corkqc.com.
2. Fortes, M.A., 1993. *Cork and Corks*. In European Review, vol. 1, pp. 189–195.
3. Haralick, R.M., Shanmugam, K., Dinstein, I., 1973. *Textural Features for Image Classification*. In IEEE Transactions on Man and Cybernetics, vol. 3, pp. 610–621.
4. ICMC, Instituto del Corcho, Madera y Carbón Vegetal, 2006. *Instituto de Promoción del Corcho (ICMC-IPROCOR)*. At http://www.iprocor.org, Spain.
5. Sahoo, P.K., Soltani, S., Wong, A.K.C., 1988. *A Survey of Thresholding Techniques*. In Computer Vision, Graphics, and Image Processing, vol. 41, pp. 233–260.
6. Shah, S.K., Gandhi, V., 2004. *Image Classification Based on Textural Features Using Artificial Neural Network (ANN)*. In IE(I) Journal-ET, vol. 84, pp. 72–77.
7. Shapiro, L.G., Stockman, G.C., 2001. *Computer Vision*, Ed. Prentice Hall, New Jersey.
8. Sonka, M., Hlavac, V., Boyle, R., 1998. *Image Processing, Analysis and Machine Vision,* Ed. PWS Publishing, USA, 2nd edition.

Visual Topological Map Building in Self-similar Environments

Toon Goedemé, Tinne Tuytelaars and Luc Van Gool

PSI - VISICS
Katholieke Universiteit Leuven, Belgium
{tgoedeme,tuytelaa,vangool}@esat.kuleuven.be

Abstract. This chapter describes a method to automatically build topological maps for robot navigation out of a sequence of visual observations taken from a camera mounted on the robot. This direct non-metrical approach relies completely on the detection of loop closings, i.e. repeated visitations of one particular place. In natural environments, visual loop closing can be very hard, for two reasons. Firstly, the environment at one place can look differently at different time instances due to illumination changes and viewpoint differences. Secondly, there can be different places that look alike, i.e. the environment is *self-similar*. Here we propose a method that combines state-of-the-art visual comparison techniques and evidence collection based on Dempster-Shafer probability theory to tackle this problem.

Keywords. Mobile robot navigation, topological map building, omnidirectional vision, Dempster-Shafer theory.

1 Introduction and Related Work

In every mobile robot application, the internal representation of the perceived environment is of crucial importance. The environment map is the basis for other tasks, like localisation, path planning, navigation, etc. The map building field can be divided in two major paradigms: geometrical maps and topological maps, even if hybrid types have been implemented [1].

In the traditional geometrical paradigm, maps are quantitative representations of the environment wherein locations are given in metrical coordinates. One approach often used is the occupancy map: a grid of evenly spaced cells each containing the information whether the corresponding position in the real world is occupied.

Because the latter is error-prone, time-consuming, and memory-demanding, we chose for the topological paradigm. Here, the environment map is a qualitative graph-structured representation where nodes represent distinct places in the environment, and arcs denote traversable paths between them. This flexible representation is not dependent on metrical localisation such as dead reckoning, is compact, allows high-level symbolic reasoning and mimics the internal map humans and animals use [2].

Several approaches for automatic topological map building have been proposed, differing in the method and the sensor(s) used. In our work, we solely use a camera as sensor. We chose for an omnidirectional system with a wide field-of-view.

Other researchers worked mainly with other sensors, such as the popular laser range scanner. A standard approach is the one of [3], who construct generalised Voronoi diagrams out of laser range data.

Very popular are various probabilistic approaches of the topological map building problem. Ranganathan et al. [4] for instance use Bayesian inference to find the topological structure that explains best a set of panoramic observations, while Shatkay and Kaelbling [5] fit hidden Markov models to the data. If the state transition model of this HMM is extended with robot action data, the latter can be modeled using a partially observable Markov decision process or POMDP, as in [6] and [7]. Zivkovic et al. [8] solve the map building problem using graph cuts.

In contrast to these global topology fitting approaches, an alternative way is detecting *loop closings*. During a ride through the environment, sensor data is recorded. Because it is known that the driven path is traversable, an initial topological representation is one long edge between start and end node. Now, extra links are created where a certain place is revisited, i.e. an equivalent sensor reading occurs twice in the sequence. This is called a loop closing. A correct topological map results if all loop closing links are added.

In natural environments, loop closing based on mere vision input can be very hard, for two reasons. Firstly, the environment at one place can look different at different time instances due to illumination changes, viewpoint differences and occlusions. A comparison technique that is not robust to these changes will overlook some loop closings. Secondly, there can be different places that look alike, i.e. the environment is *self-similar*. These would add erroneous loop closings and thus yield an incorrect topological map as well.

In [9], the authors detect loops by comparing omnidirectional images with local feature techniques, a robust technique we also adopt. But, there method suffers indeed from self-similarities, as we experienced in previous work [10].

Also in loop closing, probabilistic methods are introduced to cope with the uncertainty of link hypotheses. Chen and Wang [11], for instance, use Bayesian inference. Beevers and Huang [12] recently introduced Dempster-Shafer probability theory into loop closing, which has the advantage that ignorance can be modeled and no prior knowledge is needed. Their approach is promising, but limited to simple sensors and environments. In this chapter, we present a new framework for loop closing using rich visual sensors in natural complex environments, which is also based on Dempster-Shafer mathematics but uses it differently.

We continue the chapter with a brief summary of Dempster-Shafer theory in Sect. 2. Then we describe the details of our algorithm in Sect. 3. Section 4 details our real-world experiments. The chapter ends with a conclusion in Sect. 5.

2 Dempster-Shafer

The proposed visual loop closing algorithm relies on Dempster-Shafer theory [13] [14] to collect evidence for each loop closing hypothesis. Therefore, a brief overview of the central concepts of Dempster-Shafer theory is presented in this section.

Dempster-Shafer theory offers an alternative to traditional probabilistic theory for the mathematical representation of uncertainty. The significant innovation of this framework is that it makes a distinction between multiple types of uncertainty. Unlike traditional probability theory, Dempster-Shafer defines two types of uncertainty:

- *Aleatory Uncertainty* – the type of uncertainty which results from the fact that a system can behave in random ways (a.k.a. stochastic or objective uncertainty)
- *Epistemic Uncertainty* – the type of uncertainty which results from the lack of knowledge about a system (a.k.a. subjective uncertainty or ignorance)

This makes it a powerful technique to combine several sources of evidence to try to prove a certain hypothesis, where each of these sources can have a different amount of knowledge (*ignorance*) about the hypothesis. That is why Dempster-Shafer is typically used for sensor fusion.

For a certain problem, the set of mutually exclusive possibilities, called the *frame of discernment*, is denoted by Θ. For instance, for a single hypothesis H, where $\neg H$ is the inverse hypothesis of H about an event this becomes $\Theta = \{H, \neg H\}$. For this set, traditional probability theory will define two probabilities $P(H)$ and $P(\neg H)$, with $P(H)+P(\neg H) = 1$. Dempster-Shafer's analogous quantities are called *basic probability assignments* or *masses*, which are defined on the *power set* of Θ: $2^\Theta = \{A | A \subseteq \Theta\}$. The *mass* $m : 2^\Theta \to [0, 1]$ is a function meeting the following conditions:

$$m(\emptyset) = 0 \qquad \sum_{A \in 2^\Theta} m(A) = 1. \tag{1}$$

For the example of the single hypothesis H, the power set becomes $2^\Theta = \{\emptyset, \{H\}, \{\neg H\}, \{H, \neg H\}\}$. A certain sensor or other information source can assign masses to each of the elements of 2^Θ. Because some sensors do not have knowledge about the event (e.g. it is out of the sensor's field-of-view), they can assign a certain fraction of their total mass to $m(\{H, \neg H\})$. This mass, called the *ignorance*, can be interpreted as the probability mass assigned to the outcome 'H OR $\neg H$', i.e. when the sensor does not know about the event, or is – to a certain degree – uncertain about the outcome.[1]

Sets of masses about the same power set, coming from different information sources can be combined together using *Dempster's rule of combination*:

$$m_1 \oplus m_2(C) = \frac{\sum_{A \cap B = C} m_1(A) m_2(B)}{1 - \sum_{A \cap B = \emptyset} m_1(A) m_2(B)} \tag{2}$$

This combination rule is useful to combine evidence coming from different sources into one set of masses. Because these masses can not be interpreted as classical probabilities, no conclusions about the hypothesis can be drawn from them directly. That is why two additional notions are defined, *support* and *plausibility*. They are computed as:

$$Spt(A) = \sum_{B \subseteq A} m(B) \qquad Pls(A) = \sum_{A \cap B \neq \emptyset} m(B) \tag{3}$$

These values define a confidence interval for the real probability of an outcome: $P(A) \in [Spt(A), Pls(A)]$. Indeed, due to the vagueness implied in having non-zero ignorance, the exact probability can not be computed. But, decisions can be made based on the lower and upper bounds of this confidence interval.

[1] This means also that no prior probability function is needed, *no knowledge* can be expressed as total ignorance.

3 Algorithm

We apply this mathematical theory on loop closing detection based on omnidirectional image input. Our target application is as follows. A robot is equipped with an omnidirectional camera system and is guided through an environment, e.g. by means of a joystick. While driving around, images are captured at constant time intervals. This yields a sequence of images which the automatic method described in this chapter transforms in a topological map. Later, this map can be used for localisation, path planning and navigation, as we described in previous work [15] and [16].

3.1 Omnidirectional Camera System

The visual sensor we use is a catadioptric system composed by a colour camera and an hyperbolic mirror, as shown in Fig. 1. This system is mounted on top of a robot, in our case the electric wheel chair Sharioto. Typical images are shown in Fig. 7.

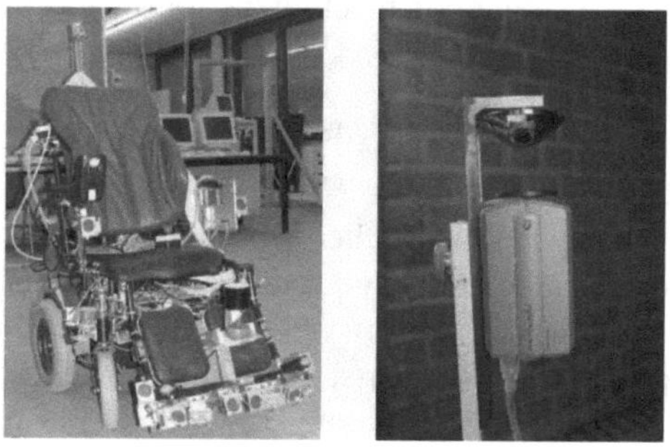

Fig. 1 Left: the wheel chair test platform. Right: the omnidirectional camera system.

3.2 Image Comparison

Because we do not have any other kind of information, the entire topological loop closing is based on images. The target is to find a good way to compare images, such that a second visit to a certain place can be detected as two similar images in the input sequence. As explained before, one of the main challenges is the appearance variation of places. At different time instances (i.e. during different visits), the images acquired at a certain place can vary a lot. This is mostly due to three reasons:

- **Illumination differences**: The same place is illuminated with a different light source (e.g. somebody switched on a light, the sun emerges from behind the clouds, ...).
- **Occlusions**: Part of the image can be hidden because of e.g. people passing by.
- **Viewpoint differences**: It can never be guaranteed that the robot comes back to exactly the same position. Even for small viewpoint changes the image looks different.

We want to recognise a place despite these factors, requiring the image comparison to be invariant or at least robust against them. Our proposed image comparison technique makes use of *fast wide baseline features*, namely SIFT [17] and Vertical Column Segments [18]. These techniques compute local feature matches between the two images, invariant to the illumination. Such a local technique is also robust to occlusions. Figure 2 shows an example of the found correspondences using these two kinds of features. These matches were found in less than half a second on up-to-date hardware and 640×480 images.

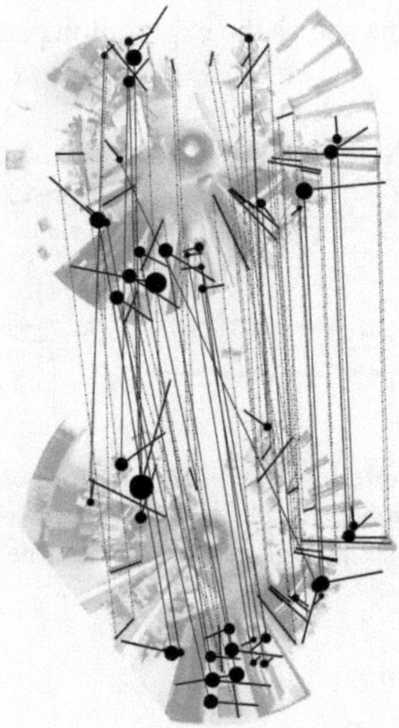

Fig. 2 A pair of omnidirectional images, superimposed with corresponding column segments (radial lines, matches indicated with dotted line) and SIFT features (circles with tail, matches with continues line). The images are rotated for optimal visibility of the matches.

The required image comparison measure (*visual distance*) must be inverse proportional to the *number of matches*, relative to the average number of features found in the images. Hence the first two factors in Eq. (4). But, also the difference in relative configuration of the matches must be taken in account. Therefore, we first compute a global angular alignment of the images by computing the average angle difference of the matches. The visual distance is now also made proportional to the average angle difference of the features after this global alignment.

$$d_V = \frac{1}{N} \cdot \frac{n_1 + n_2}{2} \cdot \frac{\sum |\Delta\alpha_i|}{N} \tag{4}$$

where N corresponds to the number of matches found, n_i the number of extracted features in image i, $\Delta\alpha_i$ the angle difference for one match after global alignment.

3.3 Image Clustering

We define a topological map as a graph; a set of *places* is connected by links denoting possible transitions from one place to another. Each place is represented by one prototype image, and transitions between places can be done by *visual servoing* from one place towards the prototype of the neighbouring place. Such a visual servoing step (as we described in [16]) imposes a maximum *visual distance* between places.

The dots in the sketch Fig. 3 denote places where images are taken. Because they were taken at constant time intervals and the robot did not drive at a constant speed, they are not evenly spread. We perform an agglomerative clustering with complete linkage based on the visual distance (Eq. (4)) on all the images, yielding the ellipse shaped clusters in Fig. 3. The black line shows the exploration path as driven by the robot.

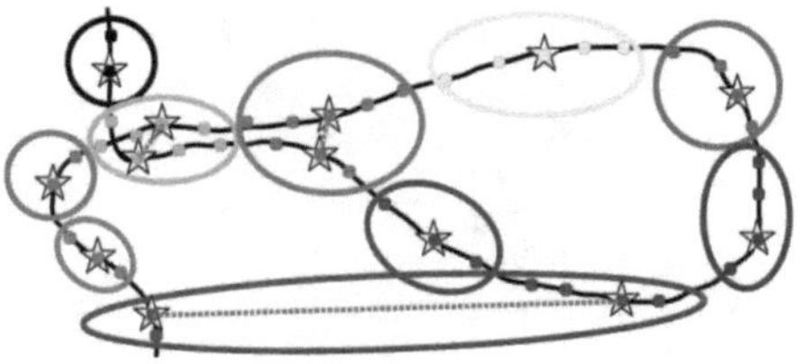

Fig. 3 Example for the image clustering and hypothesis formulation algorithms. Dots are image positions, black is exploration path, clusters are visualised with ellipses, prototypes of (sub)clusters with a star. Hypotheses are denoted by a dotted line between two stars.

3.4 Hypothesis Formulation

As can be seen in the example (Fig. 3), not all image groups nicely cover one distinct place. This is due to *self-similarities*, or distinct places in the environment that are different but look alike and thus yield a small visual distance between them.

For each of the clusters, we can define one or more subclusters. Images within one cluster who are linked by exploration path connections are grouped together. For each of these *subclusters* a prototype image is chosen as the *medoid*[2] based on the visual distance, denoted as a star in the figure.

As can be seen in the example, clusters containing more than one subcluster can be one of two possibilities:

- real *loop closings*, i.e. the robot is revisiting a place and detected that it looks alike.
- erroneous *self-similarities*, i.e. distinct places that look alike to the system.

For each pair of these subclusters within the same cluster, we define a *loop closing hypothesis* H, which states that if $H = true$, the two subclusters describe the same physical place and must be merged together. We will use Dempster-Shafer theory to collect evidence about each of these hypotheses.

[2] The medoid of a cluster is analogous to the centroid, but uses the median operator instead of the average.

3.5 Dempster-Shafer Evidence Collection

For each of the hypotheses defined in the previous step, a decision must be made if it was correct or wrong. Figure 4 illustrates four possibilities for one hypothesis. We observe that a hypothesis has more chance to be true if there are more hypotheses in the neighbourhood, like in case a and b. If no neighbouring hypotheses are present (c,d), no more evidence can be found and no decision can be made based on this data.

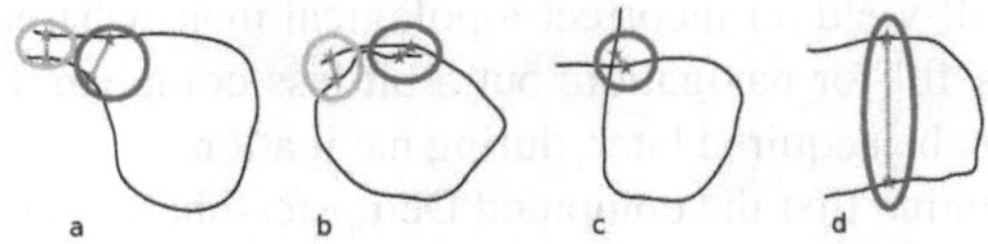

Fig. 4 Four topological possibilities for one hypothesis.

We conclude that for a certain hypothesis, a neighbouring hypothesis adds evidence to it. It is clear that, the further away this neighbour is from the hypothesis, the less certain the given evidence is. We chose to model this subjective uncertainty by means of the ignorance notion in Dempster-Shafer theory. That is why we define an *ignorance function* containing the distance between two hypotheses H_a and H_b:

$$\xi(H_a, H_b) = \begin{cases} 1 - \sin\left(\frac{d_H(H_a,H_b)\pi}{2d_{th}}\right) & (d_H \leq d_{th}) \\ 0 & (d_H > d_{th}) \end{cases} \tag{5}$$

where d_{th} is a distance threshold and $d_H(H_a, H_b)$ is the sum of the distances between the two pairs of prototypes of both hypotheses, measured in number of exploration images.

To gather *aleatory* evidence, we look at the visual similarity of both subcluster prototypes, normalised by the standard deviation of the intra-subcluster visual similarities:

$$s_V(H_a) = \frac{s_V(prot_{a1}, prot_{a2})}{\sigma_{subclus}(s_V)}, \tag{6}$$

where the visual similarity s_V is derived from the visual distance, defined in Eq. (4).

Each neighbouring hypothesis H_b yields the following set of Dempster-Shafer masses, to be combined with the masses of the hypothesis H_a itself:

$$\begin{aligned} m(\{\emptyset\}) &= 0 \\ m(\{H_a\}) &= s_V(H_b)\xi(H_a, H_b) \\ m(\{\neg H_a\}) &= (1 - s_V(H_b))\xi(H_a, H_b) \\ m(\{H_a, \neg H_a\}) &= 1 - \xi(H_a, H_b) \end{aligned} \tag{7}$$

Hypothesis masses are initialised with the visual similarity of its subcluster prototypes and a initial ignorance value (0.25 in our experiments), which models its influenceability by neighbours.

3.6 Hypothesis Decision

After combination of each hypothesis's mass set with the evidence given by neighbouring hypotheses (up to a maximum distance d_{th}), a decision must be made if this hypothesis was correct and thus if the subclusters must be united into one place or not.

Unfortunately, as stated above, only positive evidence can be collected, because we can not gather more information about totally isolated hypotheses (like c and d in Fig. 4). This not too bad, because of different reasons. Firstly, the chance for correct, but isolated hypothesis (case c) is low in typical cases. Also, adding erroneous loop closings (c and d) will yield an incorrect topological map, whereas leaving them out will keep the map useful for navigation, but a bit less complete. Of course, new data about these places can be acquired later, during navigation.

Important is to remind that the computed Dempster-Shafer masses can not directly be interpreted as probabilities. That is why we use Eq. (3) to compute the support and plausibility of each hypothesis after evidence collection. Because these values define a confidence interval for the real probability, a hypothesis can be accepted if the lower bound (the support) is greater than a threshold.

After this decision, a final topological map can be built. Subclusters connected with accepted hypotheses are merged into one place, and a new medoid is computed as prototype of it. For hypotheses that are not accepted, two distinct places should be constructed.

4 Experiments

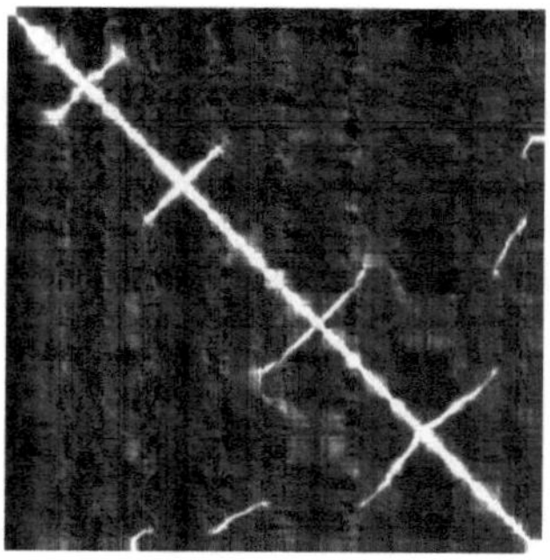

Fig. 5 Matrix showing the visual similarities between the images of the experiment.

With the camera system mounted on a electric wheel chair (see Fig. 1), we drove around in a complex natural environment, being our office floor. 463 images were recorded during this path of 275 m length. Figure 6 shows a map of the environment. It can be seen that the path visits several offices and corridors more than once, generating the possibility for a lot of loop closing hypotheses. Figure 7 gives a few typical images acquired.

Between each pair of images, fast wide baseline features are matched and the proposed visual distance measure is computed, yielding the similarity matrix visualised

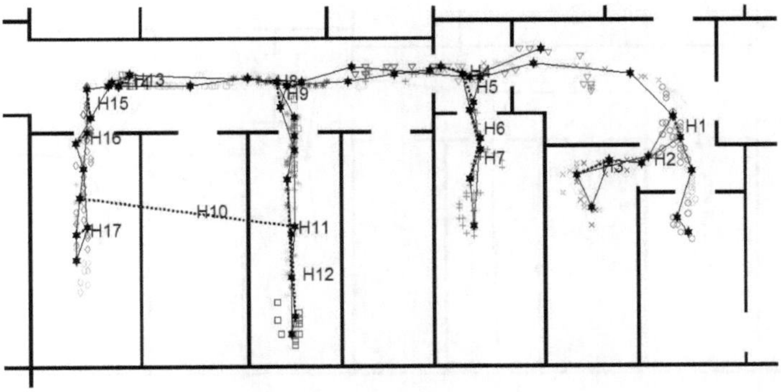

Fig. 6 Right: Map of the experiment. Scattered datapoints are image positions, shown in various symbols denoting clustering. Stars are (sub)clusters. The exploration path is visualised with thin black lines, hypotheses with thick dotted lines.

Fig. 7 Images 256, 315, 345, 388 and 430 of the experiment.

in Fig. 5. Based on this, the images are clustered as shown with different symbols in Fig. 6, resulting in 38 clusters. For each subcluster, a prototype is chosen denoted by a black star. Between subclusters within one cluster, hypotheses are formulated, denoted by thick black dotted lines.

As can be seen in the map, all but one hypothesis (number 10) are correct. This is clearly a self-similarity in the environment, the two offices do not differ enough in appearance. Figure 8 gives the Dempster-Shafer masses of each hypothesis before and after evidence collection. It is clear that after evidence combination, we have more reason to reject hypothesis 10. The other subclusters can be merged, resulting in the final topological map, shown in Fig. 9.

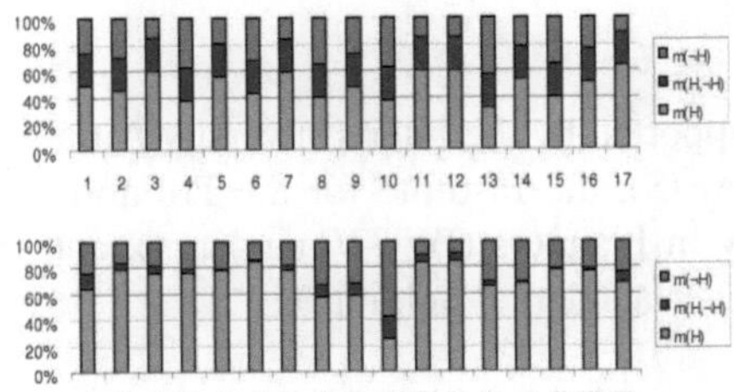

Fig. 8 Dempster-Shafer masses for each of the hypotheses, before (above) and after (below) evidence collection.

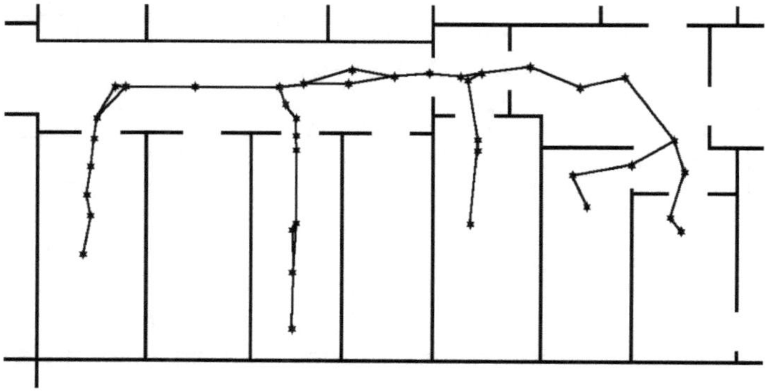

Fig. 9 Final topological map. Stars denote prototypes.

5 Conclusions

This chapter described a method to find topological loop closings in a sequence of images taken in a natural environment. The result is a vision-based topological map which can be used for localisation, path planning and navigation.

Firstly, a robust visual distance was presented. Making use of state-of-the-art wide baseline matching techniques, this enables the recognition of places despite changes in viewpoint, illumination, and the presence of occlusions.

Secondly, a mathematical model is presented to solve the problem of self-similarities, i.e. places in the environment that look alike but are different. This approach uses Dempster-Shafer probability theory to combine evidence of neighbouring loop closing hypotheses.

The real-world experiments presented illustrate the performance and robustness of the approach.

Future work planned includes the introduction of even more performant visual features such as SURF [19], and the on-line adaptation of the map while using it for navigation.

Acknowledgements

This work is partially supported by the Inter-University Attraction Poles, Office of the Prime Minister (IUAP-AMS), the Institute for the Promotion of Innovation through Science and Technology in Flanders (IWT-Vlaanderen), and the Fund for Scientific Research Flanders (FWO-Vlaanderen, Belgium).

References

1. Tomatis, N., Nourbakhsh, I.R., Siegwart, R.: Hybrid simultaneous localization and map building: closing the loop with multi-hypotheses tracking. In: ICRA, 1. (2002) 2749–2754

2. Tolman, E.C.: Cognitive maps in rats and men. In: Psychological Review (55). (1948) 189–208
3. Nagatani, K., Choset, H., Thrun, S.: Towards exact localization without explicit localization with the generalized voronoi graph. In: ICRA (1). (1998) 342–348
4. Ranganathan, A., Menegatti, E., Dellaert, F.: Bayesian inference in the space of topological maps. In: IEEE Transactions on Robotics 22(1). (2006)
5. Shatkay, H., Kaelbling, L.P.: Learning topological maps with weak local odometric information. In: IJCAI (2). (1997) 920–929
6. Koenig, S., Simmons, R.: Unsupervised learning of probabilistic models for robot navigation. In: ICRA (1). (1996) 2301–2308
7. Tapus, A., Siegwart, R.: Incremental robot mapping with fingerprints of places. In: IROS (1). (2005) 2429–2434
8. Zivkovic, Z., Bakker, B., Kröse, B.: Hierarchical map building using visual landmarks and geometric constraints. In: IROS (1). (2005) 7–12
9. Wahlgren, C., Duckett, T.: Topological mapping for mobile robots using omnidirectional vision. In: SWAR (1). (2005) 3441–3447
10. Goedemé, T., Nuttin, M., Tuytelaars, T., Van Gool, L.: Vision based intelligent wheelchair control: the role of vision and inertial sensing in topological navigation. In: Journal of Robotic Systems, 21(2). (2004) 85–94
11. Chen, C., Wang, H.: Appearance-based topological bayesian inference for loop-closing detection in cross-country environment. In: IROS (1). (2005) 322–327
12. Beevers, K., Huang, W.: Loop closing in topological maps. In: ICRA (1). (2005) 4367–4372
13. Dempster, A.P.: Upper and lower probabilities induced by a multivalued mapping. In: The Annals of Statistics (28). (1967) 325–339
14. Shafer, G.: A Mathematical Theory of Evidence. Princeton University Press, Princeton (1976)
15. Goedemé, T., Nuttin, M., Tuytelaars, T., Van Gool, L.: Markerless computer vision based localization using automatically generated topological maps. In: European Navigation Conference GNSS (1). (2004) 235–243
16. Goedemé, T., Tuytelaars, T., Vanacker, G., Nuttin, M., Van Gool, L.: Feature based omnidirectional sparse visual path following. In: International Conference on Intelligent Robots and Systems, IROS 2005. (2005) 1003–1008
17. Lowe, D.: Distinctive image features from scale-invariant keypoints. In: IJCV (60). (2004) 91–110
18. Goedemé, T., Tuytelaars, T., Van Gool, L.: Fast wide baseline matching with constrained camera position. In: Conference on Computer Vision and Pattern Recognition (1). (2004) 24–29
19. Bay, H., Tuytelaars, T., Van Gool, L.: Surf: speeded up robust features. Proceedings of the ninth European Conference on Computer Vision (1). (2006)

Image Motion Estimator to Track Trajectories Specified With Respect to Moving Objects

J. Pomares, G. J. García, L. Payá and F. Torres

Physics, Systems Engineering and Signal Theory Department
University of Alicante, Alicante, Spain
{jpomares, gjgg, laura.paya, Fernando.Torres}@ua.es

Abstract. Up to now, different methods have been proposed to track trajectories using visual servoing systems. However, when these approaches are employed to track trajectories specified with respect to moving objects, different considerations must be included in the visual servoing formulation to progressively decrease the tracking error. This paper shows the main properties of a non-time dependent visual servoing system to track image trajectories. The control action obtained integrates the motion estimation of the object from which the features are extracted. The proposed motion estimator employs information from the measures of the extracted features and from the variation of the camera locations. These variations are obtained determining the Homography matrix between consecutive camera frames.

Keywords. Motion estimation, tracking trajectories, visual control.

1 Introduction

Now, visual servoing systems are a well-known approach to guide a robot using image information. Typically, visual servoing systems are position-based and image-based classified [3]. Position-based visual servoing requires the computation of a 3-D Cartesian error for which a perfect CAD-model of the object and a calibrated camera are necessary. These types of systems are very sensitive to modeling errors and noise perturbations. In image-based visual servoing the error is directly measured in the image. This approach ensures the robustness with respect to modeling errors, but generally an inadequate movement of the camera in the 3-D Cartesian space is obtained. On the other hand, it is well known that image-based visual servoing is locally stable. This nice property ensures a correct convergence if the desired configuration is sufficiently near to the current one. This chapter shows the properties of an approach which employs image-based visual servoing to track trajectories. This method is not the main objective of the chapter and a more extensive study of the approach can be seen in our previous works [5].

Once this strategy is defined, the chapter focuses on the extension of the previous mentioned algorithms to be able to carry out the tracking of image trajectories when the visual features are in motion. Previous works, such as [3] have shown the necessity of estimating object motion and to include this estimation in the control action in order to decrease the tracking errors. The motion estimation can be solved using different algorithms like the one shown in [6] based in virtual visual servoing. In [1] an estimator which employs measurements about the camera velocity is proposed. However, this method introduces errors in the estimation due to the non accuracy measurement of the camera motion. The previous mentioned algorithms are used to achieve a given configuration of the features in the image. In this chapter, we define method to improve the estimation of the camera velocity based on visual information. This method is used to define a motion estimator to be applied during the tracking of trajectories.

This chapter is organized as follows: The main aspects of the visual servoing system to track trajectories are first described in Sect. 2. Section 3 shows a method to estimate the motion of the object from which the features are extracted. Section 4, simulation and experimental results confirm the validity of the proposed algorithms. The final section presents the main conclusions arrived at.

2 Tracking Image Trajectories

First, the notation employed and the desired trajectory to be tracked are described. In this chapter, we suppose that the robot must track a desired trajectory in the 3D space, $\gamma(t)$, using an eye-in-hand camera system (see Fig. 1). By sampling the desired trajectory, $\gamma(t)$, a sequence of N discrete values is obtained, each of which represents N intermediate positions of the camera ${}^{k}\gamma/k \in 1,...,N$. From this sequence, the discrete trajectory of the object in the image $S = \{{}^{k}\mathbf{s}/k \in 1,...,N\}$ can be obtained, where ${}^{k}\mathbf{s}$ is the set of M points or features observed by the camera at instant k, ${}^{k}\mathbf{s} = \{{}^{k}f_i/i \in 1,...,M\}$. In the next section, a non time-dependent visual controller to track the previous mentioned image trajectory is described. In this chapter we are not interested in image processing issues; therefore, the image trajectory is generated using four grey marks whose centres of gravity will be the extracted features (see Fig. 2).

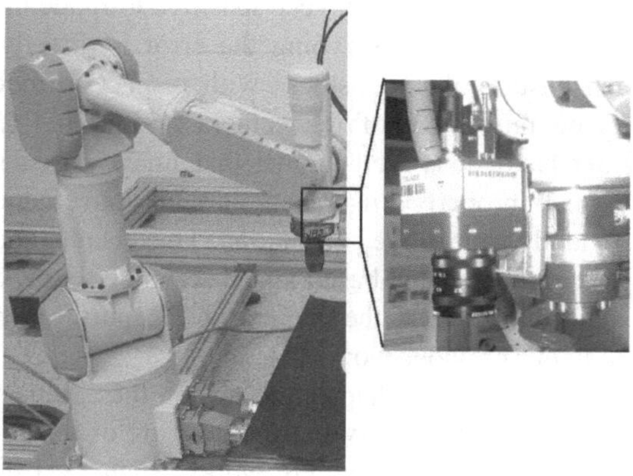

Fig. 1 Eye-in-hand camera system.

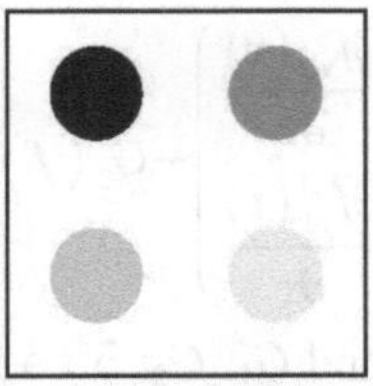

Fig. 2 Pattern from which the image features are extracted.

2.1 Visual Servo Control

Using a classical image based visual servoing system to carry out the tracking of a given trajectory, the following velocity must be applied to the robot (with respect to the coordinate frame located at the eye-in-hand camera):

$$\boldsymbol{v}_{\mathrm{V}}^{\mathrm{C}} = -\lambda \cdot \boldsymbol{e} + \hat{\mathbf{J}}_{\mathrm{f}}^{+} \cdot \frac{\partial \mathbf{s}_{\mathrm{d}}(t)}{\partial t} \tag{1}$$

where $\lambda > 0$ is the gain of the proportional controller and $\boldsymbol{e} = \hat{\mathbf{J}}_{\mathrm{f}}^{+} \cdot (\mathbf{s} - \mathbf{s}_{\mathrm{d}}(t))$ where $\mathbf{s}$ are the extracted features from the image, $\hat{\mathbf{J}}_{\mathrm{f}}^{+}$ is an estimation of the pseudoinverse of the interaction matrix [3] and $\mathbf{s}_{\mathrm{d}}(t)$ is a time dependent function. This function provides at each moment a set of desired visual features in order to allow the tracking of the desired trajectory in the image.

In order to allow the tracking in a non-time dependent way, the following control action obtained can be used:

$$\boldsymbol{v}_{\mathrm{V}}^{\mathrm{C}} = -\lambda \cdot \hat{\mathbf{J}}_{\mathrm{f}}^{+} \cdot \boldsymbol{e}_{\mathrm{f}} \tag{2}$$

where $\boldsymbol{v}_{\mathrm{V}}^{\mathrm{C}}$ is the velocity obtained with respect to the camera coordinate frame; $\lambda > 0$ is the gain of the controller; $\hat{\mathbf{J}}_{\mathrm{f}}^{+}$ is the pseudoinverse of the estimated interaction matrix; $\boldsymbol{e}_{\mathrm{f}} = \mathbf{s} - \mathbf{s}_{\mathrm{d}}$; $\mathbf{s}=[f_1, f_2,\ldots, f_M]^{\mathrm{T}}$ are the set of features extracted from the image; $\mathbf{s}_{\mathrm{d}}=[f_1+m_1\Phi_1(f_1), f_2+m_2\Phi_2(f_2),\ldots, f_M+m_M\Phi_M(f_M)]^{\mathrm{T}}$; Φ_i is the movement flow for the feature i, $\boldsymbol{m}=\{m_1, m_2,\ldots, m_M\}$ determines the progression speed.

Now, for the sake of clarity, the sub-index that indicates which feature is being considered and the super-index that indicates the instant in which these features are obtained are omitted. The movement flow, Φ, is a set of vectors converging towards the desired trajectory in the image.

We consider, for a given feature, a desired parameterized trajectory in the image $f_d:\Gamma \rightarrow \Im$ where $\Gamma \subset \Re$. The coordinates of this trajectory in the image are $f_d(\tau)=[f_{xd}(\tau), f_{yd}(\tau)]$ and $\boldsymbol{f}$ are the current coordinates of the feature in the image. The error vector $\boldsymbol{E}(f)=(E_x, E_y)$ where $E_x=(f_x-f_{xd})$ and $E_y=(f_y-f_{yd})$ is defined, where $f_d=(f_{xd}, f_{yd})$ are the coordinates of the nearest point to $\boldsymbol{f}$ in the desired trajectory. The movement flow, Φ, is defined as:

$$\Phi = G_1(\boldsymbol{f}) \cdot \begin{pmatrix} \dfrac{\partial f_{\mathrm{xd}}(\tau)}{\partial \tau} \\ \dfrac{\partial f_{\mathrm{yd}}(\tau)}{\partial \tau} \end{pmatrix} - G_2(\boldsymbol{f}) \cdot \begin{pmatrix} \dfrac{\partial \mathrm{U}}{\partial E_{\mathrm{x}}} \\ \dfrac{\partial \mathrm{U}}{\partial E_{\mathrm{y}}} \end{pmatrix} \tag{3}$$

where **U** is the potential function and G_1, G_2: $\Im \rightarrow \Re^+$ are weight functions so that $G_1 + G_2 = 1$. Details about the weight functions and the potential function **U** can be seen in our previous works [5].

3 Tracking Moving Objects

In order to track trajectories specified with respect to moving objects, the motion of the object must be included in the control action proposed in Eq. (2). Doing so, the new control action will be:

$$\boldsymbol{v}^{\mathrm{C}} = -\lambda \cdot \hat{\mathbf{J}}_{\mathrm{f}}^{+} \cdot \boldsymbol{e}_{\mathrm{f}} - \frac{\partial \hat{e}}{\partial t} \tag{4}$$

where $\partial\hat{e}/\partial t$ represents the estimation of the variations of $\boldsymbol{e} = \hat{\mathbf{J}}_{\mathrm{f}}^{+} \cdot \boldsymbol{e}_{\mathrm{f}}$ due to the movement of the object from which the features are extracted. As is shown in [1], the estimation of the velocity of a moving object tracked with an eye-in-hand camera system can be obtained from the measurements of the camera velocity and from the error function. Thus, from Eq. (2) the value of the estimation of the error variation due to the movement of the tracked object can be obtained in this way (to obtain an exponential decrease of the error it must fulfil that $\dot{\boldsymbol{e}} = -\lambda \cdot \boldsymbol{e}$):

$$\frac{\partial \hat{e}}{\partial t} = \dot{\boldsymbol{e}} - \boldsymbol{v}^{\mathrm{C}} \tag{5}$$

From Eq. (5), the value of the motion estimation can be obtained using the following expression:

$$\left(\frac{\partial \hat{e}}{\partial t}\right)_{\mathrm{k}} = \frac{\boldsymbol{e}_{\mathrm{k}} - \boldsymbol{e}_{\mathrm{k-1}}}{\Delta t} - \boldsymbol{v}_{\mathrm{k-1}}^{\mathrm{C}} \tag{6}$$

where Δt can be obtained determining the delay at each iteration of the algorithm, $\boldsymbol{e}_{\mathrm{k}}$ and $\boldsymbol{e}_{\mathrm{k-1}}$ are the error values at the instants k and k–1, and $\boldsymbol{v}_{\mathrm{k-1}}^{\mathrm{C}}$ is the camera velocity measurement at the instant k–1 with respect the camera coordinate frame. As is shown in our previous works [4] the estimations obtained from Eq. (6) depends on the measurement of the camera motion which cannot be measured without errors. Therefore, in order to improve the global behaviour of the system it is necessary to obtain a more accuracy estimation of the camera motion.

To compute the camera velocity in the previous iteration $\boldsymbol{v}_{\mathrm{k-1}}^{\mathrm{C}}$ it is necessary to obtain the rotation $\mathbf{R}_{\mathrm{k-1}}$ and translation $\boldsymbol{t}_{\mathrm{dk-1}}$ between the two last frames. To do so, first of all, we call Π the plane containing the object. Considering P a 3D point observed by the camera, the same point in the image space is p. With $p_{\mathrm{k-1}}$ we represent the position of the feature in the image captured at instant k–1. In the next image (which corresponds with the image obtained by the camera after one iteration of the control

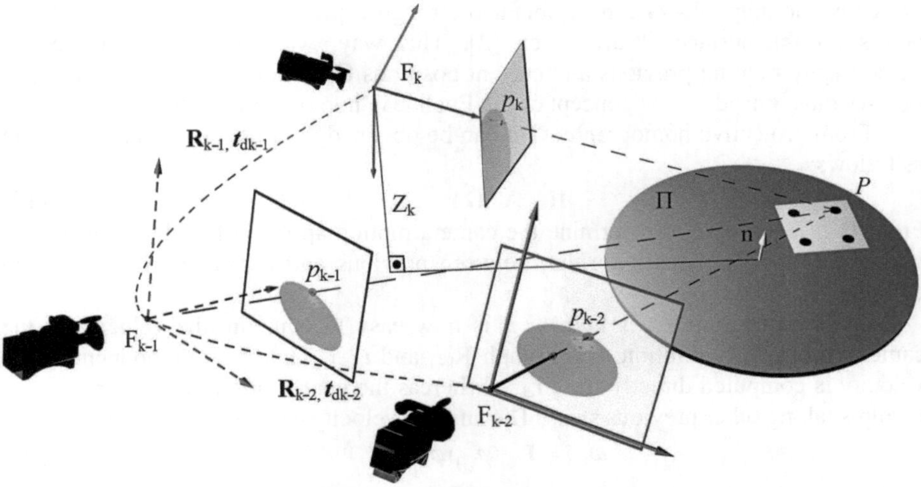

Fig. 3 Scheme of the motion estimation.

loop), the same point will be located at p_k position (see Fig. 3). The projective homography **G** is defined as:

$$\mu_{k-1} p_{k-1} = \mathbf{G} p_k + \beta \boldsymbol{\zeta} \tag{7}$$

where μ_{k-1} is a scale factor, $\boldsymbol{\zeta} = \mathbf{A}\mathbf{R}_{k-1} \boldsymbol{t}_{dk-1}$ is the projection in the image captured by the camera at time k–1 of the optical centre when the camera is in the next position, and β is a constant scale factor which depends on the distance Z_k from the contact surface to the origin of the camera placed at the current position:

$$\beta = \frac{d(P,\Pi)}{Z_k} \tag{8}$$

where $d(P,\Pi)$ is the distance from the contact surface plane to the 3D point. Although β is unknown, applying Eq. (7) between the instant k–1 (at previous iteration) and the current camera positions we can obtain that:

$$\beta = sign\left(\frac{(\mu_{k-1} p_{k-1} - \mathbf{G}_{k-1} p_k)_1}{(\mathbf{A}\mathbf{R}_{k-1} \boldsymbol{t}_{dk-1})_1}\right) \frac{\left\| \mathbf{G}_{k-1} p_k \wedge p_{k-1} \right\|}{\left\| \mathbf{A}\mathbf{R}_{k-1} \boldsymbol{t}_{dk-1} \wedge p_{k-1} \right\|} \tag{9}$$

where subscript 1 indicates the first element of the vector and **A** is a non singular matrix containing the camera internal parameters:

$$\mathbf{A} = \begin{bmatrix} f \cdot p_u & -f \cdot p_u \cdot \cot(\theta) & u_0 \\ 0 & f \cdot p_v / \sin(\theta) & v_0 \\ 0 & 0 & 1 \end{bmatrix} \tag{10}$$

where u_0 and v_0 are the pixel coordinates of the principal point, f is the focal length, p_u and p_v are the magnifications in the u and v directions respectively, and θ is the angle between these axes.

If P is on the plane Π, β is null. Therefore, from Eq. (7):

$$\mu_{k-1} p_{k-1} = \mathbf{G} p_k \tag{11}$$

Projective homography **G** can be obtained through expression Eq. (11) if at least four points on the surface Π are given [2]. This way, we can compute projective homography relating previous and current positions $\mathbf{G}_{k-1}$. In order to obtain $\mathbf{R}_{k-1}$ and $\boldsymbol{t}_{dk-1}$ we must introduce the concept of the Euclidean homography matrix **H**.

From projective homography **G** it can be obtained the Euclidean homography **H** as follows:

$$\mathbf{H} = \mathbf{A}^{-1}\mathbf{G}\mathbf{A} \tag{12}$$

From **H** it is possible to determine the camera motion applying the algorithm shown in [7]. So, applying this algorithm between previous and current positions we can compute $\mathbf{R}_{k-1}$ and $\boldsymbol{t}_{dk-1}$.

As iteration time Δt is known, it is now easy to compute the velocity of the camera at previous iteration $\boldsymbol{v}_{k-1}$ through $\mathbf{R}_{k-1}$ and $\boldsymbol{t}_{dk-1}$. The linear component of the velocity is computed directly from $\boldsymbol{t}_{dk-1}$, whereas the computation of angular velocity requires taking other previous steps. The angular velocity can be expressed as:

$$\boldsymbol{\omega}_{k-1} = \mathbf{T}_{k-1}(\boldsymbol{\varphi}_{k-1})\dot{\boldsymbol{\varphi}}_{k-1} \tag{13}$$

where $\boldsymbol{\varphi}_{k-1} = \left[\alpha_{k-1} \quad \beta_{k-1} \quad \gamma_{k-1}\right]$ are the Euler angles ZYZ obtained from $\mathbf{R}_{k-1}$, $\dot{\boldsymbol{\varphi}}_{k-1}$ are the time derivative of the previous Euler angles and:

$$\mathbf{T}_{k-1}(\boldsymbol{\varphi}_{k-1}) = \begin{pmatrix} 0 & -\sin\alpha_{k-1} & \cos\alpha_{k-1}\sin\beta_{k-1} \\ 0 & \cos\alpha_{k-1} & \sin\alpha_{k-1}\sin\beta_{k-1} \\ 1 & 0 & \cos\beta_{k-1} \end{pmatrix} \tag{14}$$

Applying Eq. (7) it is possible to compute the angular velocity of the camera to achieve current position from the previous iteration. This way linear and angular velocity of the camera between the two iterations are computed:

$$\boldsymbol{v}_{k-1} = \begin{pmatrix} \boldsymbol{v}_{tk-1} \\ \boldsymbol{\omega}_{k-1} \end{pmatrix} = \begin{pmatrix} \dfrac{\boldsymbol{t}_{dk-1}}{\Delta t} \\ \mathbf{T}_{k-1}(\boldsymbol{\varphi}_{k-1})\dot{\boldsymbol{\varphi}}_{k-1} \end{pmatrix} \tag{15}$$

4 Results

4.1 Experimental Setup

The system architecture is composed of an eye-in-hand PHOTONFOCUS MV-D752-160-CL-8 camera at the end-effector of a 7 d.o.f. Mitsubishi PA-10 robot. The camera is able to acquire and to process up to 100 frames/second using an image resolution of 320×240.

In this chapter we are not interested in image processing issues; therefore, the image trajectory is generated using four grey marks whose centres of gravity will be the extracted features.

4.2 Simulation Results

In this section simulation results are obtained from the application of the proposed visual servoing system to track trajectories specified with respect to moving objects.

To do so, the motion estimator described in Sect. 3 is used. The motion of the features extracted during the experiment is shown in Fig. 4. In this figure the evolution of the visual features observed from the initial camera position is represented (in this figure we consider that any control action is applied to the robot). We can see that the target object describes a periodical and rectangular motion. To better show the object motion, in Fig. 5 the image error evolution is represented. As previously, in Fig. 5 any control action is applied to the robot (in this figure the desired features are the ones provided by the movement flow). This motion will not end until the trajectory is completely tracked. Figure 6 shows the desired image trajectory and the ones obtained considering and without considering motion estimation. This figure shows that, using motion estimation, the system tracks the desired trajectory avoiding error due to the motion of the object from which the features are extracted.

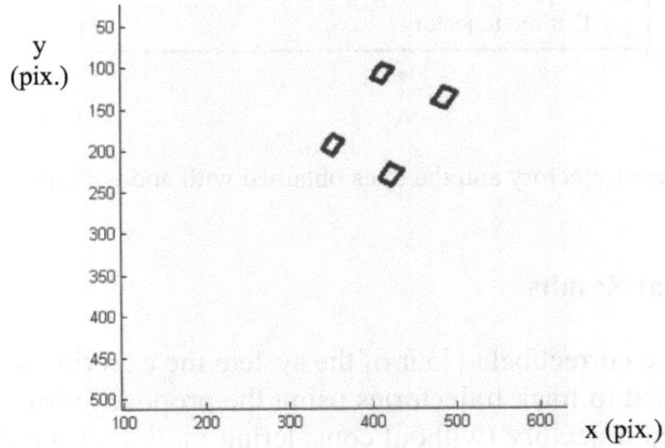

Fig. 4 Image trajectory observed from the initial camera position.

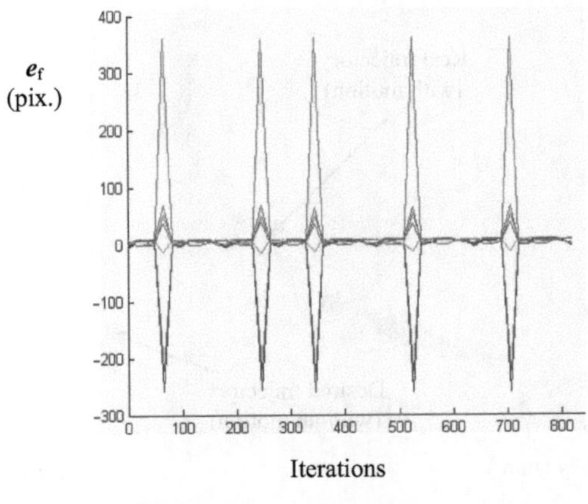

Fig. 5 Image error due to the object motion.

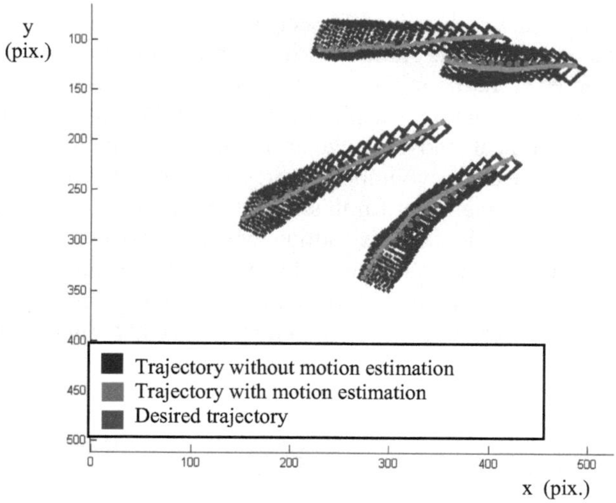

Fig. 6 Desired image trajectory and the ones obtained with and without motion estimation.

4.3 Experimental Results

In order to verify the correct behaviour of the system the experimental setup described in Sect. 4.1 is applied to track trajectories using the proposed visual servoing system. In Fig. 7 the desired trajectory (without considering motion of the object from which the trajectory is specified) is shown. In the same figure is also represented the obtained trajectory when the object from which the features are extracted is in motion. We can observe that the motion is lineal (is not equal to the one described in Sect. 4.2).

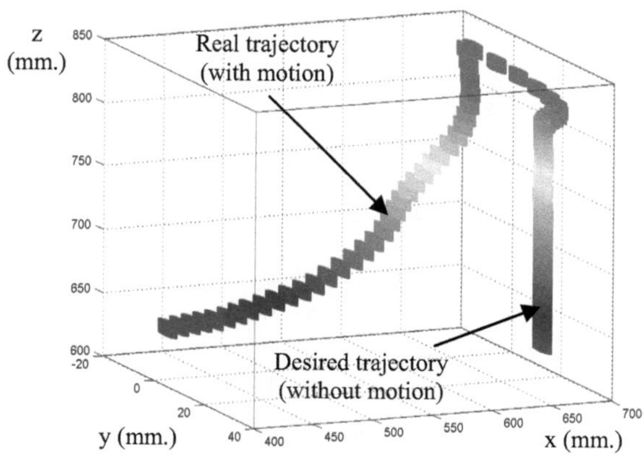

Fig. 7 Comparison between the desired 3D trajectory and the one obtained using motion estimation.

In order to observe the improvement introduced by the motion estimation in the visual servoing system, in Fig. 8 a comparison in the image space between the desired trajectory and the ones obtained with and without motion estimation is represented.

We can see that using motion estimation the system reduces rapidly the tracking error due to the motion of the object. The object motion is not constant, therefore, when the motion estimation is not carried out, the error increases quickly when the object velocity increases. This tracking error is clearly reduced using the motion estimation proposed in this chapter.

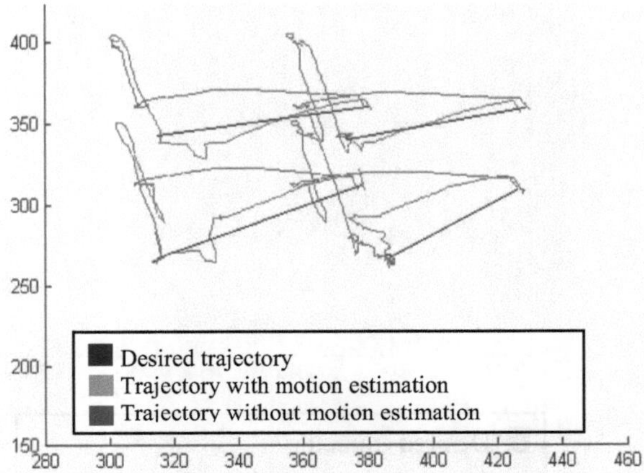

Fig. 8 Desired image trajectory and the ones obtained with and without motion estimation.

In order to show more clearly the improvement obtained with the motion estimator a new experiment will be presented. In Fig. 9 the trajectory described by the system in the image space with and without considering motion estimation, is represented. The desired trajectory is also shown. In this case the visual features are in motion. This motion is carried out by another robot. It can be observed that when we introduce the estimation in the system, the error caused by the movement of the object is considerably reduced. Thus, without estimation the system is able to arrive to the final of the trajectory when the object motion ends. This is because when the tracking reaches a point in the image, the movement flow tries to reduce the error caused by the object motion, however, this control action cannot compensate the motion, and only can reduce the error when the motion ends.

5 Conclusions

In this chapter we have shown a visual servoing system to track image trajectories specified with respect to moving objects. In order to avoid the errors introduced by the motion, it is required to include in the control action of the visual servoing system the effect of this motion. The chapter describes the necessity to obtain a good estimation of the camera motion and, to do so, a method based on visual information has been proposed. Simulation and experimental results show that using the proposed estimator a good tracking is obtained avoiding the errors introduced by the motion.

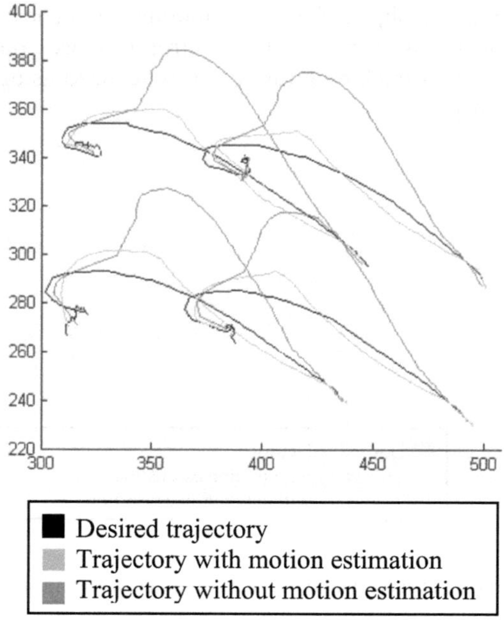

Fig. 9 Comparison between the desired trajectory in the image and the one obtained with and without motion estimation.

Acknowledgements

This work was funded by the Spanish MCYT project DPI2005-06222 "Diseño, implementación y experimentación de escenarios de manipulación inteligentes para aplicaciones de ensamblado y desensamblado automático" and by the project GV05/007: "Diseño y experimentación de estrategias de control visual-fuerza para sistemas flexibles de manipulación".

References

1. Bensalah, F., Chaumette, F. 1995. Compensation of abrupt motion changes in target tracking by visual servoing. In *IEEE/RSJ Int. Conf. on Intelligent Robots and Systems, IROS'95*, Pittsburgh, vol. 1, pp. 181–187.
2. Hartley R., Zisserman A. 2000. *Multiple View Geometry in Computer Vision.* Cambridge Univ. Press, pp. 91–92.
3. Hutchinson, S., Hager, G., Corke, P. 1996. A Tutorial on Visual Servo Control. IEEE Trans. on Robotics and Automation, vol. 12, no. 5, pp. 651–670.
4. Pomares, J., Torres, F., Gil, P. 2002. 2-D Visual servoing with integration of multiple predictions of movement based on Kalman filter. In *Proc. of IFAC 2002 15th World Congress*, Barcelona, vol. L.

5. Pomares, J., Torres, F., 2005. Movement-flow based visual servoing and force control fusion for manipulation tasks in unstructured environments. IEEE Trans. on Systems, Man, and Cybernetics—Part C, vol. 35, no. 1, pp. 4–15.
6. Pressigout, M., Marchand, E. 2004. Model-free augmented reality by virtual visual servoing. In *IAPR Int. Conf. on Pattern Recognition, ICPR'04*, Cambridge, vol. 2, pp. 887–890.
7. Zhang, Z., Hanson, A.R. 1996. Three-dimensional reconstruction under varying constraints on camera geometry for robotic navigation scenarios, Ph.D. Dissertation, University of Massachusetts Amherst.

Depth Gradient Image Based on Silhouette: A Solution for Reconstruction of Scenes in 3D Environments

Pilar Merchán[1], Antonio Adán[2] and Santiago Salamanca[3]

[1]Escuela de Ingenierías Industriales, Universidad de Extremadura, Avda. Elvas s/n, 06071Badajoz, Spain
pmerchan@unex.es
[2]Escuela Superior de Informatica, Universidad de Castilla La Mancha 13071 Ciudad Real, Spain
Antonio.Adan@uclm.es
[3]Escuela de Ingenierías Industriales, Universidad de Extremadura, Avda Elvas, s/n, 06071Badajoz, Spain
ssalaman@unex.es

Abstract. Greatest difficulties arise in 3D environments when we have to deal with a scene with dissimilar objects without pose restrictions and where contacts and occlusions are allowed. This work tackles the problem of correspondence and alignment of surfaces in such a kind of scenes. The method presented in this chapter is based on a new representation model called Depth Gradient Image Based on Silhouette (DGI-BS) which synthesizes object surface information (through depth) and object shape information (through contour). Recognition and pose problems are efficiently solved for all objects of the scene by using a simple matching algorithm in the DGI-BS space. As a result of this the scene can be virtually reconstructed. This work is part of a robot intelligent manipulation project. The method has been successfully tested in real experimentation environments using range sensors.

Keywords. 3D computer vision, reconstruction, 3D object recognition, range image processing.

1 Introduction

1.1 Statement of the Problem in a Practical Environment

The work presented in this chapter is integrated in a robot-vision project where a robot has to carry out an intelligent interaction in a complex scene. The 3D vision system takes a single range image of the scene which is processed to extract information about the identity of the objects and their pose in the scene. Figure 1a presents the real environment with the components that we are using in our work: the scene, the immobile vision sensor and the robot. In the worst case, the complexity of this scene includes: no shape-restrictions, shades, occlusion, cluttering and contact between objects. Figure 1b shows a prototype of a scene that we have dealt with.

An intelligent interaction (grasping, pushing, touching, etc.) of a robot in such a scene is a complex task which involves several and different research fields: 3D image processing, computer vision and robotics. There are three main phases in the interaction: segmentation of the scene into their constituent parts, recognition and pose of the objects and robot planning/interaction. In this chapter we specifically present an efficient recognition/pose solution that allows us to know the layout of the objects in the scene. This information is essential to carry out a robot interaction in the scene. Therefore we have integrated this work as a part of the general project which is being currently used in real applications.

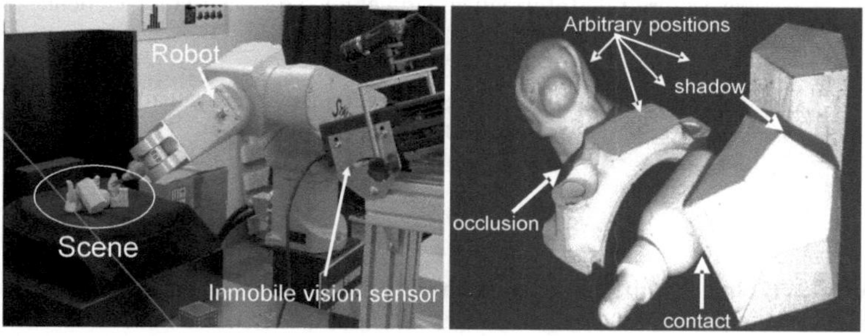

Fig. 1 (**a**) Robot interaction in the scene. (**b**) Example of complex scene.

1.2 Previous Work on Contour and Surface based Representations

Object recognition in complex scenes is one of the current problems in the computer vision area. In a general sense, the recognition involves identifying an unknown object, which is arbitrarily posed, among a set of objects in a database. In 3D environments, recognition and positioning are two different concepts that can be separately handled when complete information of the scene is available. Nevertheless, if only a part of the object is sensed in the scene, recognition and pose appear closely related. In others words, the object is recognized through the computation of the best alignment scene-model.

As we know, there is a wide variety of solutions to recognize 3D objects. Most of them use models based on geometrical features of the objects and the solution depends on the feature matching procedure.

Among the variety of methods are those that recognize 3D objects by studying sets of 2D images of the object from different viewpoints. Some of these methods rely on extracting the relevant points of a subset of canonical views of the object [16] [17]. The main drawback of these techniques is that the detection of relevant points may be sensitive to noise, illumination changes and transformations. There are also strategies that use shape descriptors instead of relevant points of the object [19] [21] but most of them need the complete image of the object to solve the recognition problem.

Matching and alignment techniques based on contour representations are usually effective in 2D environments. Thus, different techniques based on: Fourier descriptors [24], moment invariants [5], spline curves [4], wavelets [12] and contour curvature [14], [23] can be found in the literature. Nevertheless, in 3D environments these techniques have serious limitations. The main restrictions are: limited object pose, viewpoint restrictions (usually reduced to one freedom degree) and occlusions not allowed. Main troubles of the contour based techniques occur due to the fact that the silhouette's information may be insufficient and ambiguous. Thus, similar silhouettes might correspond to different objects from different viewpoints. Consequently, a representation based on the object contour may be ambiguous, especially when occlusion circumstances occur. As a result, the matching problem is usually tackled by means of techniques that use surface information instead of contour information.

Surface matching algorithms solve the recognition problem using 3D sensors. In this case two different processes arise: surface correspondence and surface registration. Surface correspondence is the process that establishes which portions of two surfaces overlap. Using the surface correspondence, registration computes the transformation that aligns the two surfaces. Obviously the correspondence phase is the most interesting part of the algorithm. Recently Planitz et al. [15] have published a survey about these techniques. Next, we will present a brief summary of the most important works that are closely related to ours.

Chua and Jarvis [9] code surrounding information at a point of the surface through a feature called *Point Signature*. Point signature encodes information on a 3D contour of a point P of the surface. The contour is obtained by intersecting the surface of the object with a sphere centered on P. The information extracted consists of distances of the contour to a reference plane fixed to it. So, a parametric curve is computed for every P and it is called point signature. An index table, where each bin contains the list of indexes whose minimum and maximum point signature values are common, is used for making correspondences. The best global correspondence produces the best registration.

Johnson and Hebert [11] have been working with polygonal and regular meshes to compare two objects through *Spin Image* concept. Spin image representation encodes information not for a contour but for a region around a point P. Two geometrical values (α,β) are defined for the points of a region and a 2D histogram is finally constructed.

In [22] a representation, that stores surface curvature information from certain points, produces images called *Surface Signatures* at these points. As a result of this, a standard Euclidean distance to match objects is presented. Surface signature representation has several points in common with spin image. In this case, surface curvature information is extracted to produce 2D images called surface signature where 2D coordinates correspond to other geometrical parameters related to local curvature. *Geometric Histogram* matching [3] is a similar method that builds another kind of 2D geometric histogram.

Zhang [25] also uses 2D feature representation for regions. In this case a curvature-based representation is carried out for arbitrary regions creating *harmonic shape images.* Model and scene representations are matched and the best local match is chosen. A. Adán and M. Adán [1] present a new strategy for 3D objects recognition using a flexible similarity measure that can be used in partial views. Through a new curvature feature, called *Cone Curvature*, a matching process yields a coarse transformation between surfaces. Campbell and Flynn [7] suggest a new local feature based technique that selects and classify the highly curved regions of the surface. These surfaces are divided up into smaller segments which define a basic unit in the surface. The registration is accomplished through consistent poses for triples of segments.

Cyr and Kimia [8] measure the similarity between two views of the 3D object by a metric that measures the distance between their corresponding 2D projected shapes. Liu et al. [20] propose the Directional Histogram Model for objects that have been completely reconstructed.

Most methods described impose some kind of restriction related to 3D sensed information or the scene itself. The main restrictions and limitations are: several views are necessary [16] [17], isolated object or few objects in the scene, shape restrictions [1] [3] [22] points or zones of the object must be selected in advance [7] [11] [22] [25] and occlusion is not allowed [20].

In this chapter we deal with the problem of correspondence and registration of surfaces by using a new strategy that synthesizes both surface and shape information in 3D scenes without restrictions. In other words: only a view of the scene is necessary, multi-occlusion is allowed, no initial points or regions are chosen and no shape restrictions are imposed. Our technique is based on a new 3D representation called *Depth Gradient Image Based on Silhouette* (DGI-BS). Through a simple DGI-BS matching algorithm, the surface correspondence is solved and a coarse alignment is yielded.

The chapter is structured as follows. DGI-BS model is presented throughout Sect. 2 including the DGI-BS representation for occlusion cases. Section 3 presents the DGI-BS based surface correspondence algorithm and Sect. 4 deals with correspondence verification and registration. Section 5 presents the main results achieved in our lab and finally conclusions and future work are set in Sect. 6.

2 DGI-BS Model

2.1 Concept

DGI-BS model is defined on the range image of the scene. Range image yields the 3D coordinates of the pixels corresponding to the intensity image of the scene. If the reference system is centred in the camera, the coordinate corresponding to the optic axis (usually Z axis) gives the depth information. So, the depth image is directly available.

Let us suppose a scene with only one object. Let Z be the depth image of the object from an arbitrary viewpoint v and let H be the silhouette of the object from v. Note that the silhouette H can be marked in Z.

After recording the list of pixels along the silhouette, we calculate (see Fig. 2 above), for every pixel j of H, a set of depth gradients corresponding to pixels that are in the 2D normal direction to P_j towards inside of the object. Formally

$$\nabla Z_{H,i}^{j} = Z^{i} - Z_{H}^{j} \tag{1}$$

where

$$\forall P_j \in H, \quad j = 1,\ldots,p \qquad d(P_j, P_i) = i \cdot s \qquad i = 1,\ldots,t \tag{2}$$

In expression (2), p is the number of pixels of the silhouette H, d is the Euclidean distance (in the depth image Z) between pixels P_j and P_i, and s is the distance between two samples. Note that the first pixel of H is selected at random.

Once $\nabla Z_{H,i}^{j}$ has been calculated for every point of H, we obtain a $t \times p$ matrix which represents the object from the viewpoint v. That is we have called *DGI-BS* representation from v.

$$DGI - BS(v) = \begin{bmatrix} \nabla Z_{H,1}^{1} & \nabla Z_{H,1}^{2} & . & . & \nabla Z_{H,1}^{p} \\ \nabla Z_{H,2}^{1} & \nabla Z_{H,2}^{2} & & & \nabla Z_{H,2}^{p} \\ . & . & & & . \\ \nabla Z_{H,t}^{1} & \nabla Z_{H,t}^{2} & . & . & \nabla Z_{H,t}^{p} \end{bmatrix} \quad (3)$$

Note that *DGI-BS*(v) is a small image where j-th column is the set of depth gradients for pixel P_j and i-th row stores the gradients for points that are $i{\cdot}s$ far from contour pixels in the corresponding normal directions. Figure 2 (right) shows an example of DGI-BS using a grey-level code where black colour pixels correspond to samples that are outside the object.

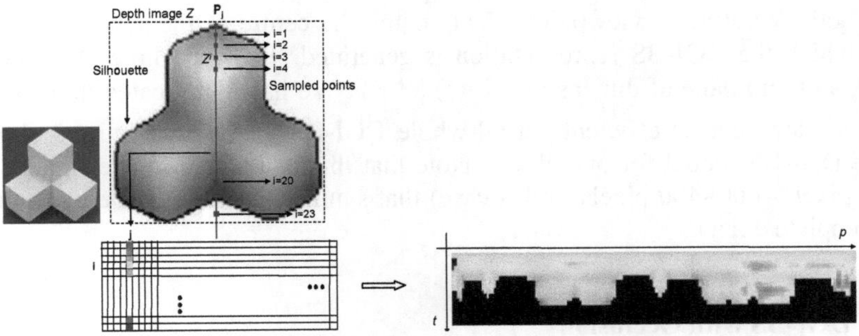

Fig. 2 Depth gradient concept (left). DGI-BS representation for a single view (right).

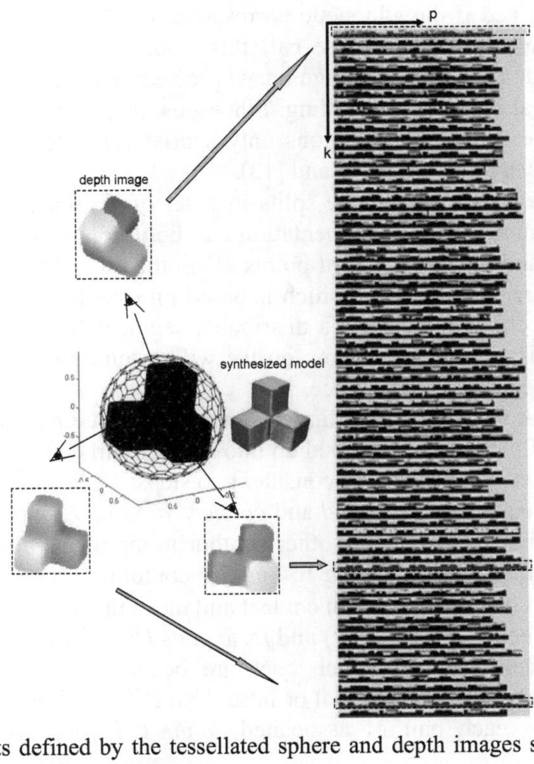

Fig. 3 Viewpoints defined by the tessellated sphere and depth images synthetically generated (left). Whole BGI-BS model (right).

In practice a factor $f_{\text{mm-pixel}}$ turns $\nabla Z_{H,i}^{j}$ into pixels. $f_{\text{mm-pixel}}$ is maintained whatever scene we work with. This change makes the DGI-BS representation invariant to camera-scene distance.

2.2 The Whole DGI-BS Model

The whole DGI-BS model of an object is an image that comprises k DGI-BS representations from k different viewpoints. In practice we use a synthesized high-resolution geometric model of the object to obtain the depth images. The viewpoints are defined by the nodes of a tessellated sphere that contains the model of the object. Each node N defines a viewpoint ON, O being the centre of the sphere (Fig. 3 left), from which the DGI-BS representation is generated. Thus, a whole DGI-BS model consists of an image of dimension $((k * t), 2 * p)$. We have duplicated dimension p in order to carry out an efficient partial-whole DGI-BS matching. Figure 3 shows the whole DGI-BS model for one object. Note that this is a single image smaller than 1 mega pixel (1600×400 pixels in this case) that synthesizes the surface information of the complete object.

2.3 DGI-BS with Occlusion

Let us suppose that an object is occluded by others in a complex scene. In this case DGI-BS can be obtained if the silhouette corresponding to the non-occluded part of the object is obtained in advance. We call this '*real silhouette*'. Therefore, it is necessary to carry out a preliminary range image processing consisting of two phases: segmentation and real silhouettes labelling. Since this chapter is focused on present the DGI-BS and due to length limitations only a brief reference of these issues is given. More details can be found in [2] and [13].

Segmentation means that the scene splits in a set of disjointed surface portions belonging to the several objects. Segmentation can be accomplished by discovering depth discontinuity and separating set of points 3D in the range image. We have used the technique of Merchán et al. [13] which is based on establishing a set of suitable data exploration directions to perform a distributed segmentation. Since this issue is not the matter of this chapter, from now on we will assume that the range and the intensity image are segmented in advance.

Concerning the real silhouettes labelling, a generic silhouette H, corresponding to a segmented portion of the scene, is divided up into smaller parts which are classified as real or false silhouettes. To do that, we consider two steps:

I. Define H as a sequence of *isolated* and *connected* parts. A part is *isolated* if all their points are far enough from every other contour in the image and *connected* if it is very close to another part belonging to another contour. In conclusion, this step finds the parts of the contours that are in contact and those that are not.

II. Define H as a sequence of *real* (R) and *false* parts (F). The goal is to find which parts occlude to others (R) and which ones are occluded parts (F). The set of *connected* parts must be classified as real or false. Using Z (depth image) we compare the mean depths for each pair of associated *connected parts* (belonging to the contours H and H') and we classify it. After comparing all *connected* pairs of H, the *real silhouettes* are finally obtained (see Fig. 4).

After carrying out the real silhouette labelling process we can obtain the DGI-BS representation for every segmented surface of the scene. In the case of an occluded object, we build DGI-BS only for the longest real silhouette of H. In practice, one or two real parts (which correspond to one and two occlusions) are expected. Of course, cases with more occlusions are possible but in any case the largest real part must be at least 30% of H. Otherwise very little information of the object would be available.

Note that in occlusions circumstances DGI-BS is a $t \times p'$ sub-matrix of the non-occluded DGI-BS version. Since a close viewpoint is available in the whole DGI-BS model, this property can be easily verified. Figure 5 illustrates an example of this property. Now dimension $p'<p$ but dimension t is image of the object is not complete more sampled points fall outside the object. That is why a few more number of black pixels appears in it.

3 Surface Correspondence through DGI-BS

The whole DGI-BS model gives complete information of an object. Consider (O,X,Y,Z) to be a reference system O being the geometric centroid of the object, v being a viewpoint of the scene defined by the polar coordinates (ρ,θ,φ) and ϕ being the camera swing angle. When a surface portion Θ is segmented in a complex scene, DGI-BS$_\Theta$ can be searched in the whole model of the object. Thus, two coordinates (k_Θ, p_Θ) in the whole DGI-BS space can be determined in this matching process: the best view (index k) and the best fitting point (index p).

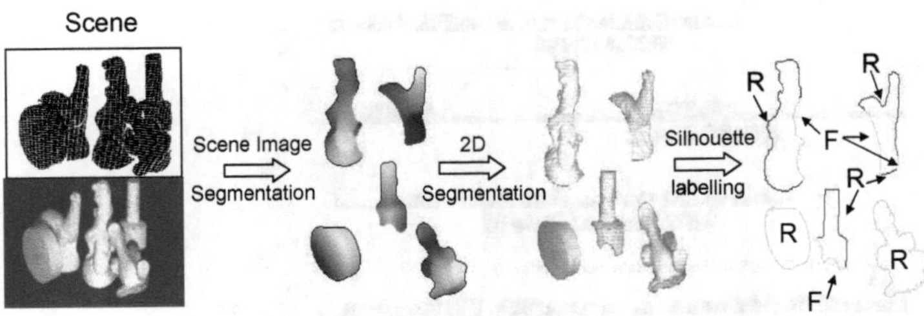

Fig. 4 Scene C: range image processing: segmentation and real silhouettes labelling (R=Real and F=false silhouette).

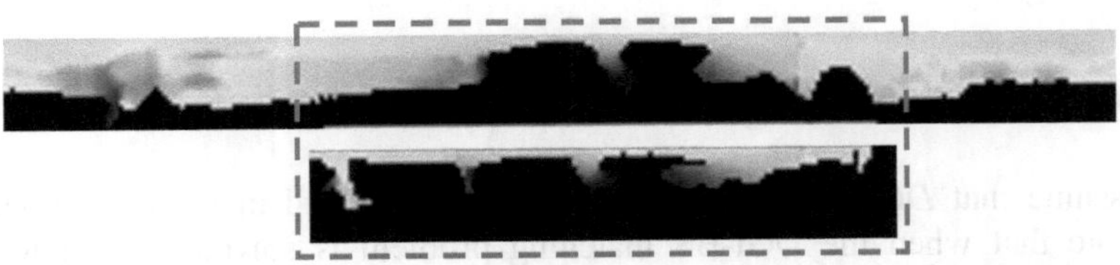

Fig. 5 DGI-BS representation of a non-occluded object (above) and DGI-BS representation of the same object under occlusion circumstances placed in the best matching position (below).

Now we will present the sequence of steps in the surface correspondence algorithm.

The segment is projected from camera viewpoint and the depth image Z_Θ and its silhouette H_Θ are obtained. After that, the real part of H_Θ is extracted according to the procedure explained in Sect. 2.3. If there are several real parts of the silhouette the longest one is chosen and the corresponding DGI-BS$_\Theta$ representation is generated. Then a scene-model matching in the DGI-BS space is carried out and two coordinates in the DGI-BS space are determined.

For a database with n objects we have n candidate views, one for each object of the database, with their respective n associated fitting points. For every pair of index $(k, p)_m$, $m=1,...,n$, a mean square error, $e_{DGI-BS}(m)$ is defined for each candidate.

Note that the surface correspondence is based on both surface and contour information. Surface information is supplied by DGI-BS representation and contour information is intrinsic to the DGI-BS definition itself. However DGI-BS representation could be ambiguous, especially on flat surfaces and hard occlusion cases. Consequently, to make the surface correspondence algorithm more efficient, we have added extrinsic information of the silhouette. This information consists of the *Function of Angles* (FA). FA contains the tangent angle for every pixel of *real* part of the silhouette. This is a typical slope representation of a contour.

Like $e_{DGI-BS}(m)$, for every pair of index $(k, p)_m$, $m=1,...,n$, FA matching error, $e_{FA}(m)$, is calculated. Finally, by minimizing the error $e = e_{DGI-BS} \cdot e_{FA}$ the best surface correspondence is finally selected. On the other hand, a sorted list of candidates is stored for further purposes. Figure 6 shows the best DGI-BS matching for several segments of scene C. It can be seen the DGI-BS of the best view (index k) and the DGI-BS of the segments at the fitting position (index p). The scene-model surface correspondences are shown on the right.

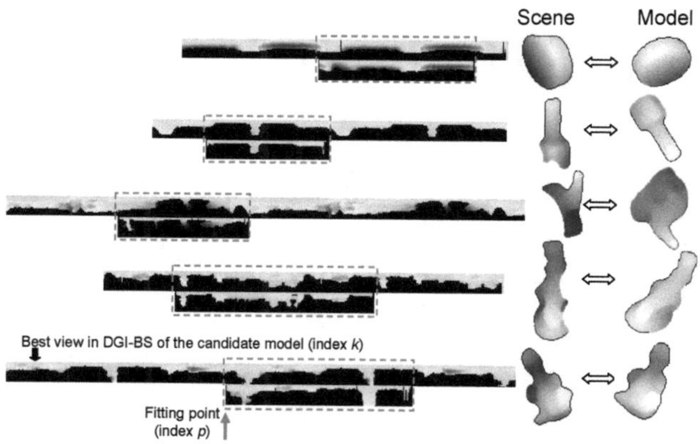

Fig. 6 DGI-BS matching for several objects of the scene C.

4 Correspondence Verification through Registration

Let us assume that $DGI\text{-}BS_M$ is the best candidate obtained in the correspondence phase. Note that when the DGI-BS matching problem is solved a point to point correspondence is established in the depth images Z_Θ and Z_M and, consequently, the same point to point correspondence is maintained for the corresponding 3D points C_Θ and C_M. Thus a coarse transformation T^* between C_Θ and C_M is easily calculated. After that we compute a refined transformation that definitively aligns the two surfaces in a common coordinate system. Thus, through the error yielded by a registration technique, we can analyse the goodness of the coarse transformation T^* and evaluate the validity of the DGI-BS matching.

We have used the well known ICP registration technique. ICP algorithm has been widely studied since the original version [6] and many researchers have proposed a multitude of variants throughout the 90's [18]. In this work, we have used the *k-d* tree algorithm and the point-to-point minimization [10].

The goodness of the registration and, indirectly, the recognition/pose result is validated taking into account the value of e_{ICP}. In other words, we want to establish if the candidate chosen in the correspondence phase was right or wrong.

In order to calculate a threshold value for e_{ICP}, an off-line ICP process is performed over a set of random viewpoints for all synthetic models of the database. As a result of that an e_{ICP} histogram is obtained and a normal distribution is fitted to it. Finally we define:

$$e_{\max} = ZS_{ICP} - \overline{e}_{ICP} \tag{4}$$

where Z is the value of the normal distribution corresponding to the 100 percentile ($Z = 3{,}99$), S_{ICP} is the standard deviation and $\overline{e}_{ICP}$ is the mean of the distribution.

After carrying out the surface correspondence and alignment phases for the best candidate, if $e_{ICP} > e_{\max}$ we consider that the surface correspondence has been wrong. In this case, the next candidate in the list is taken and a new e_{ICP} is computed and evaluated again (see Fig. 7).

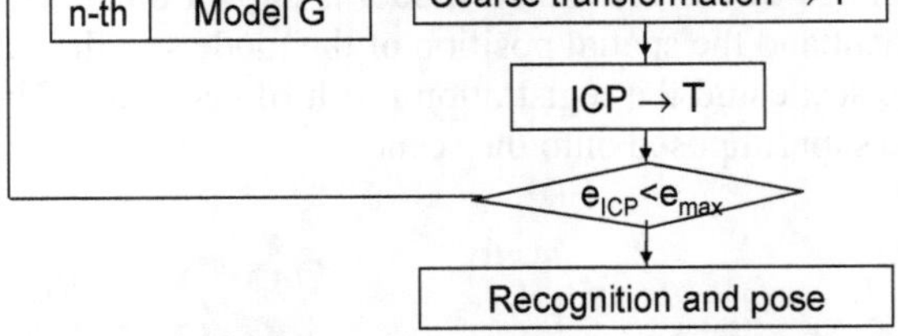

Fig. 7 Correspondence verification and registration.

5 Experimental Results

We have tested our algorithm on a set of 20 scenes captured with a *Gray Range Finder* sensor which provides coordinates (*x,y,z*) of the points of the scene. The depth map of the scene is obtained through the camera and sensor calibrations. Scenes are composed of several objects that are posed with no restrictions and with the same texture (Fig. 8). Dimension of the scene is 30×30 cm and the sensor-scene distance is 120 cm. Our database is made up of 27 objects.

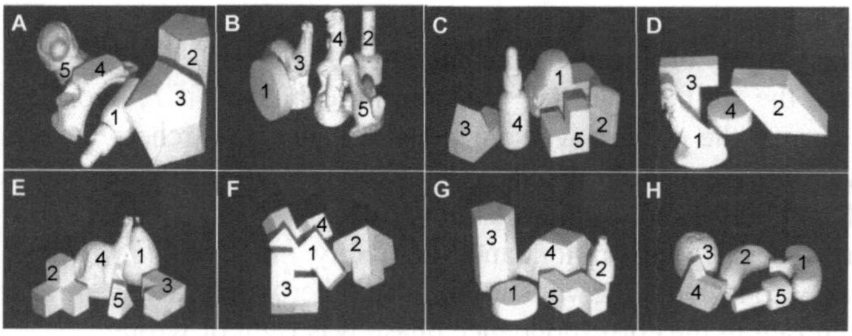

Fig. 8 Some of the scenes used in our experimentation.

The values selected for the parameters to create the whole DGI-BS representation are:

- Factor $f_{mm\text{-}pixel}$ =10 pixels/cm, so depth image dimension is about 120×120 pixels. Likewise, depth images of segmented objects are converted using the same scale factor. This means that the correspondence method is independent of the object-camera distance.
- The whole representations have been obtained by using a tessellated sphere of 80 nodes ($k = 80$). We have also used tessellated spheres with more resolution (320 nodes) but the results do not improve meaningfully and the computational time increases a lot.
- Number of sampled pixels in the 2D normal direction $t = 20$ and distance in pixels between samples $s = 5$, so the penetration inside the depth image is 100 pixels.

Figure 9 illustrates the recognition/pose process of occluded objects in scene F. The surface portions obtained after the segmentation phase appears on the left. In this case there are three portions ①, ② and ④ that are occluded. Their corresponding depth images are shown and the real silhouettes are marked onto them. Below we present the result of the scene-model matching in the DGI-BS space for the best candidate. The best view (index k) at the best fitting position (index p) can be seen for all them. Finally, the depth image of the associated model is shown of the right. On the right we show the alignment and the spatial position of the models in the scene. We can see the 3D points of the scene and the registration result of segments. The corresponding rendered models are superimposed onto the scene.

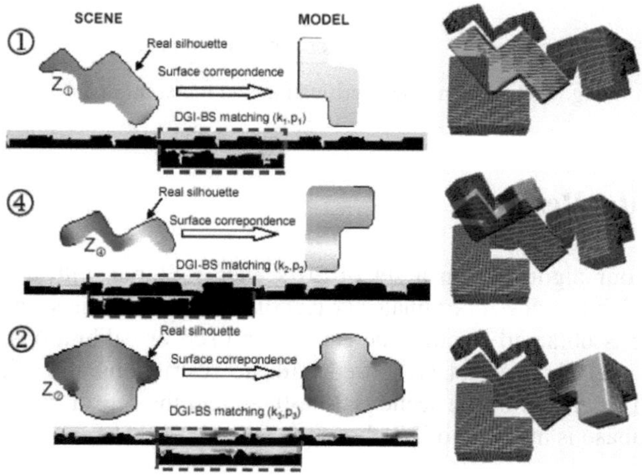

Fig. 9 Example of surface correspondence (left) and alignment of occluded surfaces (right).

Table 1 shows a summary of the results for all objects of the scenes A to H. 60.5% of the objects were occluded. We have included some important information like errors in the segmentation phase ('*seg. error*') and noise in the range image ('*noise*'). A segmentation error would imply that two objects were labelled as one object or that an object was segmented in several segments. The first case corresponds to two surfaces that are joined and where there are not depth discontinuities between both surfaces. This is quite uncommon case but of course the approach would fail. The second case occurs in self-occlusion circumstances but the method works as in a single occlusion case except that the surface portions will be smaller than in non-occlusions cases. In all cases the scenes were segmented in their constituent parts achieving 100% of effectiveness. On the other hand, the noise in the range is due to shades or highlighting regions. Coordinate k, which corresponds to the best matching view, is evaluated as right (R) or erroneous (E) in the column '*view*'. This is a visual evaluation that report us the total percentage of failures in the DGI-BS matching phase. In our experimentation we obtained 2.6% of failures in this phase (1/38).

The column labelled '*candidate*' means the first candidate of the list that verifies $e_{ICP} < e_{\max}$ and that finally is chosen as candidate. In 63.2% of the cases the first candidate was the first of the list and the average position was 2. Finally, the last column gives information about the ICP algorithm itself showing the number of iterations in the ICP algorithm. In summary we can conclude that the approach is highly effective (97.4%).

Table 1 Results for all objects of the scenes in Figure 8.

SCENE	Object	Occlusion	Seg. error	Noise	View	Candidate	e_{ICP}	Iterations ICP
A	1	Yes	No	No	R	9º	0.1108	35
	2	Yes	No	No	R	1º	0.1187	41
	3	No	No	Yes	R	1º	0.1237	43
	4	Yes	No	Yes	R	1º	0.1244	18
	5	Yes	No	No	R	8º	0.0836	51
B	1	No	No	No	R	1º	0.0693	35
	2	Yes	No	No	R	1º	0.0997	21
	3	Yes	No	No	R	1º	0.0924	45
	4	Yes	No	Yes	R	5º	0.1163	36
	5	No	No	No	R	1º	0.0800	35
C	1	Yes	No	No	E	2º	0.1593	79
	2	Yes	No	No	R	2º	0.1339	18
	3	Yes	No	No	R	6º	0.1979	30
	4	No	No	No	R	1º	0.1393	22
	5	No	No	Yes	R	3º	0.1427	25
D	1	No	No	Yes	R	1º	0.1640	29
	2	No	No	No	R	1º	0.1545	23
	3	Yes	No	No	R	1º	0.1617	38
	4	Yes	No	No	R	3º	0.1559	35
E	1	Yes	No	No	R	1º	0.1235	32
	2	No	No	No	R	1º	0.1337	18
	3	No	No	Yes	R	2º	0.1421	44
	4	Yes	No	No	R	1º	0.1515	47
	5	No	No	Yes	R	1º	0.1145	34
F	1	Yes	No	No	R	1º	0.1328	24
	2	Yes	No	No	R	3º	0.0090	35
	3	No	No	No	R	1º	0.1000	53
	4	Yes	No	Yes	R	1º	0.0939	45
G	1	No	No	No	R	1º	0.1154	35
	2	Yes	No	Yes	R	1º	0.1359	12
	3	Yes	No	No	R	1º	0.1353	28
	4	Yes	No	No	R	1º	0.1526	11
	5	No	No	Yes	R	1º	0.1252	39
H	1	Yes	No	No	R	3º	0.1227	67
	2	Yes	No	No	R	2º	0.1953	46
	3	Yes	No	Yes	R	4º	0.1676	78
	4	No	No	Yes	R	2º	0.1381	19
	5	No	No	No	R	1º	0.1335	50

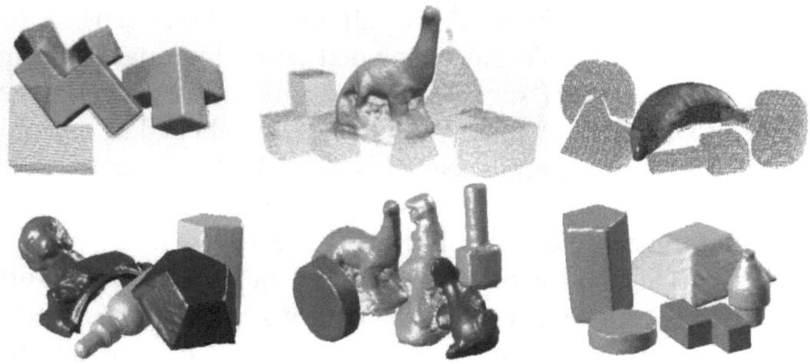

Fig. 10 Example of recognition and pose of multi-occluded objects (above) and rendered representation of the complete scene (below).

Figure 10 shows examples of pose for multi-occluded objects (scenes E, F and H) and the complete reconstruction of the scenes A, B and G.

6 Conclusions

This work deals with the problem of correspondence and registration of surfaces by using a strategy that synthesizes both surface and shape information in 3D scenes with occlusions. The method provides at the same time recognition and pose of segments of the scene. This work is part of a robot intelligent manipulation project.

DGI-BS (Depth Gradient Image Based on Silhouette) representation is a complete and discrete representation suitable in 3D environments. Moreover, DGI-BS representation can be applied to obtain a complete model as well as a partial model of an object. This property allows us to use it in complex scenes with multi-occlusion.

Recognitions and pose problems are solved by using a simple matching algorithm in the DGI-BS space that yields a coarse transformation. Afterwards, registration is sorted out by computing ICP alignment of corresponded surfaces.

We are currently working on the detection and correction of wrong correspondences. Part of these problems may be due to noise and errors in 3D data acquisition stages.

Acknowledgements

This work has been supported by the Spanish projects PBI05-028 JCCLM and DPI2006-14794-C02.

References

1. Adán, A., Adán, M., 2004. A flexible similarity measure for 3D shapes recognition. In *IEEE TPMI*, 26(11), 1507–1520.
2. Adán, A., Merchán, P., Salamanca, S., Vázquez, A., Adán, M., Cerrada, C., 2005. Objects Layout Graph for 3D Complex Scenes. In *Proc. of ICIP'05*.

3. Ashbrook, A., Fisher, R., Werghi, N., Robertson, C., 1998. Aligning arbitrary surfaces using pairwise geometric histograms. In *Proc. Noblesse Workshop on No-linear Model Based Image Analysis*, 103–108.
4. Alferez, R., Wang,Y.F., 1999. Geometric and illumination invariants for object recognition. In *IEEE TPAMI* 21, 505–535
5. Bamieh, B., De Figuereido, R.J.P., 1986. A general moment-invariants/attributed graph method for three-dimensional object recognition from single image. In *IEEE T. Rob. Aut.*, RA-2, 31–41
6. Besl, P. McKay, N., 1992. A method for registration of 3-D shapes. In *IEEE TPAMI*, 14(2), 239–256.
7. Campbell, R., Flynn, P. J., 2002. Recognition of free-form objects in dense range data using local features. In *16th International Conference on Pattern Recognition.*
8. Cyr, C. M., Kimia, B. B., 2004. A similarity-based aspect-graph approach to 3D object recognition. In *Int. Journal of Computer Vision*, 57(1), 5–22
9. Ching S. C., Jarvis R., 1997. Point signatures: a new representation for 3D object recognition. In *International J. of Comp. Vision*, 25(1), 63–85
10. Horn, B., 1988. Closed-form solution of absolute orientation using orthonormal matrices. In *Journal of the Optical Society A.*, 5 (7)
11. Andrew E. Johnson and Martial Hebert. 1999. Using spin images for efficient object recognition in cluttered 3d scenes. In *IEEE TPAMI*, 21(5), 433–449
12. Lee, J.D., 2000. A new scheme for planar shape recognition using wavelets. *In Comp. Mach. Appl.* 39, 201–216
13. Merchán, P., Adán, A., Salamanca, S., Cerrada, C., 2002. 3D complex scenes segmentation from a single range image using virtual exploration. In *LNAI*, 2527, 923–932
14. Mokhararian, F.S., 1997. Silhouette-based occluded object recognition through curvature scale space. In *Mach. Vis. Appl.* 10, 87–97
15. Planitz, B.M., Maeder, A.J., Williams, J.A., 2005. The correspondence framework for 3D surface matching algorithms. In *Comp. Vis. Image Underst.* 97, 347–383
16. Roh, K.C., Kweon, I.S., 2000. 3-D object recognition using a new invariant relationship by single view. In *Pattern Recognition* 33, 741–754
17. Rothwell, C.A., Zisserman, A., Forsyth, D.A., Mundy, J.L., 1995. Planar object recognition using projective shape representation. In *Int. J. Com. Vision* 12, 57–59
18. Rusinkiewicz, S., Levoy, M., 2001. Efficient variant of the ICP algorithm. In *Proc. of 3DIM'01*. 145–152.
19. Trazegnies, C., Urdiales, C., Bandera, A., Sandoval, F., 2003. 3D object recognition based on curvature information of planar views. In *Pattern Rec.* 36, 2571–2584.
20. Liu, X., Sun, R., Kang, S.B., Shum, H.Y., 2003. Directional histogram for three-dimensional shape similarity. In *Proc. of the IEEE Conference on CVPR*, I, 813–820.
21. Xu, D., Xu,W., 2005. Description and recognition of object contours using arc length and tangent orientation. In *Pattern Rec. Letters*, 26, 855–864
22. Yamany, S., Farag, A., 2002. Surfacing signatures: An orientation independent free-form surface representation scheme for the purpose of objects registration and matching. In *IEEE TPAMI*, 24(8),1105–1120.
23. Zabulis, X., Sporring, J., Orphanoudakis, S.C., 2005. Perceptually relevant and piecewise linear matching of silhouettes. In *Pattern Recognition*, 38, 75–93
24. Zahn, C.T., Roskies, R.C., 1972. Fourier descriptors for plane closed curve. In *IEEE T. Comp.*, 21, 195–281
25. Zhang, D. 1999. Harmonic Shape Images: A 3D free form representation and its applications in surface matching. In *PhD thesis CMU.* Pennsylvania. USA.

Tracking Multiple Objects using the Viterbi Algorithm

Andreas Kräußling

Research Establishment for Applied Sciences (FGAN)
Neuenahrer Straße 20, 53343 Wachtberg
a.kraeussling@fgan.de

Abstract. Tracking multiple targets is a great challenge for most tracking algorithms, since these algorithms tend to loose some of the targets when they get close to each other. Hence, several algorithms like the MHT, the JPDAF and the PMHT have been developed for this task. However, these algorithms are specialised on punctiform targets, whereas in mobile robotics one has to deal with extended targets. Therefore, in this chapter an algorithm is proposed that can solve this problem. It uses the Viterbi algorithm and some geometrical characteristics of the problem. The proposed algorithm was tested with real world data.

Keywords. Mobile robotics, tracking, multiple targets, Viterbi algorithm, Kalman filter.

1 Introduction and Related Work

For many real world applications it is essential that a robot is able to interact with its environment. This is true for multi–robot systems where a group of robots has to solve a given task or where robots are supposed to support people. For such situations, the awareness of the position of people and other robots is a fundamental ability for a mobile unit to be able to interact with its environment in an appropriate way.

This problem can be analysed under the superordinate concept of tracking. Tracking denotes the estimation of the position of an object based on consecutive sensor measurements. It is well studied in the field of aerial surveillance with radar devices [1]. In the area of mobile robots tracking is also a well established research topic [2] [5] [13] [16]. In mobile robotics laser range scanners are one of the preferred sensor devices. A Sick laser range scanner for example can measure the distance to the next reflecting obstacle with a high angular resolution of e.g. 1°. Lasers have rapidly gained popularity for mobile robotic applications such as collision avoidance, navigation, localisation and map building [22].

The problem of tracking people and other objects in densely populated environments with a robot–borne laser scanner can be characterised in the following way: most of the readings are from obstacles like walls or other objects and only a few measurements come from the tracked object itself. The problem of allocation of data obtained from the presently accounted target is called the data association problem [1]. As a solution to this problem, a tracking algorithm might use a validation gate which separates the signals belonging to the current target from other signals. A second characteristic of tracking people with laser range scanners is the occurrence of several measurements from the same object. In contrast to common radar based tracking sensors the Sick laser scanner has a much higher resolution. This leads to the fact that the tracked object generates several measurements. Therefore, we have to deal with what we call extended objects instead of punctiform objects like in the common radar tracking literature. Thereby, punctiform targets are those ones, which are the origin of just one measurement. A third characteristic of tracking in the field of mobile robotics is the occurrence of crossing targets. This means that two or more targets get very close to each other, so that they cannot be separated by common tracking algorithms [4] [6] [7] [10] [11]. This situation can appear e.g. when two humans meet, talk to each other and split again and is a well known problem in mobile robotics [13].

There are several methods for tracking punctiform crossing targets in clutter:

1. the MHT (Multi Hypothesis Tracker) introduced by Reid in 1979 [15].
2. the JPDAF (Joint Probabilistic Data Association Filter) introduced by Fortmann, Bar–Shalom and Scheffe in 1983 [4].
3. the PMHT (Probabilistic Multi Hypothesis Tracker) introduced by Streit and Luginbuhl in 1994 [21].

Of course, an extension of these algorithms for tracking extended crossing objects is straightforward. But there are several reasons, why such approaches might fail:

- in most cases there are several measurements from the same target.
- the crossing can last for a longer time period.
- one of the objects might be occluded by the other object for some time.
- the objects can accomplish very abrupt manoeuvres during the crossing or especially at the end of the crossing.

As an example we give the results for an EM based method [20], which is an extension of the PMHT [21] to extended targets. Figure 1 shows the results when applying this algorithm to real data.

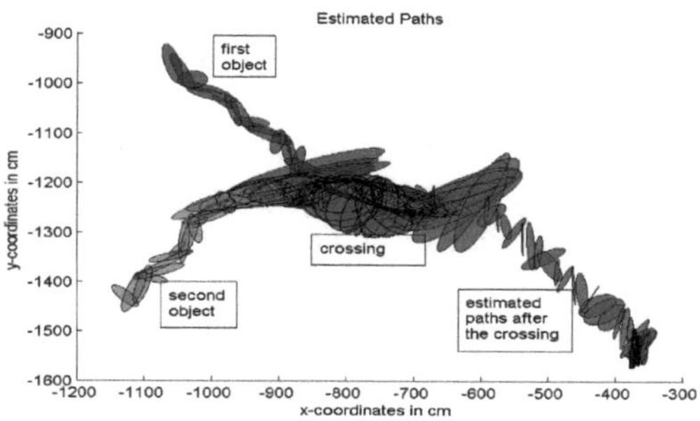

Fig. 1 Crossing of two objects, real data, EM based method.

It shows the estimates for the position of the objects by use of ellipses. Thereby the estimated position is the centre of the ellipse, whereas the shape of the ellipse represents the actual geometry of the tracked object. The objects start in the left and move to the right. Obviously, the algorithm looses the one of the targets, which moves to the upper right after the crossing (see Fig. 2). Moreover, the computational burden for applying these algorithms is very high when applied to extended targets, since these objects can be the origin of up to ten measurements. These difficulties are well known in the mobile robotics community:

- Tracking moving objects whose trajectories cross each other is a very general problem ... Problems of this type cannot be eliminated even by more sophisticated methods ... [13].
- Tracks are lost when people walk too closely together ... [18].

With respect to these circumstances our research group has developed a new method that solves the problem of tracking two extended crossing targets [9] [10] [11]. Unfortunately this algorithm is not able to deal with more than two crossing objects in its current state. Therefore, recently our research group has developed a more general algorithm, which is able to deal with an arbitrary number of crossing objects, i.e. objects that get very close to each other. This algorithm is the main contribution of this work.

The remainder of this chapter is organized as follows. In Sect. 2 the mathematical background including the used model, the validation gate, and two basic tracking algorithms is given. In Sect. 3 the new algorithm for tracking multiple interacting objects is introduced. Furthermore, in Sect. 4 the experimental results are presented. Finally, Sect. 5 contains the conclusions.

2 The Mathematical Background of the Algorithms

2.1 The Model

The dynamics of the objects to be observed and the observation process itself are modeled by a hidden Gauß–Markov chain with the equations

$$x_k = Ax_{k-1} + w_{k-1} \tag{1}$$

and

$$z_k = Bx_k + v_k. \tag{2}$$

x_k is the object state vector at time k, A is the state transition matrix, z_k is the observation vector at time k and B is the observation matrix. Furthermore, w_k and v_k are supposed to be uncorrelated zero mean white Gaussian noises with covariances Q and R.

Since the motion of a target in the plane has to be described a two dimensional kinematic model is used. Therefore, it is

$$x_k = \begin{pmatrix} x_{k1} & x_{k2} & \dot{x}_{k1} & \dot{x}_{k2} \end{pmatrix}^\top \tag{3}$$

with x_{k1} and x_{k2} the Cartesian coordinates of the target and $\dot{x}_{k1}$ and $\dot{x}_{k2}$ the corresponding velocities. z_k just gives the Cartesian coordinates of the target. For the coordinates the equation of a movement with constant velocity is holding, i.e.

$$x_{k+1,j} = x_{kj} + \Delta T \dot{x}_{kj}. \tag{4}$$

ΔT is the time interval between two consecutive measurements. For the progression of the velocities we use the equation

$$\dot{x}_{k+1,j} = e^{-\Delta T/\Theta} \dot{x}_{kj} + \Sigma \sqrt{1 - e^{-2\Delta T/\Theta}} u(k) \tag{5}$$

from [23] with the zero mean white Gaussian noise $u(k)$ with $E[u(m)u(n)^\top] = \delta_{mn}$. Thus, the velocity is supposed to decline exponentially. The term

$$\Sigma \sqrt{1 - e^{-2\Delta T/\Theta}} u(k) \tag{6}$$

models the process noise and the accelerations.

2.2 The Validation Gate

The validation gate is realized using the Kalman filter. The Kalman filter calculates a prediction $y(k+1|k)$ for the measurements $z_{k+1,l}$ from the actually handled target at time step $k+1$ via the formula

$$y(k+1|k) = B \cdot A \cdot x(k|k). \tag{7}$$

$x(k|k)$ is the estimate for the position of the target at time step k. For every sensor reading $z_{k+1,l}$ of the time step $k+1$ ($l = 1, \ldots, 360$) the Mahalanobis distance λ [12] with

$$\lambda = (z_{k+1,l} - y(k+1|k))^\top \cdot [S(k+1)]^{-1} \cdot$$

$$\cdot (z_{k+1,l} - y(k+1|k)) \tag{8}$$

is computed. Then all measurements with $\lambda > \lambda_{max}$ with a given threshold λ_{max} are excluded. See [1] for further details. This procedure results in a set $\{\hat{z}_{k+1,i}\}_{i=1}^{m_{k+1}}$ of m_{k+1} selected measurements $\hat{z}_{k+1,i}$. The matrix $S(k+1)$ is the innovations covariance from the Kalman filter. In common filter applications this matrix is calculated from the predictions covariance $P(k+1|k)$ with the equation

$$S(k+1) = BP(k+1|k)B^\top + R \tag{9}$$

with the given covariance matrix R of the measurement noise. But for tracking extended objects this approach is not sufficient, since there is an additional influence of the extendedness of the object to the deviation of the measurements from the prediction $y(k+1|k)$. To take care of this feature an accessory positive definite matrix E should be added in (10). Otherwise a lot of the measurements from the target would be excluded by the gating process. Because the lateral dimension of people usually shows a radius in the range of 30 cm, the entries of E should be in the range of 900. Thus, after some optimization process we used

$$E = \begin{pmatrix} 780 & 0 \\ 0 & 780 \end{pmatrix} \tag{10}$$

and

$$S(k+1) = BP(k+1|k)B^{\top} + R + E. \tag{11}$$

The values of the entries of the matrix E vastly exceed the values of the entries of the matrix R, so that the main contribution in (13) comes from the matrix E.

2.3 The Kalman Filter Algorithm with Equal Weights

This algorithm first calculates an unweighted mean z_{k+1} of the m_{k+1} measurements $\{\hat{z}_{k+1,i}\}_{i=1}^{m_{k+1}}$, that have been selected by the gate, i.e.

$$z_{k+1} = \frac{1}{m_{k+1}} \sum_{i=1}^{m_{k+1}} \hat{z}_{k+1,i}. \tag{12}$$

This mean is used as the input for the update equation of the Kalman filter, i.e.

$$x(k+1|k+1) = x(k+1|k) + K_{k+1}(z_{k+1} - y(k+1|k)) \tag{13}$$

with the predictions $x(k+1|k)$ and $y(k+1|k)$ and the Kalman gain K_{k+1} derived from the Kalman filter. Finally, the estimates $x(k+1|k+1)$ are further improved by the use of the Kalman smoother [19]. The corresponding algorithm is called Kalman filter algorithm (KFA). As has been shown in [6] [7] [10] [11] it is very fast and gives good information about the position of the target, but cannot reproduce multimodal probability distributions. Thus it is not able to handle multiple interacting targets.

2.4 The Viterbi Based Algorithm

The Viterbi algorithm has been introduced in [24]. A good description is also given in [3]. It has been recommended for tracking punctiform targets in clutter in [14] and for tracking extended targets in [8]. It calculates for each selected measurement $\hat{z}_{k+1,i}$ a separate estimate $x(k+1|k+1)_i$. For the calculation of the estimates $x(k+1|k+1)_i$ in the update (15) the measurement $\hat{z}_{k+1,i}$ and the predictions $x(k+1|k)_j$ and $y(k+1|k)_j$ from the predecessor j are used. When tracking punctiform targets in clutter, the predecessor is determined by minimizing the length of the paths ending in $\hat{z}_{k+1,i}$. When regarding extended targets in most cases all measurements in the validation gate are from the target, so that it is not meaningful to consider the lengths of the paths ending in the possible predecessors $\hat{z}_{k,j}$ when determining the predecessor. Therefore, a better choice for the predecessor is that one for which the Mahalanobis distance [12]

$$\nu_{k+1,j,i}^{\top}[S(k+1)]^{-1}\nu_{k+1,j,i} \tag{14}$$

is kept to a minimum. There, $\nu_{k+1,j,i}$ is the innovation

$$\nu_{k+1,j,i} = z_{k+1,i} - y(k+1|k)_j \tag{15}$$

and $S(k+1)$ is the innovations covariance. This procedure is similar to a nearest neighbour algorithm. When applying the Viterbi algorithm the application of the validation gate is performed in the following way. At first for every selected measurement $\hat{z}_{k,j}$ the gate is applied to the measurements at time $k+1$. That results in the sets $Z_{k+1,j}$ of measurements which have passed the particular gate for the measurement $\hat{z}_{k,j}$ successfully. The set of all measurements $\hat{z}_{k+1,i}$, that are associated with the target, is then just the union of these sets. By this procedure it is ensured, that the corresponding tracking algorithms can deal with multimodal probability distributions to some extend, which is a major improvement when dealing with multiple interacting targets.

Because the Viterbi algorithm computes a set of track estimates for the different measurements obtained from an extended target, the user finally has to decide, which estimate to use as the "true" track estimate for the target. In the case of tracking punctiform objects in clutter, one generally chooses the estimate with minimum path length, because this is most likely the best hypothesis. In case of tracking extended objects, where all measurements are likely to originate from the target, the path lengths of the tracks corresponding to individual measurements are not meaningful. Instead, we currently nondeterministically choose one of the estimates belonging to one object. The corresponding algorithm is called Viterbi based algorithm (VBA) and has been introduced in [8]. The disadvantages of this algorithm are a great computational burden and a poor information about the position of the tracked objects [6] [7] [10] [11].

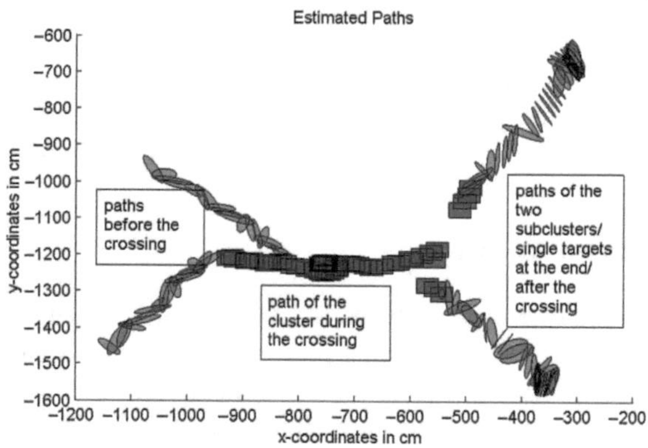

Fig. 2 Two crossing targets.

3 An Algorithm for Tracking Multiple Interacting Objects

Our new algorithm uses two classes of objects:

– single targets.

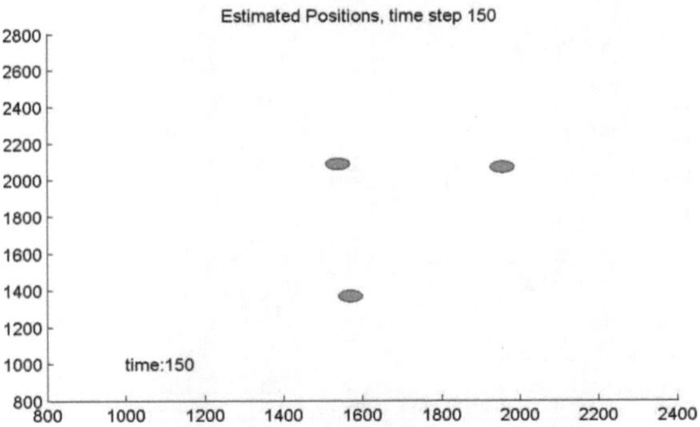

(**a**) The algorithm tracks three distinct, movingtargets.

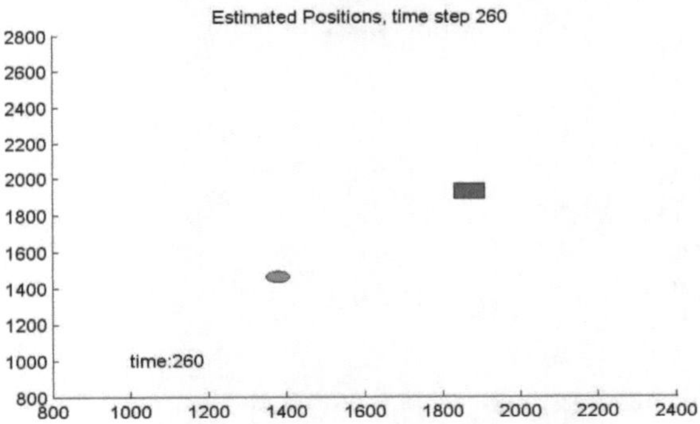

(**b**) Two of the targets get in close proximity of each other, and are therefore represented by one cluster.

Fig. 3 First experiment.

– clusters, which represent at least two interacting objects, i.e. objects that are moving very close to each other.

Single targets are tracked with the KFA, since there is no need for representing multimodal probability distributions. Furthermore, this algorithm is very fast and gives very accurate information about the position of the tracked object.

Clusters are tracked with the VBA, since multimodal distributions have to be represented. This approach guarantees that none of the objects that are associated with the cluster is lost. This fact is important especially when the objects split and start to move separately again.

Three different events have to be regarded when tracking multiple interacting objects:

1. The merging of two single targets. This means that two single targets get very close to each other. This is the case, if at least one measurement is located in the validation gates of both targets. Then the algorithm stops to track the two single targets with the KFA and starts tracking a cluster, which contains both targets, using the VBA. Therefore, it uses the measurements located in the validation gates of at least one target.

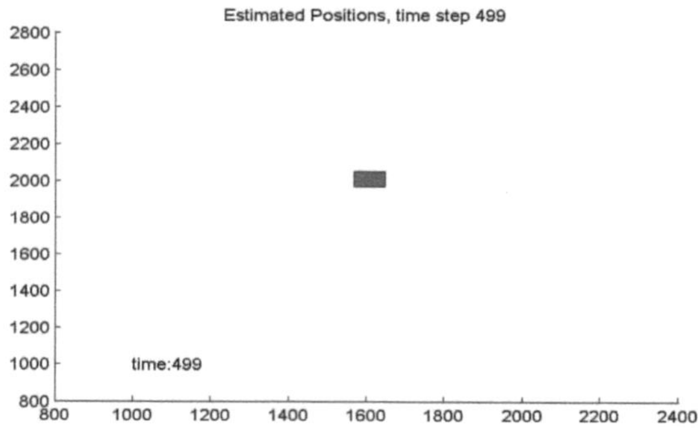

(**a**) The third target joins the other two and is also associated with the cluster.

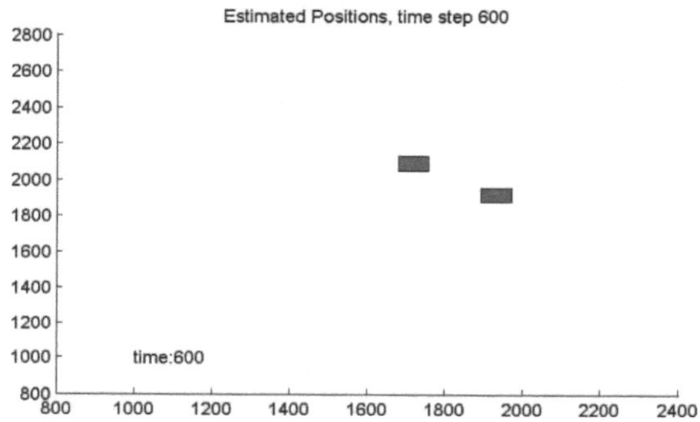

(**b**) One of the targets separates from the others, thus we can track subclusters.

Fig. 4 First experiment.

2. The merging of a single target and a cluster. This means that a single target and a cluster get very close to each other. This happens, if at least one measurement is located in the validation gates of the target and the cluster. In this case the algorithm stops to track the single target and the cluster separately. Instead it starts tracking a combined cluster. Therefore, it uses the measurements located in the validation gates of either the single target or the previously considered cluster or both.
3. The merging of two clusters. This means that two clusters get very close to each other. This is the case, if at least one measurement is located in the validation gates of both clusters. If this is true, the algorithm stops to track the two clusters and starts tracking a combined cluster. Therefore, it uses the measurements located in the validation gates of at least one of the previously considered clusters.

Note, that whenever a merging takes place, the algorithm remembers the single targets which correspond to the new combined cluster.

For each tracked cluster, we have to detect if it splits into its single targets. For this, we define three conditions, that are checked by the algorithm:

1. The position estimates corresponding to the measurements in the validation gates are separated into subclusters. For this purpose, we select the first estimate, which

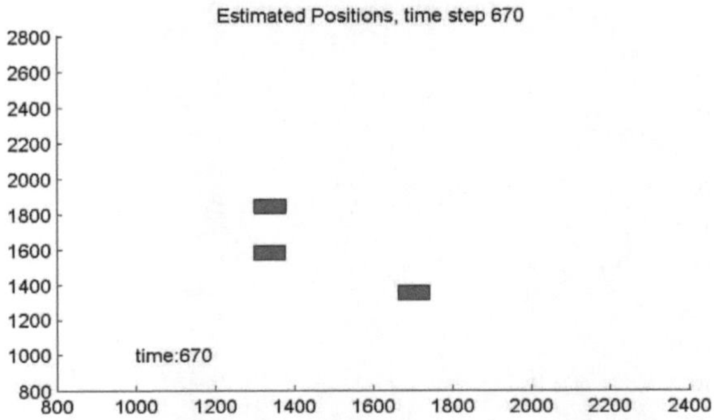

(**a**) The other two targets also separate and therefore three subclusters are tracked.

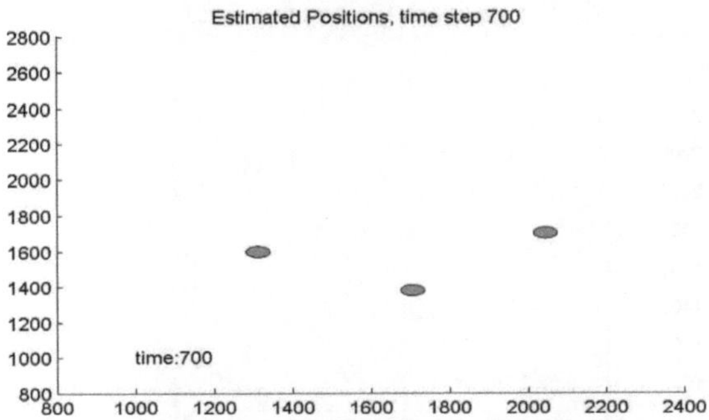

(**b**) Once these subclusters leave each others proximity, the single targets can be tracked.

Fig. 5 First experiment.

then is associated with the first subcluster. For all other estimates associated with the cluster, the Euclidian distance to the first estimate is calculated. If this distance is below a certain threshold, the estimate is associated with the first subcluster. In our experiments, we set the threshold to 150 cm, which corresponds to the maximum distance between the legs of a walking person. We then have to consider the estimates, for which the Euclidean distance to the first subcluster exceeds this manually chosen threshold. Using the same procedure we applied for building the first subcluster, we now construct subclusters until all estimates are associated with one of these smaller clusters. If the number of subclusters equals the number of targets that were merged into this cluster, the first condition for the dispersion of the cluster is fulfilled. Then, we proceed with step 2.

2. We now check the distance between the subclusters pairwise. If the distance is above a manually chosen bound, we regard these clusters as separated. We chose the value of that bound to be 300 cm. The second condition is fulfilled, if the number of pairs of separated subclusters equals $\frac{n(n-1)}{2}$. Thereby, n is the number of single targets associated with the cluster. Hence, we are checking if all subclusters are pairwise separated.

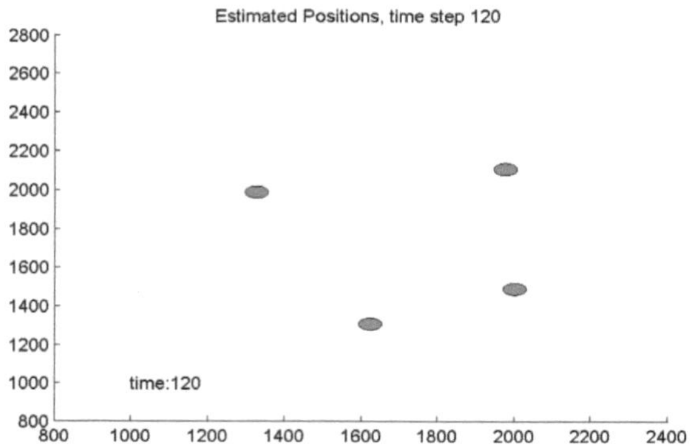

(**a**) The algorithm tracks four distinct, moving targets.

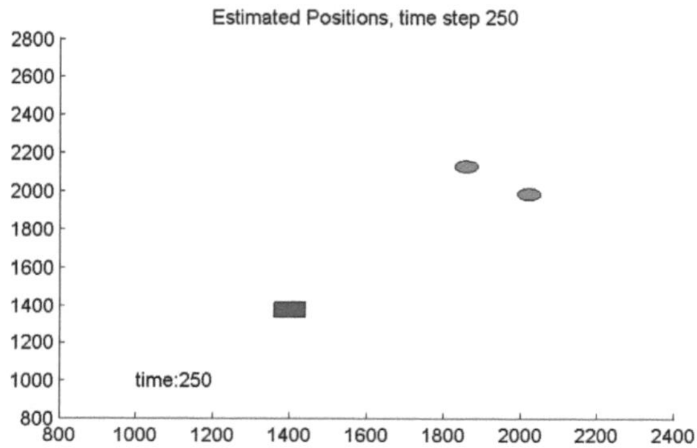

(**b**) Two of the targets get in close proximity of each other, and are therefore represented by one cluster.

Fig. 6 Second experiment.

3. Above this, we can separate single subclusters from the cluster. This follows the same logic as in step 1. Note, the algorithm is not able to determine, how many targets are represented by a single subcluster.

If conditions 1 and 2 are met, the n subclusters are associated with the n single targets, that are therefrom tracked with the KFA. When separating targets from clusters, we cannot guarantee if the target association is the same as before merging the targets into the cluster. We believe that there is no solution to this problem, if we exclusively use anonymous sensors like laser range finders.

4 Experiments

The presented algorithm was tested with real world data, recorded in our laboratory. In this section we show the data of two experiments. Before this, Fig. 2 illustrates how our algorithm works in the case of two targets using the data from Fig. 1. In the figures in this section single targets are indicated by ellipses, whereas clusters are indicated by

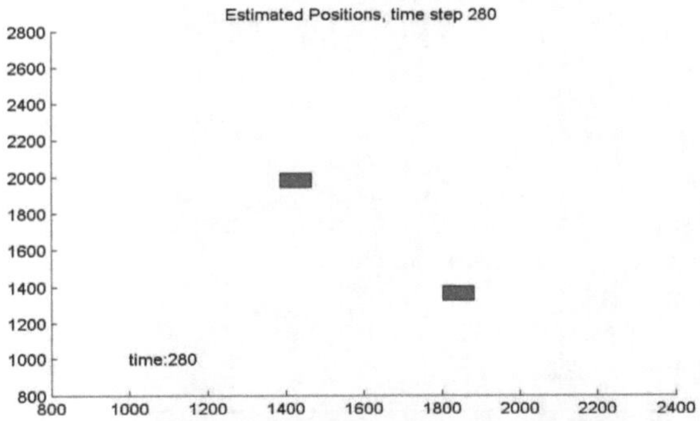

(**a**) The third and fourth target get close to each other and are also represented by a cluster.

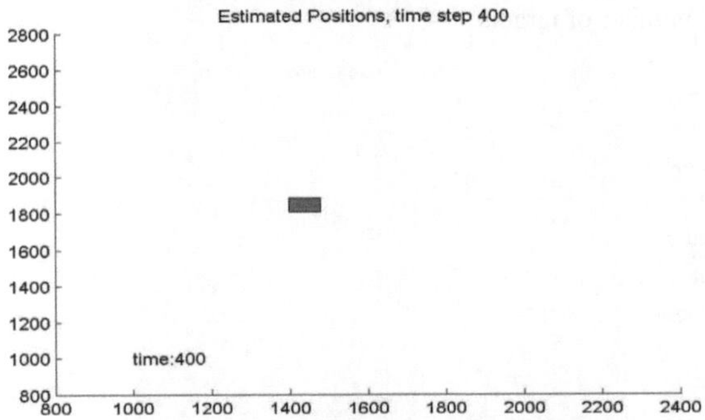

(**b**) The two clusters join and thus all four targets are associated with one cluster.

Fig. 7 Second experiment.

rectangles. As soon as the above mentioned conditions are met, we are able to track the two targets separately again.

In the first experiment, three persons moved around the observer in counterclockwise direction. At first, these three persons are tracked separately and thus, the algorithm uses the KFA (Fig. 3a). Then, two of the single targets get into close proximity of each other and are therefore merged into one cluster, which is tracked using the VBA (Fig. 3b). In the next figure, the third person joins the group and is also associated with the cluster. This combined cluster is still tracked by the VBA (Fig. 4a). Next, the combined cluster split after a while into two subclusters which are indicated in Fig. 4b. Then, the cluster split into three subclusters, which are indicated in Fig. 5a. At this stage of the experiment, condition one is met, but not condition two. Finally, the three persons split up fulfilling condition two and the persons are tracked as single targets again (Fig. 5b). Since we are not able to tell how many targets are associated with a subcluster in general, only the total number of targets is known. Thus, we have to wait until all targets are separated according to the defined conditions before the algorithm tracks the individual targets using the KFA.

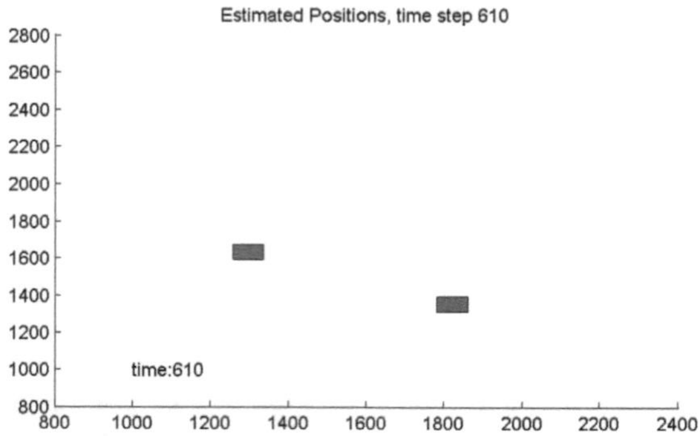

(**a**) The combined cluster splits after a while into two subclusters, each of them consisting of an unknown number of targets.

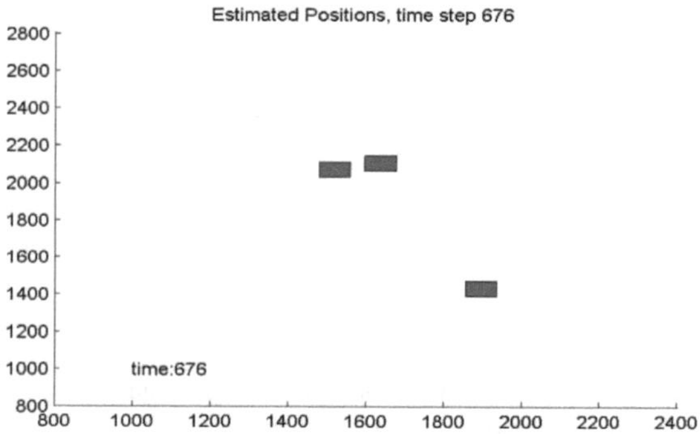

(**b**) One of the subclusters splits into two subclusters.

Fig. 8 Second experiment.

In the second experiment, we let the algorithm track four people. All four persons started walking alone. Consequently, they are tracked independently using the KFA (Fig. 6a). Then two of them are walking together and are therefore tracked in a single cluster using the VBA (Fig. 6b), whereas the other two people are still walking separately. In the next step, the other two persons are also walking together. The result is that these two are also tracked using a single cluster and thus, the algorithm tracks two distinct clusters, each representing two persons (Fig. 7a). Then, in Fig. 7b all four of them are walking together, what results in the merging of the two clusters to one combined cluster, representing all four persons. Next, this combined cluster splits into two subclusters, each consisting of an unknown number of targets (Fig 8a). In Figs. 8b and 9a these subclusters split up into four subclusters. Finally, condition two is met and therefore the four targets are tracked individually again (Fig. 9b).

In the experiments, we showed that our algorithm is able to keep track of all targets for all situations that can occur when tracking multiple targets:

- the merging of two single targets to a cluster.
- the merging of a single target and a cluster.

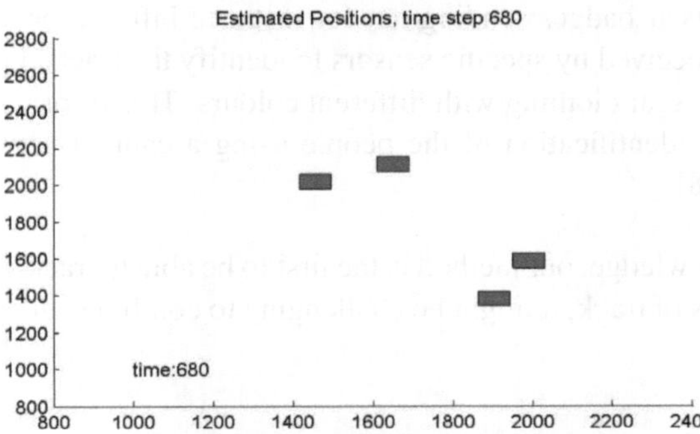

(**a**) The targets split after a while and are therefrom tracked as subclusters.

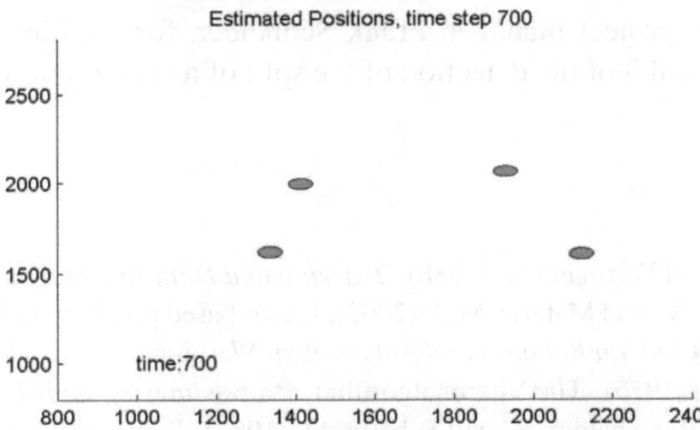

(**b**) Then, the targets can be tracked individually.

Fig. 9 Second experiment.

- the merging of two clusters to a combined cluster.
- the splitting of a cluster into subclusters which are indicated separately in the graphics.
- the splitting of several targets tracked in a cluster to single targets.

5 Conclusions

In this work we presented a novel algorithm for tracking multiple extended interacting objects. It consists of a Kalman filter tracking procedure for tracking single targets with high accuracy and low computational effort, and a Viterbi based method for tracking objects that are close to each other. The latter part facilitates clustering to be able to separate the targets correctly if they split up.

In experiments we showed this algorithm to be capable of tracking multiple crossing targets without any restriction. Of course, since we only use laser range finders, the target association after a crossing may be interchanged. There are several approaches that might be investigated to eliminate this drawback:

- different colours of the pairs of trousers people wear might result in different intensities of the reflected laser beams.

- people could wear badges sending out for instance infrared or ultrasound signals which can be received by specific sensors to identify the tracked people [17].
- usually people wear clothing with different colours. This information could be exploited for the identification of the people using a camera network as has been proposed in [18].

Since, up to our knowledge, our method is the first to be able to track several interacting objects without loss of track, it might be challenging to combine our approach with one of these attempts.

Acknowledgements

I want to thank my project manager, Frank Schneider, for the hint to use clustering methods for step 1 and 3 of the detection of the split of a cluster into its single targets.

References

1. Bar-Shalom, Y. and Fortmann, T. (1988). *Tracking and Data Association.* Academic Press.
2. Fod, A., Howard, A., and Mataric, M. J. (2002). Laser–based people tracking. In *Proceedings of the IEEE Intl. Conf. on Robotics and Automation*, Washington, DC, USA, pp. 3024–3029.
3. Forney, Jr., G.-D. (1973). The Viterbi algorithm. *Proceedings of the IEEE*, 61(3):268–278.
4. Fortmann, T. E., Bar-Shalom, Y., and Scheffe, M. (1983). Sonar tracking of multiple targets using joint probabilistic data association. *IEEE Journal of Oceanic Engineering*, OE–8(3).
5. Fuerstenberg, K. C., Linzmeier, D. T., and Dietmayer, K. C. J. (2002). Pedestrian recognition and tracking of vehicles using a vehicle based multilayer laserscanner. In *Proceedings of IV 2002, Intelligent Vehicles Symposium*, Versailles, France, volume 1, pp. 31–35.
6. Kräußling, A. (2006). Tracking extended moving objects with a mobile robot. In *Proceedings of the 3rd IEEE Conference on Intelligent Systems, London, UK.*
7. Kräußling, A. (2008). Tracking extended moving objects with a mobile robot. In Chountas, P., Petrounias, I., and Kacprzyk, J. (Eds.), *Intelligent Techniques and Tools for Novel System Architectures,* volume 109, Studies in Computational Intelligence Series, Springer.
8. Kräußling, A., Schneider, F. E., and Wildermuth, D. (2004a). Tracking expanded objects using the Viterbi algorithm. In *Proceedings of the 2nd IEEE Conference on Intelligent Systems, Varna, Bulgaria.*
9. Kräußling, A., Schneider, F. E., and Wildermuth, D. (2004b). Tracking of extended crossing objects using the Viterbi algorithm. In *Proceedings of the 1st International Conference on Informatics in Control, Automation and Robotics (ICINCO)*, Setubal, Portugal.
10. Kräußling, A., Schneider, F. E., and Wildermuth, D. (2005). A switching algorithm for tracking extended targets. In *Proceedings of the 2nd International Conference on Informatics in Control, Automation and Robotics (ICINCO), also to be published in the Springer book of selected papers of ICINCO 2005*, Barcelona, Spain.
11. Kräußling, A., Schneider, F. E., Wildermuth, D., and Sehestedt, S. (2007). A switching algorithm for tracking extended targets. In Filipe, J., Ferrier, J.-L., Cetto, J.A., Carvalho, M. (Eds.), *Informatics in Control, Automation and Robotics II,* Springer.
12. Mahalanobis, P. C. (1936). On the generalized distance in statistics. *Proceedings of the National Institute of Science*, 12:49–55.

13. Prassler, E., Scholz, J., and Elfes, E. (1999). Tracking people in a railway station during rush–hour. In Christensen, H. I., (ed.), *Computer Vision Systems*, volume 1542, pp. 162–179. Springer, lecture notes in computer science edition.
14. Quach, T. and Farooq, M. (1994). Maximum likelihood track formation with the Viterbi algorithm. In *Proceedings of the 33rd Conference on Decision and Control, Lake Buena Vista,* Florida, USA.
15. Reid, D. B. (1979). An algorithm for tracking multiple targets. *IEEE Trans. Automatic Control*, AC–24:843–854.
16. Schulz, D., Burgard, W., Fox, D., and Cremers, A. B. (2001). Tracking multiple moving objects with a mobile robot. In *Proceedings of the 2001 IEEE Computer Society Conference on Computer Vision and Pattern Recognition (CVPR 2001)*, Kauai, Hawaii.
17. Schulz, D., Fox, D., and Hightower, J. (2003). People tracking with anonymous and Id–sensors using Rao–Blackwellised particle filters. In *Proceedings of the 18th International Joint Conference on Artificial Intelligence (IJCAI 2003),* Acapulco, Mexico.
18. Schumitch, B., Thrun, S., Bradski, G., and Olukotun, K. (2006). The information–form data association filter. In *Proceedings of the 2005 Conference on Neural Information Processing Systems (NIPS),* MIT Press, Vancouver, BC, Canada.
19. Shumway, R. H. and Stoffer, D. S. (2000). *Time Series Analysis and Its Applications.* Springer.
20. Stannus, W., Koch, W., and Kräußling, A. (2004). On robot–borne extended object tracking using the EM algorithm. In *Proceedings of the 5th Symposium on Intelligent Autonomous Vehicles,* Lisbon, Portugal.
21. Streit, R. L. and Luginbuhl, T. E. (1994). Maximum likelihood method for multi–hypothesis tracking. *Signal and Data Processing of Small Targets, SPIE*, 2335.
22. Thrun, S. (1998). Learning metric-topological maps for indoor mobile robot navigation. *Artificial Intelligence*, 99(1):21–71.
23. van Keuk, G. (1971). Zielverfolgung nach Kalman–Anwendung auf elektronisches radar. Technical Report 173, Forschungsinstitut für Funk und Mathematik, Wachtberg–Werthhoven, Germany.
24. Viterbi, A. J. (1967). Error bounds for convolutional codes and an asymptotically optimum decoding algorithm. *IEEE Transactions on Information Theory*, IT–13(2).

Reactive Simulation for Real-Time Obstacle Avoidance

Mariolino De Cecco[1], Enrico Marcuzzi[2], Luca Baglivo[2] and Mirco Zaccariotto[2]

[1]Dep. Mech Struct Eng, University of Trento, Via Mesiano 77, Trento, Italy
mariolino.dececco@unipd.it
[2]CISAS, Centre of Studies and Activities for Space, Via Venezia 1, 35131 Padova, Italy
{enrico.marcuzzi,luca.baglivo}@unipd.it

Abstract. This chapter provides a new approach to the dynamic path planning and obstacle avoidance in unknown and dynamic environments. The system is based on the interaction between four different modules: the Path Planner, the Graph which memorizes all the local target, the "Sentinel", and the module which computes the Reactive Simulation every time an obstacle is detected along the path. The Reactive Simulation takes in account the kinematics model of the vehicle and the actual state conditions to make a real-time simulation in order to predict the trajectory of the differential drive robot that would allow the safe reaching of the local target.

Keywords. AGV, reactive simulation, obstacle avoidance.

1 Introduction

The main aspects in the field of Autonomous Guided Vehicles are the safety of motion planning and accurate pose measurement. Both of these aspects are still challenging problems.

Improving the degree of system autonomy makes possible to use AGV in unknown and dynamic environments, where the main problem is to avoid obstacles and trap situations. Main aspects which have to be taken into consideration are dynamic obstacles, accurate relative pose estimation, kinematics and dynamical parameters variations.

Past studies concerning obstacle avoidance [1] [2] were based on potential fields where obstacles exert repulsive forces, while the target applies an attractive force to the robot. A resultant force vector is calculated for a given robot position and it gives the current motion direction. These methods have strong limitations in approaching doors, U-shape obstacles and in avoiding trap situations. Virtual Force Histogram (VFH) was developed to improve and make the potential field method more robust: it uses a two-dimensional Cartesian histogram grid as a world model to reduce sensors data and to compute the desired control motion for the vehicle [3] [8] [9].

Based on potential field methods, other studies were focused on a dynamical building of figure (Dynamic Force Fields) where the magnitude in each point can be seen as proportional to the probability of collision at that point [5].

All these methods, when operating in unknown environment cannot really solve trap-situations and can't find an optimal path (it can only be found when complete environmental information is given).

In order to search for optimality it is possible to build a map during motion with the application of Kalman filter [6], but this can become useless in very dynamic environments, like factories or ports with a lot of vehicles in motion, thus wasting computational time.

To cope with dynamic environments, sensor-based motion generation techniques were developed: environmental changes or moving obstacles detected by sensors imply a “reactive path planning”, adapting robot motions to every new event, and sensorial measures are used to create a local model of the environment exploited to drive the robot safely, also in dense and cluttered scenarios [4].

A different approach [7] is a complete trajectory generation: based on real-time perceptual information, a feasible nonholonomic motions plan (from a given initial posture to a given final posture) is generated using a parametric representation. With this approach the main problem is the computational time, and the existence of a trajectory which satisfies all the boundaries and conditions, namely the solution of a constrained optimization problem. But this approach doesn’t take into account the model of the vehicle and so possible changes in kinematics and dynamical parameters.

In fact the environment could change, as well as kinematics parameters, affecting vehicle’s path. This second aspect is usually not taken into account in path planning and obstacle avoidance.

In recent times simulation techniques have been applied to real-time systems optimization [14].

In this chapter we describe a reactive simulation starting every time that a new target position is planned during motion due to the detection of an obstacle: if the output of the simulation is safe obstacle avoidance the robot continues smoothly its path, otherwise it stops to avoid dangerous collisions and to plan a safe path. The output is a more robust real time obstacle avoidance algorithm that would allow navigation in unknown and dynamically changing environment.

An important advantage of this approach is that the use of a model for simulation permits to estimate on-line the kinematics parameters and this allows to take into account parametric variations like different diameters of wheels, inertia of masses, etc., that could affect sensibly vehicle’s motion [10].

The algorithm was implemented on an autonomous vehicle with differential drive kinematics (Fig. 1). A PXI (National Instruments) with an embedded real-time operating system (RTOS) was used to control the robot and implement the Reactive Simulation.

2 Kinematics Model

The vehicle used in the experiment has a differential drive kinematics (Fig. 2):

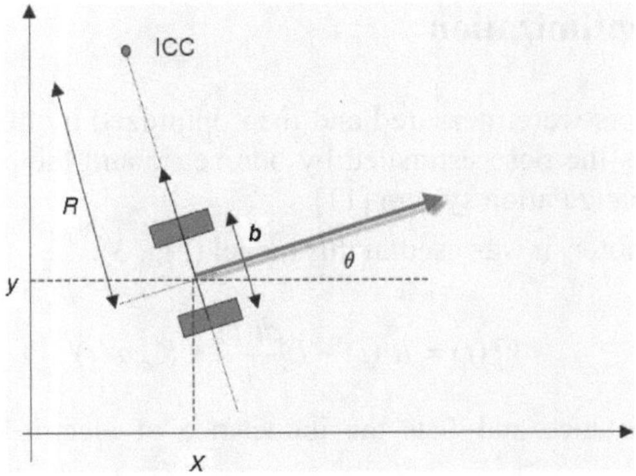

Fig. 1 Differential drive kinematics.

The discrete form of the Inertial-Odometric navigation equations is the following:

$$\begin{cases} \delta_k = \dfrac{(v_{rk} - v_{lk})}{b} T_c + \delta_{k-1} \\ x_k = (v_k \cos \delta_k) T_c + x_{k-1} \\ y_k = (v_k \sin \delta_k) T_c + y_{k-1} \end{cases} \tag{1}$$

where b is the distance between the centers of the two wheels, v_{rk} and v_{lk} are the linear velocities of right wheel and left wheel, v_k is the velocity of vehicle relative to the mid-point of axis b:

$$v_k = \frac{1}{2} * (v_{rk} + v_{lk}) \tag{2}$$

T_c is the period of the task which estimates the pose.

Fig. 2 The prototype of AGV used in the experiment.

3 Simulation Optimization

Kinematics parameters were measured and then optimized by minimizing a function cost which considers the pose estimated by odometric and the one estimated by the reference infrared triangulation system [11].

As regards the CC motor, it was used in this model (Fig. 3):

- Electrical part:

$$V_a(t) = Ri(t) + L\frac{di(t)}{dt} + K_\Phi \omega(t) \quad (3)$$

where R is the resistance and L is the inductance of electric circuit, K_Φ is the constant of torque, V_a is the input voltage, ω is the angular velocity,

- Mechanical part:

$$\tau_m(t) = K_\Phi i(t) = J\dot{\omega}(t) + B\omega(t) \quad (4)$$

where τ_m is the motor torque, J is the rotor inertia, B is the viscous friction. Combining the two parts:

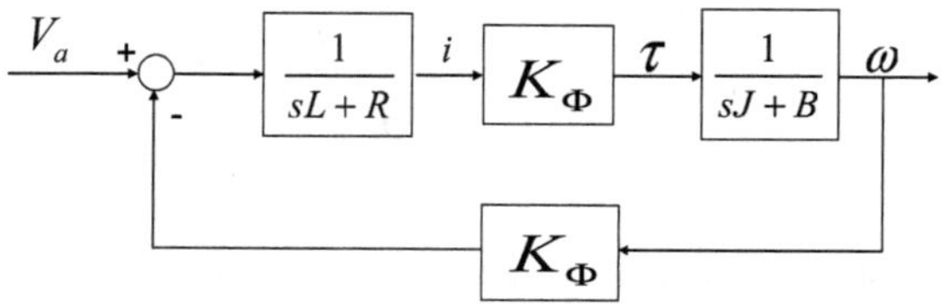

Fig. 3. Scheme of CC motor.

Concerning to Fig. 4, the trajectory is tracking projecting the center of vehicle on the segment P1P2 (where P1 is the target just reached and P2 is the new target planned) obtaining Pp, computing the point Pi:

$$Pi = Pp + d_0 L_{12} \quad (5)$$

where d_0 is distance between Pp and Pi, and imposing the angular velocity of the vehicle according to:

$$\omega = k\delta \quad (6)$$

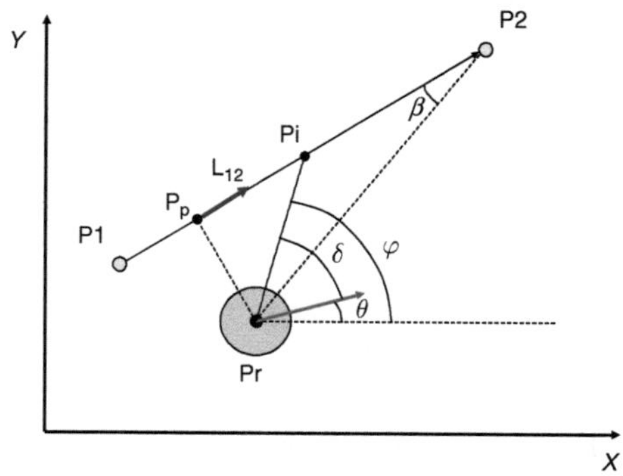

Fig. 4 Scheme of trajectory tracking.

The step of integration in simulation was chosen as a compromise between computational time (it is a real time simulation) and accuracy of trajectory. Errors between simulations with a step of integration of 1 ms and simulations with increasing steps are shown in Table 1, which summarizes a set of tests with different velocities and trajectories, and with different initial heading with respect to the point to reach:

Table 1 Maximum simulated errors increasing step of integration in simulation.

Step of integration	Maximum simulated error
25 ms	8 mm
50 ms	16 mm
100 ms	32 mm
200 ms	61 mm

A maximum error of 16 mm enters in boundaries of safe robot motion, on the contrary a step of integration of 100 ms would be too inaccurate. A good compromise was considered a maximum step of 50 ms. In Fig. 5 it is clear as the simulated trajectories get worse increasing the step of integration.

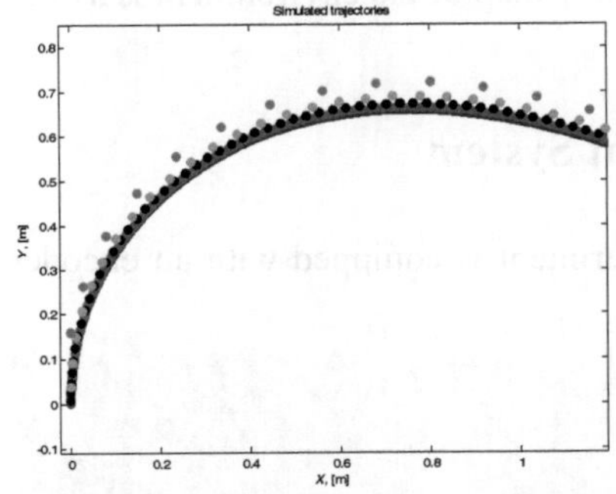

Fig. 5 A detail of simulated trajectories with increasing steps of integration: 1 ms, 25 ms (the continuous internal lines), 50 ms (the black dots), 100 ms, 200 ms (the external dots).

The computational time of Reactive Simulation is a very important aspect in real time applications. It depends on various influence quantities: actual velocity of vehicle, length of the trajectory that has to be simulated, initial pose of vehicle, step of integration. Table 2 shows computational times of Reactive Simulation: a trajectory of 2 m at the velocity of 0.4 m/s was simulated with different steps of integration, and for every step with different heading with respect to the point to reach.

Table 2. Computational time of Reactive Simulation: every time is the mean of simulations with different heading with respect to the point to reach.

Step of integration	Mean Time
10 ms	69 ms
20 ms	36 ms
30 ms	25 ms
40 ms	19 ms
50 ms	16 ms

It is clear as faster simulation has to be preferred for real time applications and so a step of integration of 50 ms was chosen

4 Dynamical Path Planning

The Planner makes a dynamical path planning to reach the goal with no a priori information about the environment. The only information are the initial position of the vehicle and the final target position. Planner searches for open spaces and for doors: the so called “openings”. Dynamically the planner makes a representation of the environment based on a graph, which memorizes all the openings. In this way the problem of dead-ends can be solved: when no opening is found, the vehicle rotates on its own axes by 180°, then it searches again and if actual openings are yet memorised in the graph, the planner understands that it is probably in a dead-end. So based on the openings memorised in the graph, the new path is planned: the vehicle goes to a point memorised but not yet visited, selected by A* search algorithm. A* search is a common algorithm based on a heuristic function which permits to make a local optimal choose (because any map of the environment is available) [12] [13].

5 The Measurement System

The vehicle used in experiment is equipped with an encoder for every wheel, and a laser range finder.

Fig. 6 Experimental apparatus used for laser characterisation.

In order to characterise the laser range finder it has been mounted on a calibration setup (see Fig. 6) composed of a bar for optical alignment, a panel hinged in the axis orthogonal to the bar, whose rotation is measured with an incremental encoder with 14,400 ppr.

By means of the above system it was first characterised the noise standard deviation that come out to be about 4 mm, then the effect of the following influence parameters on laser accuracy: temperature drift, distance, surface colour, surface material, angle of incidence with respect to the object.

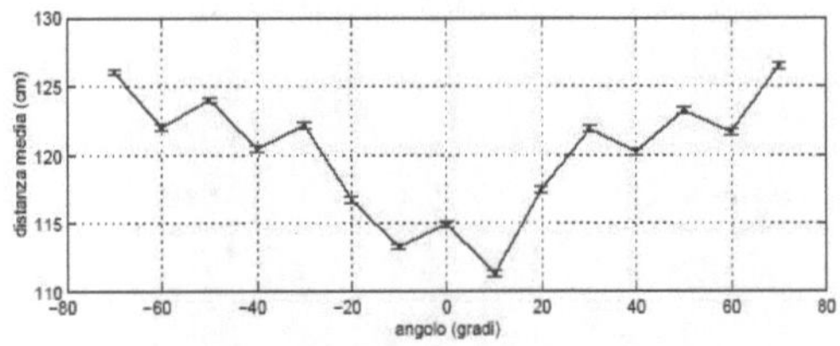

Fig. 7 Measurement as a function of the incidence angle. Measurements taken with the target at fixed distance of 115 cm with the apparatus of Fig. 6.

Above all influence quantities the angle of incidence plays a major role in degrading the measurement accuracy. The effect of the incidence angle over range estimation is shown in Fig. 7. It is evident that uncertainties of about 200 mm can arise only because of rotation or perspective effects. Uncertainty of range data was taken to apply a margin to the output of the Reactive Simulation.

The above characterisation is important also because laser scan data are now starting to be used for pose estimation [15].

6 Reactive Simulation

Reactive Simulation is an algorithm based on the vehicle's model which compute a simulation of the trajectory to the local target.
This is the logical scheme of interaction between modules (as it can be seen in Fig. 8):

1. "Planner" chooses a local target, and vehicle moves towards it (see first frame of Fig. 9a, where a simulation of obstacle avoidance is shown).
2. "Sentinel" checks environment to detect moving obstacles or something like these in robot's trajectory (in second frame of Fig. 9a, on the left an obstacle detected by the Sentinel).
3. If obstacles are detected, Sentinel alerts the Planner which find a new local target (in second frame of Fig. 9a, on the right the new local target planned).
4. Then Reactive Simulation starts, simulates the trajectory to cover to reach the actual local target, and communicates to Planner if the computed trajectory crashes into some obstacles or not (see Fig. 12c and 12d).
5. Finally planner decides to reach or not the local target. In the first case vehicle continues smoothly its path, otherwise it stops to avoid dangerous collision and to plan a safe path (in the first frame of Fig. 9b, the vehicle continues to move towards the local target).

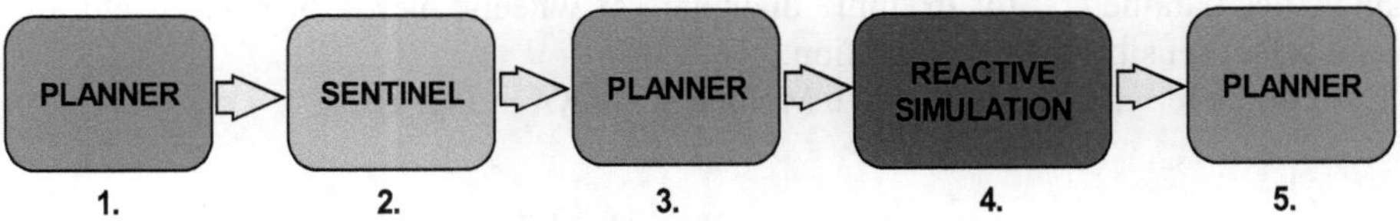

Fig. 8 Logical scheme of interaction between different modules.

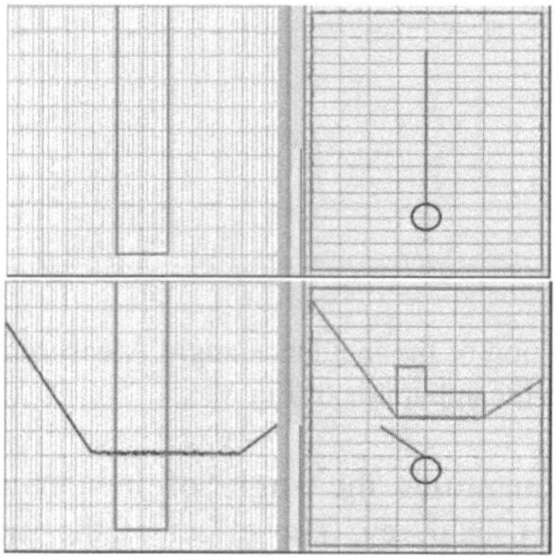

Fig. 9a Example of simulation of obstacle avoidance. On the right side is shown the trajectory of vehicle. On the left side it's shown the Sentinel, which checks a virtual corridor from the actual position of vehicle to the local target to detect any obstacle, as happens in second frame, where an obstacle suddenly appears.

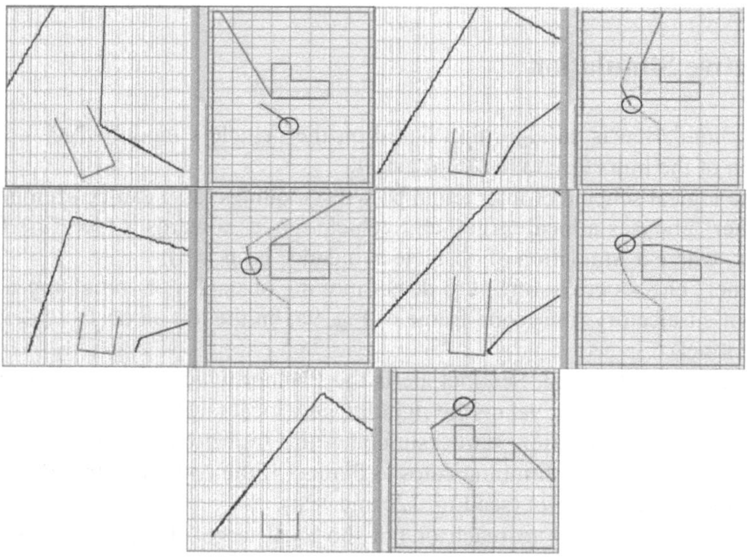

Fig. 9b Example of simulation of obstacle avoidance.

By this way, and integrating the reactive simulation with an on-line identification parameters algorithm, it could be possible to take into account variations in kinematics parameters, for example diameters of wheels, inertia of masses, etc, that could affect sensibly vehicle's motion.

A complete scheme of the Drive Module is shown in Fig. 10.

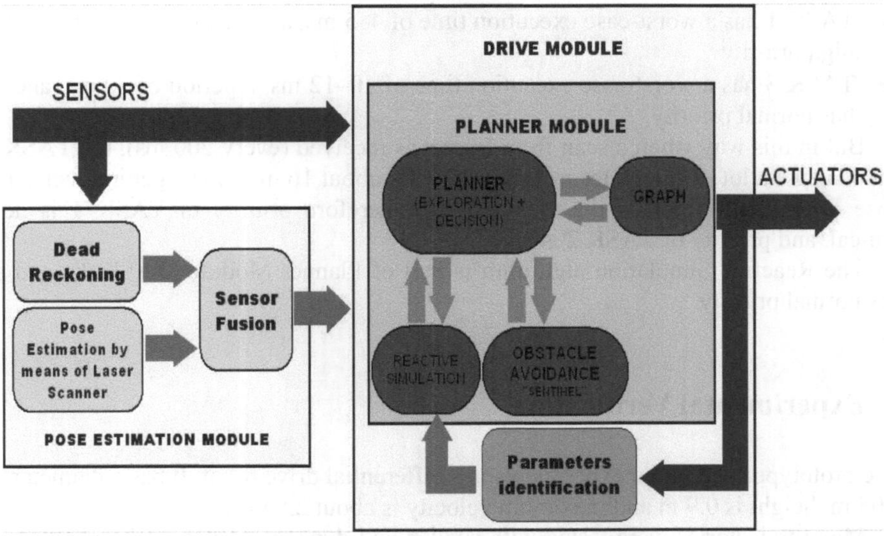

Fig. 10 Complete scheme of Drive Module, and its interaction with Sensors Fusion Algorithm and on-line Parameters Identification Algorithm.

7 Real-Time Implementation

Reactive Simulation has been implemented in real-time applications using the prototype of autonomous vehicle. It is equipped with a 333 MHz PXI (National Instruments) with an embedded real-time operating system (RTOS). The software has three main tasks (see Fig. 11):

- TASK 1 which estimates the pose with data from odometers sensors and executes the trajectory tracking
- TASK 2 of communication with laser
- TASK 3 which makes the path planning, obstacle avoidance and Reactive Simulation.

The priority (static-priority) of the tasks was initially assigned according to rate monotonic algorithm which assign the priority of each task according to its period, so that the shorter the period the higher the priority.

In order of priority:

- TASK 2 has a worst-case execution time of 2–3 ms (except when a scan is received), a period of 6 ms, and so has critical priority

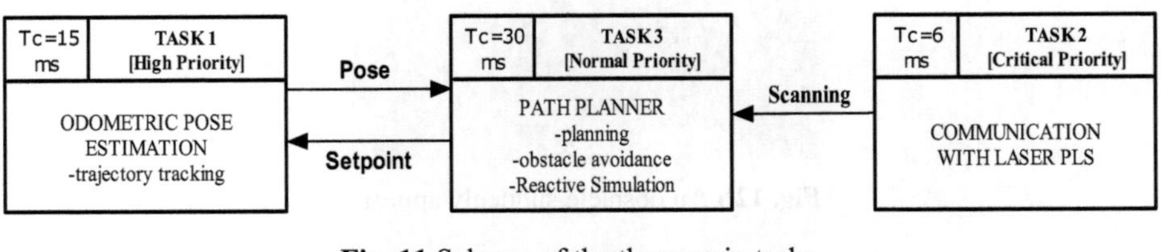

Fig. 11 Scheme of the three main tasks.

- TASK 1 has a worst-case execution time of 4–5 ms, a period of 15 ms, and so has high priority
- TASK 3 has a worst-case execution time of 10–12 ms, a period of 30 ms, and so has normal priority.

But in this way when a scan from laser was received (every 200 ms), the TASK 2 has to make a lot of calculates utilizing CPU for about 16 ms and so getting worse the pose estimation, which is the main aspect. Therefore priority of TASK 1 is now critical, and priority of TASK 2 is high.

The Reactive Simulation algorithm is part of Planner Module (TASK 3), and so has normal priority.

8 Experimental Verification

The prototype used in the experiments is a differential drive robot. It has a diameter of 1.05 m, height is 0.9 m and maximum velocity is about 2.5 m/s.

Many trajectories were tested with sudden and dynamic obstacles, in which were taken into account the standard deviation of laser's scans and the linear and angular velocities of the vehicle when the scans were received to perform a safe robot motion. In fact, the trajectory is planned based on scans closer to vehicle respect to the received scans (see pink line in Fig. 12c and 12d which represents the closer scan). The distance between the received scan and "the safe" one depends on actual linear and angular velocities of the vehicle.

An example of obstacle avoidance is shown in Fig. 12.

Fig. 12a Example of obstacle avoidance with Reactive Simulation: the vehicle starts to move toward target.

Fig. 12b An obstacle suddenly appears.

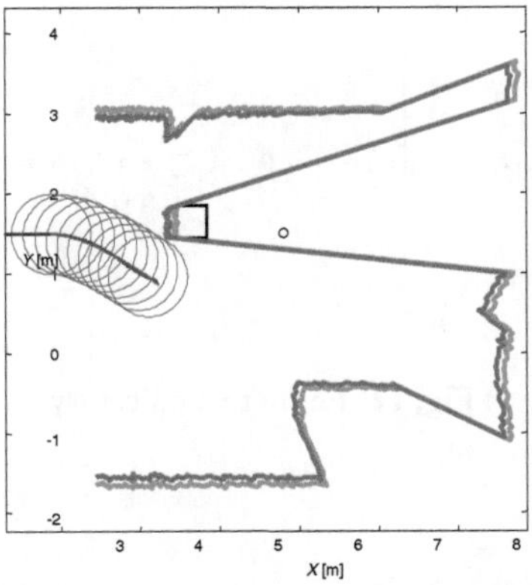

Fig. 12c A new local target was planned and Reactive Simulation computes the trajectory. The lightest grey line is the received scan and the darkest one is the calculated closer scan.

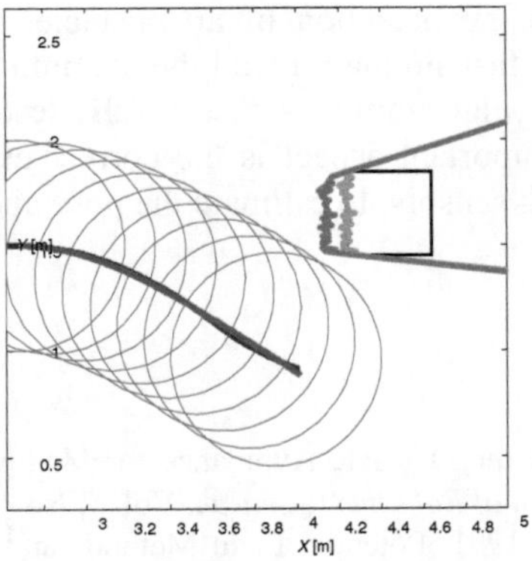

Fig. 12d A detail of Reactive Simulation where the darkest line inside the circles is the real trajectory made by vehicle after the Reactive Simulation.

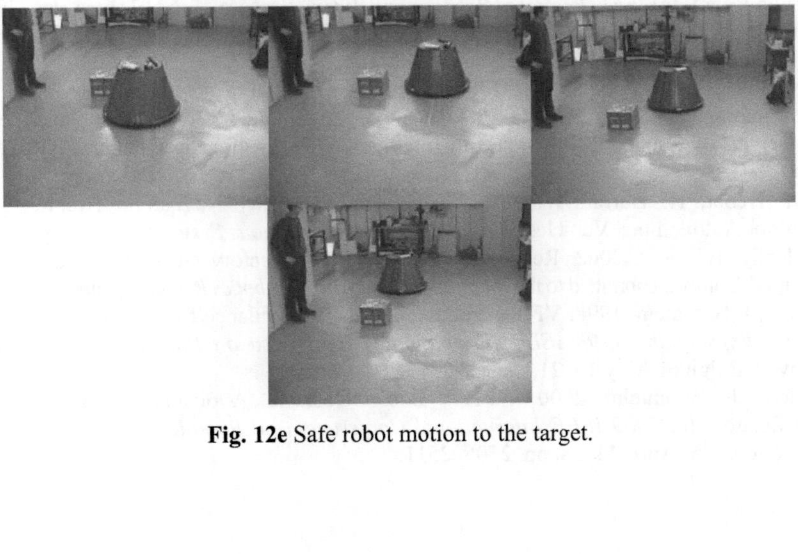

Fig. 12e Safe robot motion to the target.

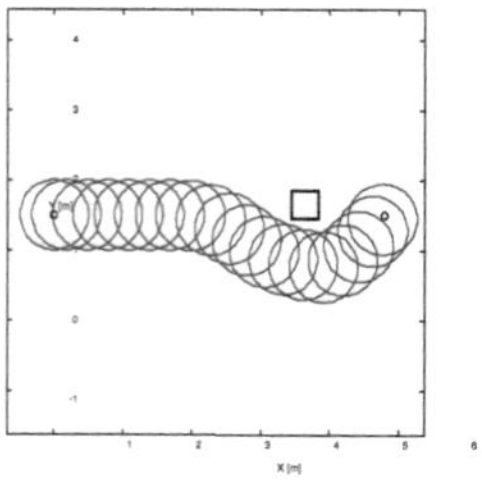

Fig. 12f Entire real trajectory.

9 Conclusions

This chapter presents a novel technique called "Reactive Simulation" for real-time obstacle avoidance. A vehicle's trajectory simulation starts every time a new local target is planned due to the detection of an obstacle. It was verified that 50 ms integration step permits a fast simulation and the maximum error enters in boundaries of safe robot motion. The algorithm was successfully tested on a vehicle in real-time applications, where an important aspect is the correct execution of the tasks which have to communicate with sensors, to estimate the pose and to plan a safe path.

References

1. O. Khatib, 1986. Real-Time Obstacle Avoidance for Manipulators and Mobile Robot. In: *The International Journal of Robotics Research*, Vol. 5, No. 1.
2. Y. Koren, J. Borentein, 1991. Potential Field Methods and their Inherent Limitations for Mobile Robot Navigation. In: *Proceedings of the IEEE conference on Robotics and automation*, Sacramento, California, April 7–12, pp. 1398–1404.
3. J. Borentein, Y. Koren, 1991. The Vector Field Histogram – Fast Obstacle Avoidance for Mobile Robot. In: *IEEE Journal of Robotics and Automation*, Vol. 7, No. 3, June, pp. 278–288.
4. Minguez, J., Montano, L., 2005. Sensor-Based Motion Generation in Unknown, Dynamic and Troublesome Scenarios. In: *Robotics and Automation Systems*. 52, pp. 290–311.
5. R.M. Planas, J.M. Fuertes, A.B. Martinez, 2002. Qualitative Approach for Mobile Robot Path Planning Based on Potential Field Methods, Automatic Control Department Technical University of Catalonia. In: Sixteenth International Workshop on Qualitative Reasoning, June 10–12.
6. E.M. Nebot, H. Durrant-Whyte, 1999. A High Integrity Navigation Architecture for Outdoor Autonomous Vehicles, In: *Robotics and Autonomous Systems*, Vol. 26, pp. 81–97.
7. A. Kelly, B. Nagy, 2002. Reactive Nonholonomic Trajectory Generation via Parametric Optimal Control, submitted to the *International Journal of Robotics Research*, Summer.
8. Ulrich, J. Borentein, 1998. VFH+: Reliable Obstacle Avoidance for Fast Mobile Robot. In: *Proceedings of the 1998 IEEE International Conference on Robotics and Automation*, Leuven, Belgium, May 16–21, pp. 1572–1577.
9. Ulrich, J. Borentein, 2000. VFH*: Local Obstacle Avoidance with Look-Ahead Verification. In: *2000 IEEE International Conference on Robotics and Automation*, San Francisco, CA, April 24–28, pp. 2505–2511.

10. M. De Cecco, 2002. Self-Calibration of AGV Inertial-Odometric Navigation Using Absolute-Reference Measurement. In: *IEEE Instrumentation and Measurement Technology Conference*, Anchorage, AK, USA, 21–23. May
11. M. De Cecco, 2000. A New Concept for Triangulation Measurement of AGV Attitude and Position. In: *Measurement Science and Technology,* Vol. 11, pp. 105–110
12. J. Chestnutt, J. Kuffner, K. Nishiwaki, S. Kagami, 2003. Planning Biped Navigation Strategies in Complex Environments. In: *IEEE Int'l Conf. on Humanoid Robotics, Humanoids.*
13. Stentz, 1994. Optimal and Efficient Path Planning for Partially-Known Environments. In: *Proc. of the IEEE International Conference on Robotics and Automation (ICRA '94)*, Vol. 4, pp. 3310–3317, May.
14. L. Baglivo, M. De Cecco, F. Angrilli, F. Tecchio, A. Pivato, 2005. An Integrated Hardware/Software Platform for Both Simulation and Real-Time Autonomous Guided Vehicle Navigation. In: *Proceedings 19th European Conference on Modelling and Simulation,* Riga, Latvia, June 1st–4th, pp. 227–232.
15. M. De Cecco, L. Baglivo, E. Ervas, E. Marcuzzi, 2006. Asynchronous and Time-Delayed Sensor Fusion of a Laser Scanner Navigation System and Odometry. In: *XVIII Imeko World Congress Metrology for a Sustainable Development, Rio de Janeiro, Brazil, in publication,* September, 17–22.

A Gain-Scheduling Approach for Airship Path-Tracking

Alexandra Moutinho and José Raul Azinheira

IDMEC - Instituto Superior Técnico, TULisbon
Av. Rovisco Pais, 1047-001 Lisbon, Portugal
moutinho,jraz@dem.ist.utl.pt

Abstract. In this chapter a gain scheduled optimal controller is designed to solve the path-tracking problem of an airship. The control law is obtained from a coupled linear model of the airship that allows to control the longitudinal and lateral motions simultaneously. Due to the importance of taking into account wind effects, which are rather important due to the airship large volume, the wind is included in the kinematics, and the dynamics is expressed as function of the air velocity. Two examples are presented with the inclusion of wind, one considering a constant wind input and the other considering in addition a 3D turbulent gust, demonstrating the effectiveness of this single controller tracking a reference path over the entire flight envelope.

Keywords. UAV, gain-scheduling, path-tracking, optimal control, airship.

1 Introduction

The range of civilian and military applications of Unmanned Aerial Vehicles (UAV's) is driving the growth of the rapidly changing market for UAVs globally. The list of applications and opportunities in this domain is already large and is continually growing. Among these are inspection oriented applications that cover different areas such as mineral and archaeologic prospecting, land use surveys in rural and urban regions, inspection of man-made structures such as pipelines, power transmission lines, dams and roads.

Most of the applications cited above have profiles that require maneuverable low altitude, low speed airborne data gathering platforms. The vehicle should also be able to hover above an observation target, present extended airborne capabilities for long duration studies, take-off and land vertically without the need of runway infrastructures, have a large payload to weight ratio, among other requisites. For this scenario, lighter-than-air (LTA) vehicles, like airships, are often better suited than balloons, airplanes and helicopters [5], mainly because: they derive the largest part of their lift from aerostatic, rather than aerodynamic forces; they are safer and, in case of failure, present a graceful degradation; they are intrinsically of higher stability than other platforms.

For all its advantages, airships are being chosen as platform to a variety of applications. Some examples are demining[1], fire detection [8], emergency management [9], target search [13] and even exploration of planetary bodies [4].

[1] www.mineseeker.com

In some cases, the airship is remotely maneuvered, but a wider range of applications is achievable with unmanned autonomous airships, which requires an effective control of the robot behavior. Like other aerial vehicles, the airship dynamics is highly nonlinear. It mostly varies with the relative airspeed, which influences both acting forces as well as actuators authority. In flight control design, it is an established practice to base the controller design in a linear description of the system, obtained assuming the motion of the aerial vehicle is constrained to small perturbations about a trimmed equilibrium flight condition [12]. For instance, Hygounenc and Souères consider linear decoupled models for the longitudinal and lateral motions. The vertical movement is regulated using a Lyapunov based approach, while the horizontal path-following problem is independently solved with a PI controller [6]. Xia and Corbett [13] apply the slidding mode technique to the linear decoupled systems in order to obtain a cooperative control system for two blimps. Complementing the linear model of the airship with the vector of visual signals, Silveira et al. [11] report a line following visual servo control scheme with a PID structure.

Although using different control methodologies, the above references all implicitly use the gain scheduling approach [1] [7], the prevailing flight control design methodology nowadays [10]. As mentioned, the linear model of the system is an approximation only valid for small perturbations around an equilibrium condition, which leads to the necessity of repeating the linearization process for several trim points over the desired operation range.

In this chapter a gain scheduled optimal controller is designed to solve the path-tracking problem for the airship of the Aurora (Autonomous Unmanned Remote mOnitoring Robotic Airship) project [3]. The control law is obtained from a coupled linear model of the airship that allows to control the longitudinal and lateral motions simultaneously. Due to the importance of taking into account wind effects, which are rather important due to the airship large volume, the wind is included in the kinematics, and the dynamics is expressed as function of the air velocity. The designed controller has been tested exhaustively in simulation environment prior to real implementation, and effective results have been obtained covering the entire flight envelope with the single controller. We present here two cases concerning the same reference tracking: a nominal case and a case considering realistic wind disturbances, which the airship platform is supposed to face during its normal operation.

In this chapter we start by describing the Aurora airship platform, followed by the definition of the model used in the control design (Sect. 2). In Sect. 4 we propose an optimal gain scheduled control, applied to the path-tracking problem in Sect. 3. So as to illustrate the valuable results obtained, in Sect. 5 we show examples of the proposed control acting over the entire flight envelope in the presence of wind disturbances. Finally, conclusions are summarized in Sect. 6.

2 Airship Description

2.1 Airship Platform

In this section we briefly describe the airship platform of the Aurora project [3].

The LTA robotic prototype has been built as an evolution of the Airspeed Airships' AS800. It is a nonrigid airship with 10.5 m long, 3.0 m diameter, and 34 m^3 of volume, whose payload capacity is approximately 10 kg and maximum speed is around 50 km/h (see Fig. 1). As main control actuators, the airship has: (i) four deflection surfaces at the tail in the "×" configuration; (ii) the thrust provided by two propellers driven by two-stroke engines; (iii) the vectoring of this propulsion group.

Fig. 1 The Aurora airship.

The airship system state variables are measured by a GPS with differential correction, an inertial measurement unit and a wind sensor.

2.2 Airship Model

Consider an underactuated airship modeled as a rigid body subject to forces and torques. Let $\{i\}$ represent the inertial frame (which, for simplicity, is considered coincident with the geographical north-east-down (NED) frame), $\{l\}$ be the body-fixed coordinate frame whose origin is located in the airship center of volume, and $\mathbf{S}(\boldsymbol{\Phi}) \in SO(3) := \{\mathbf{S} \in \mathbb{R}^{3\times3} : \mathbf{S}\mathbf{S}' = \mathbf{I}, \det(\mathbf{S}) = +1\}$ is the rotation matrix from $\{i\}$ to $\{l\}$ frame, and where $\boldsymbol{\Phi} \in \mathbb{R}^{3\times1}$ is the attitude of $\{l\}$ in relation to $\{i\}$ expressed as the Euler angles roll ϕ, pitch θ and yaw ψ.

The airship dynamic equations of motion [2] may be condensed in the form

$$\mathbf{M}\dot{\mathbf{V}} = -\boldsymbol{\Omega}_6\mathbf{M}\mathbf{V} + \mathbf{E}\mathbf{S}\mathbf{g} + \mathbf{F}_a + \mathbf{f}(\mathbf{U}_a, V_t) \tag{1}$$

$$\dot{\mathbf{P}} = \mathbf{J}\mathbf{V} \tag{2}$$

where $\mathbf{V} = [\mathbf{v}', \boldsymbol{\omega}']' \in \mathbb{R}^{6\times1}$ includes the airship linear and angular velocities relative to $\{i\}$ expressed in $\{l\}$, $\mathbf{P} = [\mathbf{p}', \boldsymbol{\Phi}']' \in \mathbb{R}^{6\times1}$ contains the airship cartesian and angular positions expressed in the $\{i\}$ frame, $\mathbf{J} = \mathrm{diag}\{\mathbf{S}', \mathbf{R}\} \in \mathbb{R}^{6\times6}$, $\mathbf{R} = \mathbf{R}(\boldsymbol{\Phi}) \in \mathbb{R}^{3\times3}$ is a coefficient matrix, and $\mathbf{M} \in \mathbb{R}^{6\times6}$ is the symmetric inertia matrix. $\boldsymbol{\Omega}_6 = \mathrm{diag}\{\boldsymbol{\Omega}(\boldsymbol{\omega}), \boldsymbol{\Omega}(\boldsymbol{\omega})\} \in \mathbb{R}^{6\times6}$, $\boldsymbol{\Omega}(\boldsymbol{\omega}) \in \mathbb{R}^{3\times3}$ is the skew-symmetric matrix equivalent to the cross-product $\boldsymbol{\omega}\times$, $\mathbf{E} \in \mathbb{R}^{6\times3}$ is the gravity matrix input, and

$\mathbf{g} \in \mathbb{R}^{3\times1}$ represents the gravity acceleration in the $\{i\}$ frame. $\mathbf{F}_a \in \mathbb{R}^{6\times1}$ stands for the aerodynamic forces and moments.

The input forces and torques $\mathbf{f}(\mathbf{U}_a, V_t) \in \mathbb{R}^{6\times1}$ are a nonlinear function of the airship airspeed V_t and of the actuators input $\mathbf{U}_a \in \mathbb{R}^{6\times1}$, which includes the elevator δ_e, the total left and right engines thrust X_T, the engines vectoring angle δ_v, and the rudder δ_r (see Fig. 2).

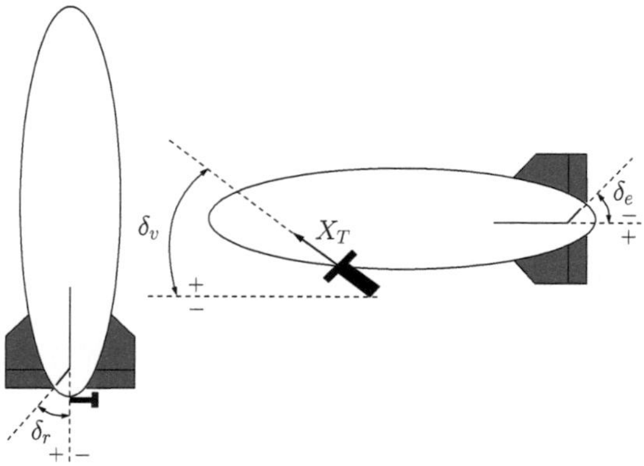

Fig. 2 Airship actuators.

Inclusion of Wind. We assume that the inertial wind velocity vector $\dot{\mathbf{p}}_w = \mathbf{S}'\mathbf{v}_w$ is constant over a region much larger than the size of the airship. This means we do not consider wind shearing effects and torques exerted on the airship ($\boldsymbol{\omega}_w = 0$).

The velocity of the airship center of volume with respect to the air represented in the $\{l\}$ frame is given by

$$\mathbf{v}_a = \mathbf{v} - \mathbf{v}_w \tag{3}$$

In order to include the wind influence in the airship equations of motion, the wind components must be supplied as inputs. Then $\mathbf{v}_a$, rather than $\mathbf{v}$, must be used in the calculation of the aerodynamic forces and moments. The airship dynamics and kinematics can then be rewritten as

$$\dot{\mathbf{V}}_a = \mathbf{M}^{-1}\left[-\boldsymbol{\Omega}_6\mathbf{M}\mathbf{V}_a + \mathbf{ESg} + \mathbf{F}_a + \mathbf{f}(\mathbf{U}_a)\right] \tag{4}$$

$$\dot{\mathbf{P}} = \mathbf{J}\mathbf{V}_a + \mathbf{S}'\mathbf{V}_w \tag{5}$$

with $\mathbf{V}_a = [\mathbf{v}_a', \boldsymbol{\omega}']'$ and $\mathbf{V}_w = [\mathbf{v}_w', 0]'$.

As mentioned before, the wind must be provided as input to (4) and (5). Since the wind is not directly measurable, we will need to compute an estimate. Knowing that

$$\mathbf{v}_a = \begin{bmatrix} V_t\cos(\alpha)\cos(\beta) \\ V_t\sin(\beta) \\ V_t\sin(\alpha)\cos(\beta) \end{bmatrix} \tag{6}$$

where the airspeed V_t, the angle of attack α and the sideslip angle β are measurable quantities, we can compute the wind velocity vector $\mathbf{v}_w$ using (3).

Linear Model. The complexity of the nonlinear dynamic equations justifies the search for a linear simplified version, also necessary if to use the gain-scheduling approach.

The linearization of the dynamic equations (4) and (5) is made for trimmed conditions around equilibrium, which is commonly an horizontal straight flight, without wind incidence.

Considering only the dynamic or perturbed part, and for the conditions stated, the motion equations are written for a perturbation vector $\mathbf{x}$ of the states around the equilibrium value $\mathbf{X}_o$, and the perturbed input $\mathbf{u}$ around the trimmed value $\mathbf{U}_o$, resulting in the matricial dynamic equation:

$$\dot{\mathbf{x}} = \mathbf{A}\mathbf{x} + \mathbf{B}\mathbf{u} \tag{7}$$

in the absence of disturbances (deterministic case).

The linearized model (7), i.e., the dynamic and input matrices $\mathbf{A}$ and $\mathbf{B}$, depend on the trim point chosen for the linearization, and in particular of the chosen airspeed V_{t_o} (we consider here low altitude flight, where the altitude variation is insufficient to significantly change the envelope pressure). The existence of a constant wind component is also to be considered.

In flight control, and as a result of the application of the small perturbations theory, two independent (decoupled) linear models are usually obtained, corresponding to the lateral and longitudinal motions [12]. Here, we chose to work with a single linear model, which allows us to design a single controller for the lateral and longitudinal movements.

Considering that all variables now correspond to the perturbation term, the state and input vectors of the dynamics equation (7) are $\mathbf{x} = [\mathbf{v}_a', \boldsymbol{\omega}', \mathbf{p}', \boldsymbol{\Phi}']'$ and $\mathbf{u} = [\delta_e, X_T, \delta_v, \delta_r]'$.

2.3 Airship Behavior over the Flight Envelope

The airship dynamics is highly nonlinear flight-dependent, with a very different behavior as the airspeed varies from the hovering or low airspeed flight, to the cruise or aerodynamic flight. At low airspeeds, the airship behaves like a balloon with two badly damped oscillation modes (pitch and roll pendulum modes), which are greatly damped as the airspeed increases. In addition, the abrupt and continuous transition between hover and aerodynamic flight in the dynamics implies a different use of actuators for each situation. When in aerodynamic flight, the most important actuators are the propellers thrust and the aerodynamic elevator/rudder control surfaces, whereas when hovering, the effective actuators are the propellers total and differential thrust, and vectoring.

Moreover, in the low airspeed region, the system may be considered underactuated as:

- the tail surfaces have reduced authority, leaving the airship to be controlled by the force inputs mostly;

- the main propellers provide four coupled force components (longitudinal and vertical forces, pitching and rolling torques) with only three inputs (total thrust, vectoring angle and differential thrust);

It is important to note that at low airspeeds, the weighting mass of the airship needs to be compensated for with the vertical forces coming from the propellers vectoring, whereas in the aerodynamic flight the vertical forces come from the aerodynamic lift resulting from the airship angle of attack.

3 Path-Tracking Problem

Let us start by define the vehicle mission: to force the output to follow a reference signal, a given function of time, and to drive to zero the tracking error:

$$\mathbf{e}(t) = \mathbf{y}(t) - \mathbf{y}_r(t) \tag{8}$$

This *path-tracking problem* differs from the path-following one by the fact that the path reference is time dependent.

We define here the cartesian position error $\mathbf{e}_p = [\eta, \delta, \gamma]'$ as expressed in the reference path frame and computed by (see Fig. 3)

$$\mathbf{e}_p = \mathbf{S}(\mathbf{\Phi}_r)(\mathbf{p} - \mathbf{p}_r) \tag{9}$$

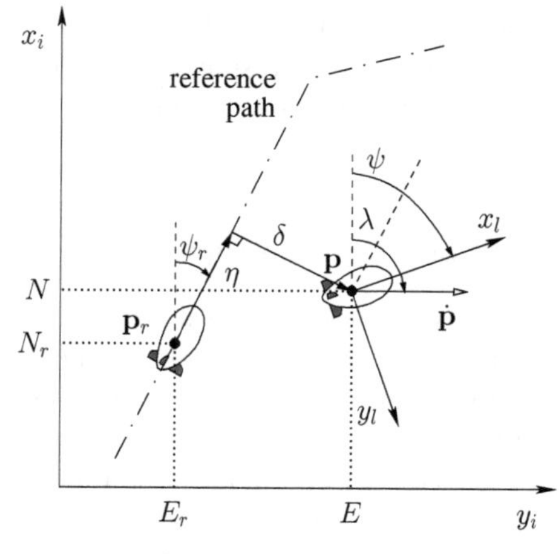

Fig. 3 Position errors definition (2D).

The reference path attitude $\mathbf{\Phi}_r = [\phi_r, \theta_r, \psi_r]'$ is obtained considering $\phi_r = 0$, and θ_r and ψ_r as the angles the reference trajectory does respectively with the horizontal plane, and the north direction.

The attitude error is the difference between the airship and reference attitudes:

$$\mathbf{e}_\Phi = \mathbf{\Phi} - \mathbf{\Phi}_{a_r} \tag{10}$$

If we were not taking into account the wind input, $\mathbf{\Phi}_{a_r} \equiv \mathbf{\Phi}_r$. However, due to the sideslip, the airship orients itself with the relative air direction (see Fig. 4). Therefore, the attitude reference $\mathbf{\Phi}_{a_r}$ corresponds to the estimated attitude of the reference velocity influenced by the wind, $\hat{\dot{\mathbf{p}}}_{a_r}$.

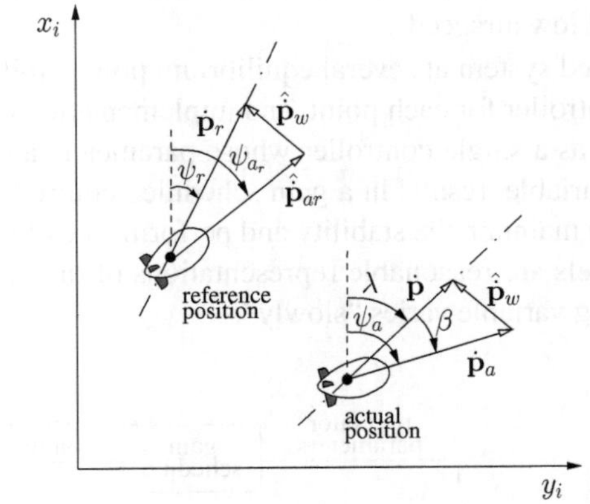

Fig. 4 Reference heading estimation (2D).

We compute the reference attitude $\mathbf{\Phi}_{a_r}$ following these three steps:

1. with the airship inertial velocity $\dot{\mathbf{p}}$, its attitude $\mathbf{\Phi}$ and the aerodynamic variables V_t, β and α (all measured variables), estimate the wind inertial velocity vector $\hat{\dot{\mathbf{p}}}_w$ using (6) and (3), and knowing that $\dot{\mathbf{p}} = \mathbf{S}(\mathbf{\Phi})'\mathbf{v}$;
2. compute the velocity in the aerodynamic frame $\hat{\dot{\mathbf{p}}}_{a_r}$ as the inertial vectors difference between the airship reference velocity $\dot{\mathbf{p}}_r$ and the wind velocity estimation $\hat{\dot{\mathbf{p}}}_w$;
3. consider $\phi_{a_r} = 0$, and θ_{a_r} and ψ_{a_r} as the angles $\hat{\dot{\mathbf{p}}}_{a_r}$ does respectively with the horizontal plane, and the north direction.

The objective of the control design proposed in the next section is then to drive to zero the errors given by (9) and (10).

4 Gain-Scheduling

The linearized system model (7) presented before is only valid for small regions around trim conditions. This reveals the basic limitation of the design via the linearization approach, the fact that the controller is guaranteed to work only in the neighborhood of a single operating (equilibrium) point. The gain scheduling technique [7] addressed here allows to extend the validity of the linearization approach to a range of operating points, in this case over the entire flight envelope.

It is sometimes possible to find auxiliary variables that correlate well with the changes in the process dynamics. In the airship case, these variables mostly correspond to the airspeed V_t and altitude h. Still, the altitude influence may be disregarded for low altitude flights where the envelope pressure is kept practically constant. The airspeed, however, has a major influence over the airship behavior. It not only determines the magnitude of the aerodynamic forces acting on the airship, but also rules the influence of the actuators on the airship motion. As example, the action of the control surfaces, corresponding to the standard inputs δ_e and δ_r, is a function of the dynamic pressure and varies as the square of the airspeed [12], which leads to a reduced authority of these actuators when flying at low airspeeds.

Obtaining a linearized system at several equilibrium points, followed by the design of a linear feedback controller for each point, and implementation of the resulting family of linear controllers as a single controller whose parameters are changed by monitoring the scheduling variable, results in a gain scheduled controller (see Fig. 5). This controller is expected to maintain the stability and performance of the linear systems as long as the design models are reasonable representations of the system dynamics and as long as the scheduling variable varies "slowly".

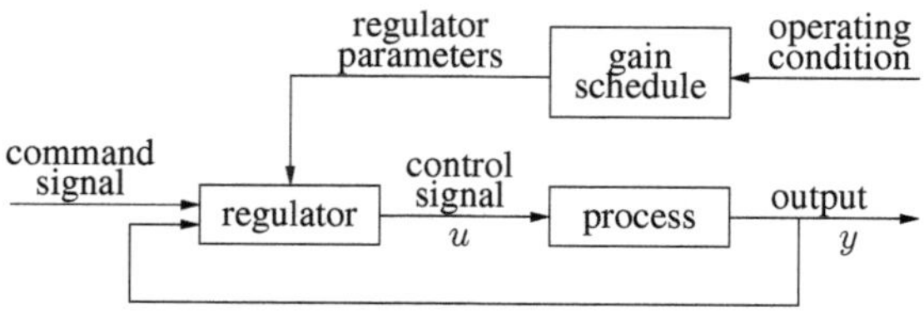

Fig. 5 Gain-scheduling block diagram.

4.1 Optimal Control

Here we discuss the design of a servo control system whose purpose is to keep the tracking error small, while the airship flight condition is kept near the equilibrium state.

Consider the linear system defined by (7), assume the complete state $\mathbf{x}$ is measurable, and that the output variable

$$\mathbf{y} = \mathbf{C}\mathbf{x} \tag{11}$$

is to track a reference input $\mathbf{y}_r$. The tracking error is then defined as

$$\mathbf{e} = \mathbf{y} - \mathbf{y}_r \tag{12}$$

Considering that the model represents the deviations from an equilibrium state $\mathbf{X}_o$, it is required that the state variables $\mathbf{x}_c$ that do not form the output vector be kept null, in order to have $\mathbf{X}_c = \mathbf{X}_{c_o}$.

The admissible control is a proportional output feedback of the form

$$\mathbf{u} = \mathbf{K}_y\mathbf{e} - \mathbf{K}_c\mathbf{x}_c = -\mathbf{K}\mathbf{z} \tag{13}$$

where $\mathbf{z} = [\mathbf{e}', \mathbf{x}_c']'$ and $\mathbf{K} = [\mathbf{K}_y\ \mathbf{K}_c]$.

We use an optimal Linear Quadratic regulator to obtain the control effort (13), that results from the minimization of the cost function

$$J = \int_0^\infty (\mathbf{z}'\mathbf{Q}\mathbf{z} + \mathbf{u}'\mathbf{R}\mathbf{u})dt \tag{14}$$

subject to the system dynamics (7). The state and control weighting matrices $\mathbf{Q} \geq 0$ and $\mathbf{R} > 0$ are the designer tools to balance the state errors $\mathbf{z}$ against the control effort $\mathbf{u}$. In the airship control case, the control weighting matrix $\mathbf{R}$ is a specially important tool in the sense that it allows the designer to change the control effort of the different actuators over the flight envelope.

The gain matrix $\mathbf{K}$ is obtained from

$$\mathbf{K} = \mathbf{R}^{-1}\mathbf{B}'\mathbf{P} \tag{15}$$

solving first the algebraic Riccati equation for the positive definite matrix $\mathbf{P}$

$$\mathbf{PA} + \mathbf{A}'\mathbf{P} - \mathbf{PBR}^{-1}\mathbf{B}'\mathbf{P} + \mathbf{Q} = 0 \tag{16}$$

The actuators request

$$\mathbf{U} = \mathbf{U}_o + \mathbf{u} \tag{17}$$

has to be computed for each linearized model, which means the procedure has to be repeated for different airspeeds. Figure 6 illustrates the closed-loop diagram for a determined equilibrium condition.

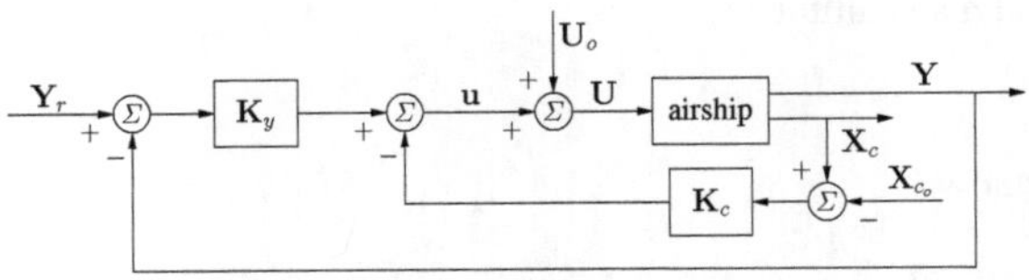

Fig. 6 Linear control block diagram.

5 Simulation Results

From the various simulations carried out, two examples are illustrated here concerning the same reference tracking: a nominal case with constant wind input and a case considering realistic wind disturbances, i.e., with an aleatoric component, which the airship platform is supposed to face during its normal operation.

In both cases the airship is to follow a reference path $\mathbf{p}_r(t)$ at constant altitude $h_r = -D_r = 50\,$m, starting deviated from the initial position at $\mathbf{p}_i = [-20, -10, -45]$m and with the initial orientation $\mathbf{\Phi}_i = [10, 10, 10]$deg.

5.1 Nominal Case

This nominal case considers a constant wind input of 4 m/s coming from north. The airship horizontal north-east trajectory with the correction from the initial deviation point and the cartesian position errors $\mathbf{e}_p = [\eta, \delta, \gamma]$ are shown in Fig. 7.

After the initial deviation is corrected, the airship position errors stabilize to zero after passing the reference path corners.

So as to avoid saturation of the thrusters when the controller is correcting the airship position, the longitudinal position error η was limited. This can be noticed by the constant rate at which the north position is corrected (η curve in Fig. 7).

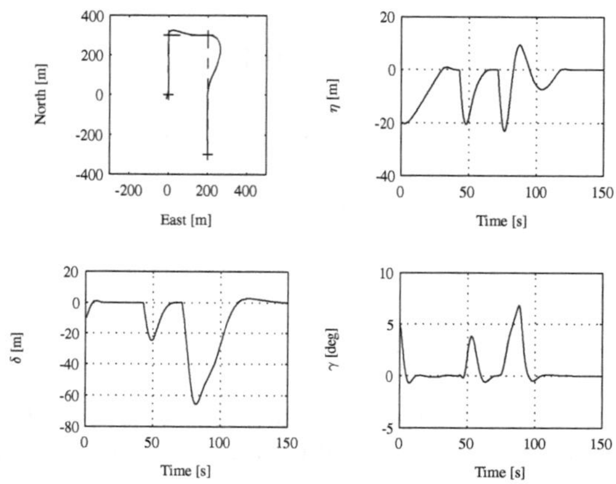

Fig. 7 Nominal case: north-east trajectory (reference - dashed, output - solid), and position errors: longitudinal η, lateral δ and altitude γ.

5.2 Realistic Case

In order to exemplify the controller robustness when in the presence of disturbances, further simulation tests included a 3D turbulent gust (simulated here by a Dryden model) with an intensity of 3 m/s in addition to a constant wind blowing at 4 m/s from north. The remaining setup values were the same used in the previous case. Figure 8 shows the horizontal north-east trajectory and the cartesian position errors $\mathbf{e}_p = [\eta, \delta, \gamma]$. The airship behavior is similar to the nominal case, with the turbulent gust noticeable by the curves oscillation. Again, the position errors are corrected to zero, having the longitudinal error η been limited to avoid the thrusters saturation.

The evolution of the airspeed V_t, represented in Fig. 9, defines the variations of the linear model used in the control design. The spam of the airspeed values over the flight envelope, between 2 and 12 m/s, is easily seen. This implies not only that the airship dynamics suffer a severe alteration, but also the actuators authority varies enormously. This can be confirmed comparing the airspeed curve with the graphics of the actuators input (elevator δ_e, total thrust X_T, vectoring angle δ_v and rudder δ_r) represented

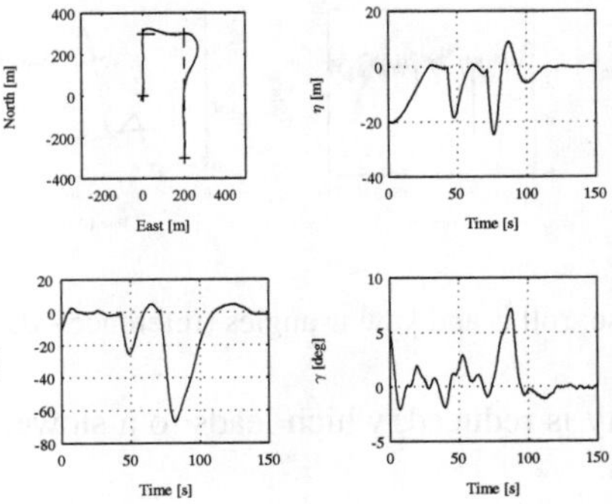

Fig. 8 Realistic case: north-east trajectory (reference - dashed, output - solid), and position errors: longitudinal η, lateral δ and altitude γ.

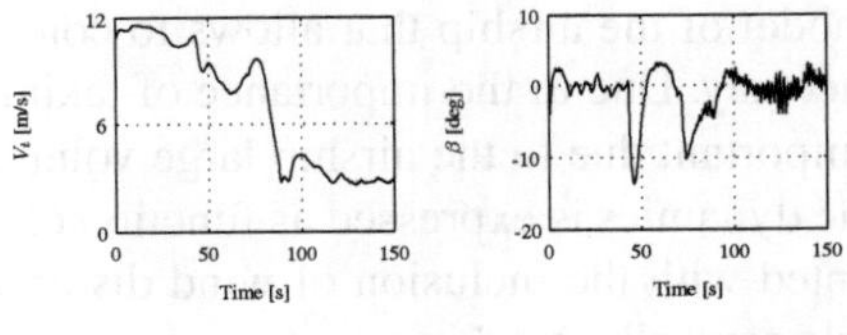

Fig. 9 Realistic case: airspeed V_t and sideslip angle β.

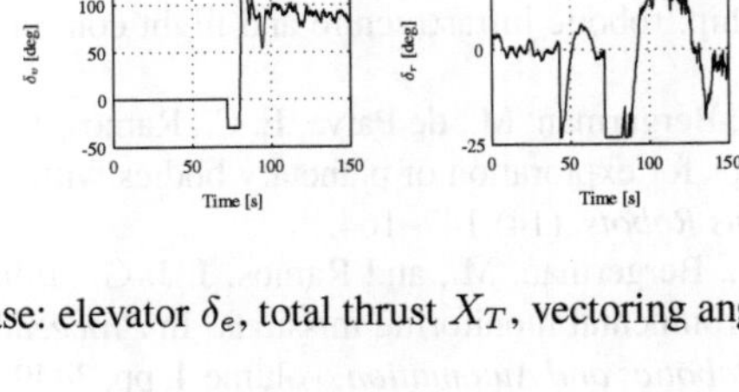

Fig. 10 Realistic case: elevator δ_e, total thrust X_T, vectoring angle δ_v and rudder δ_r.

in Fig. 10. Mostly we can observe the vectoring angle δ_v change between 0° for aerodynamic flight, and 90° at low airspeeds. The sideslip angle β is also represented in Fig. 9. We can observe that its value oscillates around zero even though the airship is submitted to a wind disturbance. This achievement is due to the fact that the wind was included in the system model used to obtain the actuators request. Finally we can observe in Fig. 11 that the rolling angle ϕ oscillates around zero, and the variation of the airship yaw ψ following the reference-path heading. As expected for low airspeeds, the

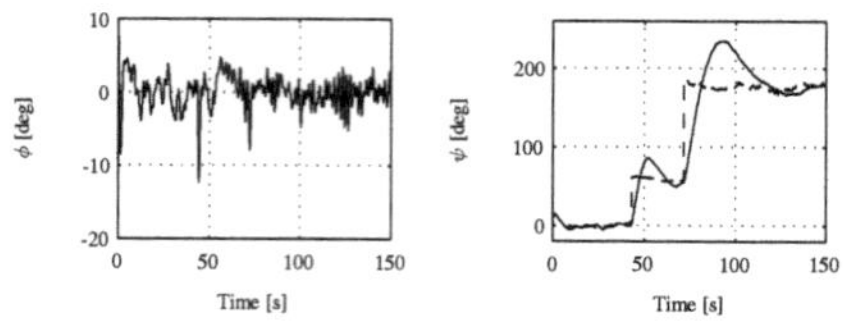

Fig. 11 Realistic case: roll ϕ and yaw ψ angles (reference - dashed, output - solid).

control surfaces authority is reduced, which leads to a slower correction of the lateral errors in these situations.

6 Conclusions

In this chapter a gain scheduled optimal controller is designed to solve the path-tracking problem of an airship, valid over the entire flight envelope. The control law is obtained from a coupled linear model of the airship that allows to control the longitudinal and lateral motions simultaneously. Due to the importance of taking into account wind effects, which are rather important due to the airship large volume, the wind is included in the kinematics, and the dynamics is expressed as function of the air velocity.

The examples presented with the inclusion of wind disturbances, demonstrate the effectiveness of this single controller tracking a reference path over the entire flight envelope. The implied variation of airspeed represents a significant problem in an airship control due to its influence to the system dynamics, as well as to the actuators authority.

References

1. Åström, K. J. and Wittenmark, B. (1989). *Adaptive Control*, Chapter 9: Gain Scheduling, pp. 343–371. Addison-Wesley Publishing Company, 1st edition.
2. Azinheira, J. R., de Paiva, E. C., and Bueno, S. S. (2002). Influence of wind speed on airship dynamics. *Journal of Guidance, Control, and Dynamics*, 25(6):1116–1124.
3. de Paiva, E. C., Azinheira, J. R., Ramos, J. J. G., Moutinho, A., and Bueno, S. S. (2006). Project AURORA airship: robotic infrastructure and flight control experiments. *Journal of Field Robotics*.
4. Elfes, A., Bueno, S. S., Bergerman, M., de Paiva, E. C., Ramos, J. J.G., and Azinheira, J. R. (2003). Robotic airships for exploration of planetary bodies with an atmosphere: autonomy challenges. *Autonomous Robots*, (14):147–164.
5. Elfes, A., Bueno, S. S., Bergerman, M., and Ramos, J. J. G. (1998). A semi-autonomous robotic airship for environmental monitoring missions. In *Proceedings of the IEEE International Conference on Robotics and Automation*, volume 4, pp. 3449–3455, Leuven, Belgium. IEEE Press.
6. Hygounenc, E. and Souères, P. (2003). Lateral path-following GPS-based control of a small-size unmanned blimp. In *Proceedings of the IEEE International Conference on Robotics and Automation*, volume 1, pp. 540–545, Taipei, Taiwan. IEEE Press.
7. Khalil, H. K. (2000). *Nonlinear Systems*. Prentice-Hall, Inc., 3rd edition.
8. Merino, L., Caballero, F., de Dios, J. R. M., and Ollero, A. (2005). Cooperative fire detection using unmanned aerial vehicles. In *Proceedings of the IEEE International Conference on Robotics & Automation*, Barcelona, Spain.

9. Rao, J., Gong, Z., Luo, J., and Xie, S. (2005). Unmanned airships for emergency managment. In *Proceedings of the IEEE International Workshop on Safety, Security and Rescue Robotics*, pp. 125–130, Kobe, Japan. IEEE Press.
10. Rugh, W. J. and Shamma, J. S. (2000). Research on gain scheduling – survey paper. *Automatica*, 36(10):1401–1425.
11. Silveira, G. F., Azinheira, J. R., Rives, P., and Bueno, S. S. (2003). Line following visual servoing for aerial robots combined with complementary sensors. In *Proceedings of the IEEE 11th International Conference on Advanced Robotics*, pp. 1160–1165, Coimbra, Portugal. IEEE Press.
12. Stevens, B. L. and Lewis, F. L. (1992). *Aircraft Control and Simulation*. John Wiley and Sons, Inc., USA.
13. Xia, G. and Corbett, D. R. (2005). Cooperative control systems of searching targets using unmanned blimps. In *Proceedings of the 5th Worth Congress on Intelligent Control and Automation*, volume 2, pp. 1179–1183, Hangzhou, P.R.China. IEEE Press.

9. [illegible] on 12 [illegible]
10. R[illegible] paper [illegible]
11. Sh[illegible] following visual [illegible] Pacific [illegible]
12. [illegible] Wiley and Sons [illegible]
13. M. G[illegible] targets using [illegible]

Semiotics and Human-Robot Interaction

João Silva Sequeira and Maria Isabel Ribeiro

Institute for Systems and Robotics, Instituto Superior Técnico Av. Rovisco Pais 1, 1049-001 Lisbon, Portugal
{jseq,mir}@isr.ist.utl.pt

Abstract. This chapter describes a robot control architecture supported on a human-robot interaction model obtained directly from semiotics concepts.The architecture is composed by a set of objects defined after a well known model of semiotic signs. These objects form an algebraic system suitable to develop formal models of human-robot interactions.They are easily mapped into motion skills and can be associated with basic forms of meaning representation.This characteristic makes them suitable to use as basis objects in high level interactions involving mobile robots. Real and simulation experiments using unicycle robots illustrate the operation of the architecture.

Keywords. Semi-autonomous robot, human-robot interaction, semiotics.

1 Introduction

This chapter describes part of an ongoing project aiming at developing new methodologies for semi-autonomous robot control that, in some sense, mimic those used by humans in their relationships. The chapter focus (i) on the use of semiotic signs to control robot motion and (ii) their use in the interaction with humans. The approach followed defines (i) a set of objects that capture key features in human-robot and robot-robot interactions, and (ii) an algebraic system with operators to work on this new space of objects.

In a wide variety of applications of robots, such as surveillance in wide open areas, rescue missions in catastrophe scenarios, and working as personal assistants, the interactions among robots and between humans and robots are a key factor for the success of a mission. In such missions contingency situations may lead a robot to request external help, often from a human operator, and hence it seems natural to search for interaction languages that can be used by both. Furthermore, a unified framework avoids the development of separate competences to handle human-robot and robot-robot interactions.

The numerous robot control architectures proposed in the literature account for human-robot interactions (HRI) either explicitely, through interfaces to handle external commands, or implicitely, through task decomposition schemes that map high level mission goals into motion commands. Examples of current architectures developed under such principles can be found in [1] [2] [3] [4]. If the humans are assumed to have enough knowledge on the robots and the environment, imperative computer languages can be used for HRI, easily leading to complex communication schemes. Otherwise, declarative, context dependent, languages, like Haskell, [5] and FROB, [6], have been proposed to simulate robot systems and also as a mean to interact with them. BOBJ was used in [7] to illustrate examples on human-computer interfacing. RoboML, [8], supported on XML, is an example of a language explicitly designed for HRI, accounting for low complexity programming, communications and knowledge representation. In [9] the robots are equipped with behaviors that convey information on their intentions to the outside environment. These behaviors represent a form of implicit communication between agents, such as the robot following a human without having been told explicitly to do so. This form of communication, without explicit exchange of data is an example of interaction that can be modeled using semiotics.

Semiotics studies the interactions among humans. These are often characterized by the loose specification of objectives, for instance as when a sentence expresses an intention of motion instead of a specific path. Capturing some of these features, typical of natural languages, is the aim of this project.

The chapter is organised as follows. Section 2 presents key concepts in semiotics to model human-human interactions and motivates their use to model human-robot interactions. These concepts are used in Section 3 to define the building blocks of the proposed architecture. Section 4 presents a set of basic experiments that demonstrate the use of the semiotic model developed along the chapter. Section 5 presents the conclusions and future work.

2 A Semiotic Perspective to Model HRI

In general robots and humans work at very different levels of abstraction. Humans work primarily at high levels of abstraction whereas robots are commonly programmed to follow trajectories, hence operating at a low level of abstraction.

Semiotics is a branch of general philosophy which studies the interactions among humans, such as the linguistic ones (see [10] for an introduction to semiotics) and hence it is a natural framework where to search for new methodologies to model human-robot interactions. In addition, this should be extended to robot-robot interactions yielding a unifying framework to handle any interaction between a robot and the external environment.

An algebraic formulation for semiotics and its use in interface design has been presented in [11]. An application of hypertext theory to world wide web was developed in [12]. Machine-machine and human-human interactions over electronic media (such as the Web) have been addressed in [13]. Semiotics has been used in intelligent control as the concept encompassing the main functions related to knowledge, namely acquisition, representation, interpretation, transformation and manipulation (see for instance [14]) and often directly related to the symbol grounding concept of artificial intelligence.

The idea underlying semiotics is that humans communicate among each other (and with the environment) through signs. Roughly, a sign encapsulates a meaning, an object, a label and the relations among them. Sign systems are formed by signs and the morphisms defined among them (see for instance [11] for a definition of sign system) and hence, under reasonable assumptions on the existence of identity maps, map composition, and composition association, can be modeled using category theory framework. The following diagram, suggested by the "semiotic triangle" common in the literature on this area (see for instance [10]), expresses the relations between the three components of a sign.

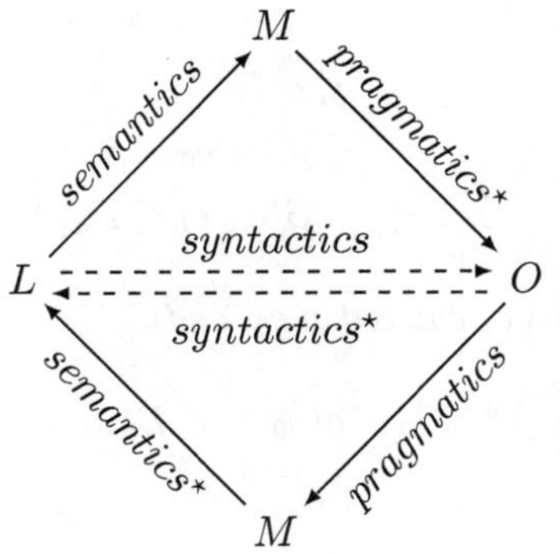

(1)

Labels, (L), represent the vehicle through which the sign is used. Meanings, (M), stand for what the users understand when referring to the sign. The Objects, (O), stand for the real objects signs refer to.

The morphisms in diagram (1) represent the different perspectives used in the study of signs. Semiotics currently considers three different perspectives: semantics, pragmatics and syntactics, [12]. Semantics deals with the general relations among the signs. For instance, it defines whether or not a sign can have multiple meanings. In human-human interactions this amounts to say that different language constructs can be interpreted equivalently, that is as synonyms. Pragmatics handles the hidden meanings that require the agents to perform some inference on the signs before extracting their real meaning. Syntactics defines the structural rules that turn the label into the object the sign stands for. The starred morphisms are provided only to close the diagrams, as the "semiotic triangle" is usually represented as an undirected graph, and are not relevant for the purpose of this work.

3 An Architecture for HRI

This section introduces the architecture by first defining a set of context free objects and operators on this set that are compatible (in the sense that they can be identified) with diagram (1). Next, the corresponding realizations for the free objects are described. The proposed architecture includes three classes of objects: motion primitives, operators on the set of motion primitives and decision making systems (on the set of motion primitives).

The first free object, named *action*, defines primitive motions using simple concepts that can be easily used to form a HRI language. The actions represent motion trends, i.e., an action represents simultaneously a set of paths that span the same bounded region of the workspace.

Definition 1 (Free action). *Let k be a time index, q_0 the configuration of a robot where the action starts to be applied and $a(q_0)|_k$ the configuration at time k of a path generated by action a.*

A free action is defined by a triple $A \equiv (q_0, a, B_a)$ *where* B_a *is a compact set and the initial condition of the action,* q_0*, verifies,*

$$q_0 \in B_a, \tag{2}$$

$$a(q_0)|_0 = q_0, \tag{2a}$$

$$\exists_{\epsilon > \epsilon_{\min}} \; : \; \mathcal{B}(q_0, \epsilon) \subseteq B_a, \tag{2b}$$

with $\mathcal{B}(q_0, \epsilon)$ *a ball of radius* ϵ *centered at* q_0*, and*

$$\forall_{k \geq 0} \;\; a(q_0)|_k \in B_a. \tag{2c}$$

□

Actions are to inclose different paths with similar (in a wide sense) objectives. Paths that can be considered semantically equivalent, for instance because they lead to a successful execution of a mission, may be enclosed within a single action.

Representing the objects in Definition 1 in the form of a diagram it is possible to establish a correspondence between (free) actions and signs verifying model (1),

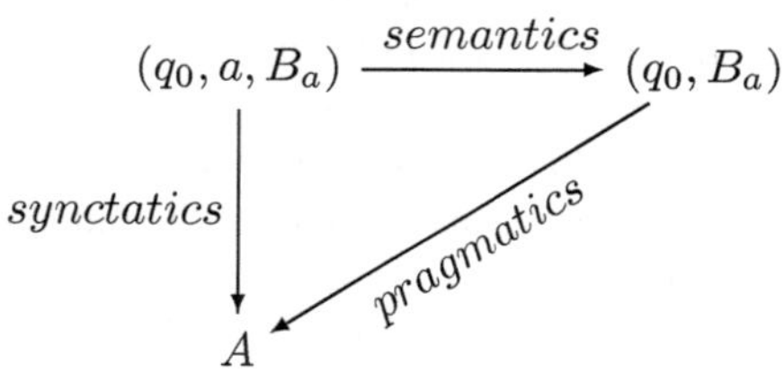

The projection map *semantics* expresses the fact that multiple trajectories starting in a neighborhood of q_0 and lying inside B_a may lead to identical results. The *pragmatics* map expresses the fact that given a bounding region and some initial condition it may be possible to infer the action being executed. The *synctatics* map simply expresses the construction of an action through Definition 1.

Following model (1), different actions can yield the same meaning, that is, two actions can produce the same net effect in a mission. Practical examples of locomotion related actions with the same meaning are easily found. For instance, passing through a door can often be done using a variety of paths each of which generated by a different action. This amounts to require that the following diagram, expressing the relations between different action spaces, commutes,

$$\begin{CD} A @>{equality}>> A' \\ @V{semantics_{(q,B)}}VV @VV{semantics_{(q',B')}}V \\ B_a @>{\mathbf{1}_M}>> B_{a'} \end{CD} \tag{3}$$

where $\mathbf{1}_M$ stands for the identity map in the space of meanings, M.

Diagram (3) provides a roadmap to define action equality as a key concept to evaluate sign semantics.

Definition 2 (Free action equality).

Two actions (a_1, B_{a_1}, q_{0_1}) *and* (a_2, B_{a_2}, q_{0_2}) *are equal, the relation being represented by* $a_1(q_{0_1}) = a_2(q_{0_2})$*, if and only if the following conditions hold*

$$a_1(q_{0_1}), a_2(q_{0_2}) \subset B_{a_1} \cap B_{a_2} \tag{4}$$

$$\begin{array}{l} \forall_{k_2 \geq 0}, \exists_{k_1 \geq 0}, \exists_{\epsilon} : \\ \quad a_1(q_{0_1})|_{k_1} \in \mathcal{B}(a_2(q_{0_2})|_{k_2}, \epsilon) \subset B_{a_1} \cap B_{a_2} \end{array} \tag{4a}$$

□

Expressions (4) and (4a) define the *equality* map in diagram (3). From (4), it suffices to choose identical bounding regions and from (4a) a goal region inside the bounding regions to where the paths generated by both actions converge.

It is worth to emphasize that Definition 1 is valid for a wide class of trajectory generation algorithms. In fact, common path planning algorithms belong to this category. For the purpose of this work, developing a formal system to support HRI modeling, the realization for the free action of Definition 1 is given by the following proposition.

Proposition 1 (Action).

Let $a(q_0)$ *be a free action. The paths generated by* $a(q_0)$ *are solutions of a system in the following form,*

$$\dot{q} \in F_a(q) \tag{5}$$

where F_a *is a Lipschitzian set-valued map with closed convex values verifying,*

$$F_a(q) \subseteq T_{B_a}(q) \tag{5a}$$

where $T_{B_a}(q)$ *is the contingent cone to* B_a *at q (see [15] for the definition of this cone).*

□

The demonstration of this proposition is just a re-statement, in the context of this chapter, of Theorem 5.6 in [15] on the existence of invariant sets for the inclusion (5).

■

The convexity of the values of the F_a map must be accounted for when specifying an action. The Lipschitz condition imposes bounds on the growing of the values of the F_a map. In practical applications this assumption can always be verified by proper choice of the map. This condition is related to the existence of solutions to (5), namely as it implies upper semi-continuity (see [15], Proposition 2.4).

Proposition 2 (Action Identity). *Two actions a_1 and a_2, implemented as in Proposition 1, are said equal if,*

$$B_{a_1} = B_{a_2} \tag{6}$$

$$\exists_{k_0} \; : \; \forall_{k>k_0}, \;\; F_{a_1}(q(k)) = F_{a_2}(q(k)) \tag{6a}$$

□

The demonstration follows by direct verification of the properties in Definition 2.

By assumption, both actions verify the conditions in Proposition 1 and hence their generated paths are contained inside $B_{a_1} \cap B_{a_2}$ which implies that (4) is verified.

Condition (6a) states that, after some time on, there are always motion directions that are common to both actions. For example, if any of the actions a_1, a_2 generates paths restricted to $F_{a_1} \cap F_{a_2}$ then condition (4a) is verified. When any of the actions generates paths using motion directions outside $F_{a_1} \cap F_{a_2}$ then condition (6a) indicates that after time k_0 they will be generated after the same set of motion directions. From 4, both actions generate paths contained inside their common bounding region and hence the generated paths verify (4a).

■

A sign system is defined by the signs and the morphisms among them. In addition to the equality map, two other morphisms are defined: action composition and action expansion. In some sense, these morphisms represent the mechanism through which actions are transformed into other actions.

Definition 3 (Free Action Composition). *Let $a_i(q_{0_i})$ and $a_j(q_{0_j})$ be two free actions. Given a compact set M, the composition $a_{j\circ i}(q_{0_i}) = a_j(q_{0_j}) \circ a_i(q_{0_i})$ verifies,*

if $B_{a_i} \cap B_{a_j} \neq \emptyset$

$$a_{j\circ i}(q_{0_i}) \subset B_{a_i} \cup B_{a_j} \tag{7}$$

$$B_{a_i} \cap B_{a_j} \supseteq M \tag{7a}$$

otherwise, the composition is undefined.

□

Action $a_{j\circ i}(q_{0_i})$ resembles action $a_i(q_{0_i})$ up to the event marking the entrance of the paths into the region $M \subseteq B_{a_i} \cap B_{a_j}$. When the paths leave the common region M the composed action resembles $a_j(q_{0_j})$. While in M the composed action generates a link path that connects the two parts.

Whenever the composition is undefined the following operator can be used to provide additional space to one of the actions such that the overlapping region is non empty.

Definition 4 (Free Action Expansion). *Let $a_i(q_{0_i})$ and $a_j(q_{0_j})$ be two actions with initial conditions at q_{0_i} and q_{0_j} respectively. The expansion of action a_i by action a_j, denoted by $a_j(q_{0_j}) \boxtimes a_i(q_{0_i})$, verifies the following properties,*

$$B_{j\boxtimes i} = B_j \cup B_i, \text{ with } B_i \cap B_j \subseteq M \tag{8}$$

where M is a compact set representing the minimal amount of space required for the robot to maneuver and such that the following property holds

$$\exists_{q_{0_k} \in B_j} \, : \, a_i(q_{0_i}) = a_j(q_{0_k}) \tag{8a}$$

meaning that after having reached a neighborhood of q_{0_k}, $a_i(q_i)$ behaves like $a_j(q_j)$.

□

Given the realization in Proposition (1) for the actions, the composition and expansion operators can be realized through the following propositions.

Proposition 3 (Action Composition). *Let a_i and a_j be two actions defined by the inclusions*

$$\dot{q}_i \in F_i(q_i) \text{ and } \dot{q}_j \in F_j(q_j)$$

with initial conditions q_{0_i} and q_{0_j}, respectively. The action $a_{j \circ i}(q_{0_i})$ is given by $\dot{q} \in F_{j \circ i}(q)$, with the map $F_{j \circ i}$ defined as

$$F_{j \circ i} = \begin{cases} F_i(q_i) & \text{if } q \ni B_i \backslash M \quad (9) \\ F_i(q_i) \cap F_j(q_j) & \text{if } q \in M \quad (9a) \\ F_j(q_j) & \text{if } q \in B_j \backslash M \quad (9b) \\ \emptyset & \text{if } B_i \cap B_j = \emptyset \quad (9c) \end{cases}$$

for some $M \subset B_j \cap B_i$.

Outside M the values of F_i and F_j verify the conditions in Proposition 1. Whenever $q \in M$ then $F_i(q_i) \cap F_j(q_j) \subset T_{B_j}(q)$.

□

The first trunk of the resulting path, given by (9), corresponds to the path generated by action $a_i(q_{0_i})$ prior to the event that determines the composition. The second trunk, given by (9a), links the paths generated by each of the actions. Note that by imposing that $F_i(q_i) \cap F_j(q_j) \subset T_{B_j}(q_j)$ the link paths can move outside the region M. The third trunk, given by (9b), corresponds to a path generated by action $a_j(q_{0_j})$.

By Proposition 1, each of the trunks is guaranteed to generate a path inside the respective bounding region and hence the overall path verifies (7).

■

The action composition in Proposition 3 generates actions that resemble each individual action outside the overlapping region. Inside the overlapping area the link path is built from motion directions common to both actions being composed. Crossing the boundary of M defines the events marking the transition between the trunks.

Whenever $F_i(q_i) \cap F_j(q_j) = \emptyset$ it is still possible to generate a link path, provided that M has enough space for maneuvering. The basic idea is to use the free action expansion to locally enlarge either $F_i(q_i)$ or $F_j(q_j)$. The following proposition provides a realization for this free operator.

Proposition 4 (Action Expansion).
Let a_i and a_j be two actions defined after the inclusions

$$\dot{q}_i \in F_i(q_i) \ \ and \ \ \dot{q}_j \in F_j(q_j)$$

The expansion $a_{j\boxtimes i}(q_{0_i})$ verifies the following properties

$$F_{i\boxtimes j} = \begin{cases} F_i & if\, q \ni B_i \backslash M \quad (10) \\ F_j \cup F_i & if\, q \in B_i \cap B_j \cup M \quad (10a) \end{cases}$$

where $M \supseteq B_i \cap B_j$ is the expansion set chosen large enough such that $F_j \cup F_i$ verifies (5a).

□

Condition (10) generates paths corresponding to the action $a_i(q_{0_i})$. These paths last until an event, triggered by the crossing of the boundary of M, is detected. At this event the overall bounding region is expanded by M and the set of paths, by F_j, as expressed by (10a).

Assuming that $F_j \cup F_i \subset T_{B_i \cap B_j \cup M}$, that is, it verifies (5a), the complete path is entirely contained inside the expanded bounding region. After moving outside M paths behave as of generated by action a_i, as required by (8a).

■

Instead of computing a priori M, the expansion operator can be defined as a process by which action a_i converges to action a_j in the sense that $F_i(q_i) \to F_j(q_j)$ and M is the space spanned by this process.
The objects and operators defined above are combined in the following diagram,

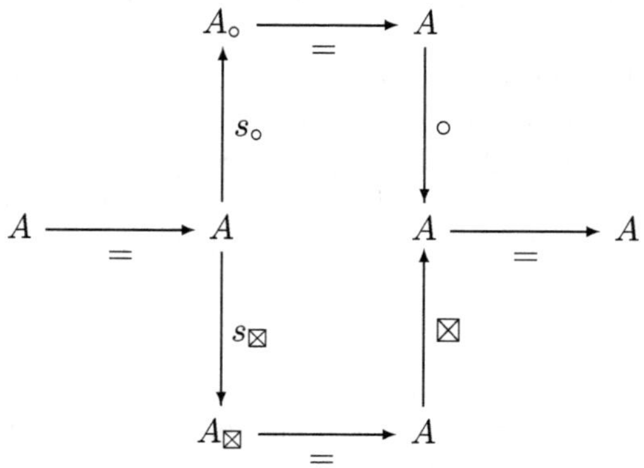

(11)

where A stands for the set of actions available to the robot, $A_\circ$ and $A_\boxtimes$ stand for the sets of actions representing primitive motions used in composition and expansion, respectively.

The product like part (in the lefthand part) of diagram (11) represents the supervision strategy that determines which of the action sets, $A_\circ$ or $A_\boxtimes$, is selected at each event. This strategy is mission dependent and is detailed in the $s_\circ$ and $s_\boxtimes$ maps.

4 Experiments

This section presents two simulation experiments using a team of three unicycle robots and two experiments with a single real unicycle robot. These experiments aim at illustrating the motion of (i) single robots under the control of the righthand side of architecture (11), and (ii) multiple robots interacting using the objects defined in the system.

Missions are specified through goal regions the robots have to reach. This resembles common human-robot interaction, with the human specifying a motion trend the robot has to follow through the goal region. Furthermore, intuitive bounding regions are easily constructed from them. The human operator is thus not explicitely present in the experiments. Nevertheless, as results clear from the experiments with a real robot, there is no lack of generality.

Interactions among the teammates are made through the action bounding regions, that is each robot has access to the bounding region being used by any teammate.

The robots use a single action, denoted $a_\circ$, defined as

$$a_\circ = \begin{cases} F(q) = (G - q) \cap H(q) \\ B(q) = \{p | p = q + \alpha G, \quad \alpha \in [0,1]\} \end{cases} \tag{12}$$

where q is the configuration of the robot, G stands for a goal set, and $H(q)$ stands for the set of admissible motion directions at configuration q (easily obtained from the well known kinematics model of the unicycle). This action simply yields a motion direction pointing straight to the goal set from configuration q. Whenever there are no admissible controls, i.e., the set of admissible motion directions that lead straight to the goal region is empty, $F(q) = \emptyset$, the bounding region is expanded using the action

$$a_\boxtimes = \begin{cases} H(q) \cup \left\{ \begin{array}{c}\text{set of motion}\\ \text{directions}\end{array} \mid d(H, G - q) \to 0 \right\} \\ \qquad\qquad\qquad\qquad\qquad\qquad \text{if } q \ni B \backslash M \\ F(q) \qquad\qquad\qquad\qquad\qquad\quad \text{otherwise} \end{cases}$$

where $d(,)$ stands for a distance between the sets in the arguments. This action corresponds to having $H(q)$ converging to $G - q$. Generic algorithms already described in the literature can be used to obtain this sort of convergence (see for instance [16]). The same set of actions and the supervisor maps are used by all the robots.

In the simulation experiments robots have access to a synthetic image representing a top view of the environment and start their mission at the lower part of the image. The irregular shape in the upper part of the images shown in the following sections represents the raw mission goal region to be reached by the robots. Basic image processing techniques are used to extract a circle goal region, centered at the centroid of the convex hull of the countour of these shapes. This circle is shown in light colour superimposed on the corresponding shape. It is worth to remark that the use of this synthetic image only simplifies the detection of a goal region, having no impact on the performance of the proposed architecture.

The simulations were implemented as a multi-thread system. Each robot simulator thread runs at an average 100 Hz whereas the architecture thread runs at an average 80 Hz. Experiments data is recorded by an independent thread at 100 Hz.

In the real experiment, the image data is used only to extract a goal region for the robot to reach. The navigation relies on odometry data.

4.1 Simulation – Independent Robots

The purpose of this experiment is to assess the behavior of the team when each robot operates isolatedly. Each robot tries to reach the same goal region (in the upper right-hand side of the image). No information is exchanged between the teammates and no obstacle avoidance behaviors are considered. In the simulation experiments robots 1 and 2 start with a 0 rad orientation whereas robot 0 starts with π rad orientation.

Figure 1 shows the trajectories generated by the robots. The symbol $\circ$ marks the position of each robot along the mission. Dashed lines connect the time related robot position marks.

The supervisor maps were chosen as,

$$\begin{aligned} s_{\circ} &\equiv a_{\circ} \;\text{ if }\; H(q) \cap G - q \neq \emptyset \\ s_{\boxtimes} &\equiv a_{\boxtimes} \;\text{ if }\; H(q) \cap G - q = \emptyset \end{aligned} \tag{13}$$

The oscilations in all the trajectories in Fig. 1 result from the algorithm used to expand the action bounding region at the initial time as the initial configuration of any of the robots does not point straight to the goal set.

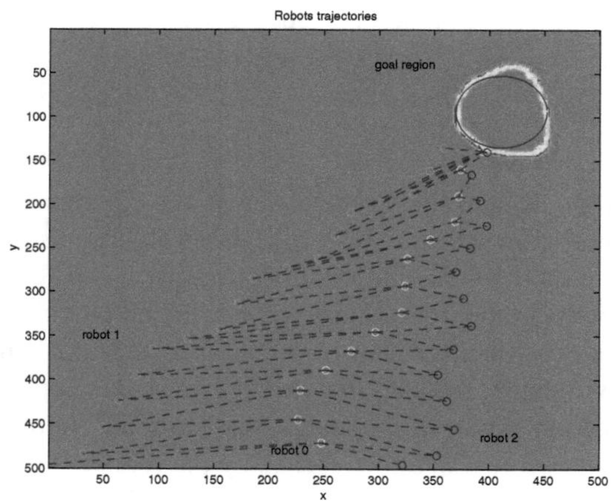

Fig. 1 Independent robots.

4.2 Simulation - Interacting Robots

This experiment illustrates the use of bounding regions as a basic form of interaction among robots, with similarities with common forms of interaction among humans.

The robots have no physical dimensions but close positions are to be avoided during the travel. This condition is relaxed as soon as the robot reaches the goal set, that is, once a robot reaches the goal region and stops the others will continue trying to reach the goal independently of the distance between them.

The supervisor maps are identical to the previous mission. However, the co-domain of $s_{\circ}$, the set $A_{\circ}$, is changed to account for the interactions among the robots.

$$A_{\circ}^{i} = \begin{cases} F^{i}(q^{i}) = \left(G^{i} - q^{i}\right) \cap H^{i}(q^{i}) \\ B^{i}(q^{i}) \backslash \cup_{j \neq i} B^{j}(q^{j}) \end{cases} \tag{14}$$

where the superscript identifies the robot the action belongs to.

Actions as (14) use only a subset of the motion directions of those in (12). The bounding region is constructed from that in (12) including the information from the motion trend of the teammates by removing any points belonging also to that of the teammates. The original mission goal is replaced by intermediate goal regions, G^{i}, placed inside the new bounding region. As the robots progress towards these intermediate goals they also approach the original mission goal. A potential drawback of this simple strategy is that the smaller bounding region constrains significatively the trajectories generated.

Figure 2 illustrates the behavior of the team when the robots interact using the actions (14). The global pattern of the trajectories shown is that of a line formation. Once the robots approach the goal region the distance between them is allowed to decrease so that each of them can reach the goal.

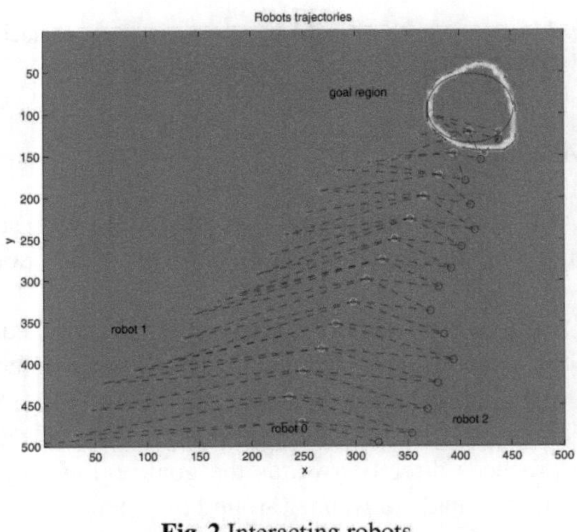

Fig. 2 Interacting robots.

Figure 3 illustrates the behavior of the team when a static obstacle is present in the environment. Similarly to common path planning schemes used in robotics, an intermediate goal region is chosen far from the obstacle such that the new action bounding region allows the robots to move around the obstacle.

From the beginning of the mission robots 0 and 1 find the obstacle in their way to the goal region and select the intermediate goal shown in the figure. After the initial stage where the robots use the expansion action aiming at aligning their trajectories with the direction of the goal. The interaction that results from the exchange of bounding regions is visible similarly as in the previous experiment. Robot 2 is not constrained by the obstacle and proceeds straight to the goal. Its influence in the trajectories of the teammates around the intermediate goal region is visible in the long dash lines connecting its position with those of the teammates.

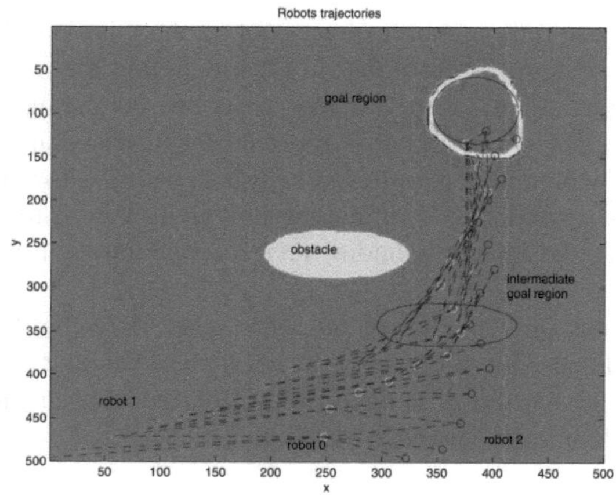

Fig. 3 Interacting robots in the presence of a static obstacle.

4.3 Real Robot Missions

These experiments, with a single robot, aims at validating the assessment of the framework done, in simulation, in Sects. 4.1 and 4.2. The supervisor maps are those in (13).

Figure 4 shows the Scout robot and the goal region as extracted using basic color segmentation procedures. No obstacles were considered in this experiment.

The corresponding trajectory in the workspace is shown in Fig. 5. The robot starts the mission with orientation $\pi/2$, facing the goal region, and hence maneuvering is not required. The robot proceeds directly towards the goal region. Figure 6 displays the trajectory when the robot is made to wander around the environment. Goal regions are defined at random instants and when detected by the robot its behavior changes to the goal region pursuing. The plot shows the positions of the robot when the targets were detected and when they are reached (within a 5 cm error distance). The visual quality of these trajectories is close to those obtained in simulation which in a sense validates the simulations.

Fig. 4 A goal region defined over an object in a laboratory scenario.

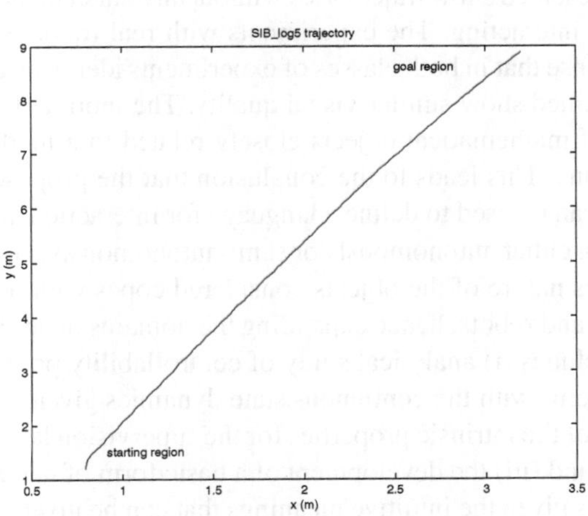

Fig. 5 Moving towards a goal region.

5 Conclusions

The chapter presented an algebraic framework to model HRI supported on semiotics principles. The framework handles in a unified way any interaction related to locomotion between a robot and its external environment.

The resulting architecture (see Diagram 11) has close connections with several other proposals in the literature namely with those where a low level control layer and a supervision layer can be identified.

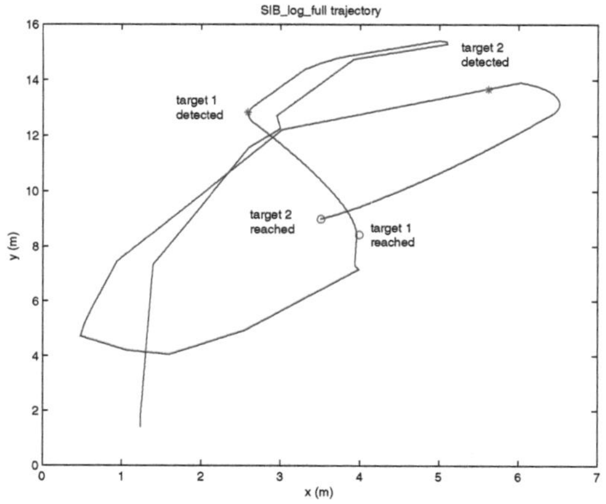

Fig. 6 Moving between multiple goals regions.

Although the initial configurations do not promote straight line motion, the simulation experiments presented show trajectories without any harsh maneuvering even when multiple robots are interacting. The experiments with real robots validate the simulation results in the sense that in both classes of experiments identical actions are used and the trajectories obtained show similar visual quality. The motion displayed is obtained using a small set of mathematical objects closely related to a model of the semiotics of human interactions. This leads to the conclusion that the proposed architecture and the objects therein can be used to define a language for interaction and control of multiple robots, operating either autonomously or semi-autonomously. For semi-autonomous robots, the semiotics nature of the objects considered copes with interactions between non skilled humans and robots, hence expanding the domains of application of robotics.

Future work includes (i) analytical study of controllability properties in the framework of hybrid systems with the continuous state dynamics given by differential inclusions, (ii) the study of the intrinsic properties for the supervision layer that may simplify design procedures, and (iii) the development of a basic form of natural language for interaction with robots given the intuitive meanings that can be given to the objects in the framework.

Acknowledgements

This work was supported by the FCT project POSI/SRI/40999/2001 - SACOR and Programa Operacional Sociedade de Informação (POSI) in the frame of QCA III.

References

1. Aylett, R., Barnes, D.: A multi-robot architecture for planetary rovers. In: Proceedings of the 5th European Space Agency Workshop on Advanced Space Technologies for Robotics & Automation. (1998) ESTEC, The Netherlands.

2. Huntsberger, T., Pirjanian, P., Trebi-Ollennu, A., Nayar, H., Aghazarian, H., Ganino, A., Garrett, M., Joshi, S., Schenker, P.: Campout: a control architecture for tightly coupled coordination of multi-robot systems for planetary surface exploration. IEEE Trans. Systems, Man & Cybernetics, Part A: Systems and Humans **33** (2003) 550–559. Special Issue on Collective Intelligence.
3. Albus, J.: A control system architecture for intelligent systems. In: Proceeding of the 1987 IEEE Intl. Conf. on Systems, Man and Cybernetics. (1987) Alexandria, VA.
4. Kortenkamp, D., Burridge, R., Bonasso, R., Schrekenghost, D., Hudson, M.: An intelligent software architecture for semi-autonomous robot control. In: Proceedings 3rd Int. Conf. on Autonomous Agents – Workshop on Autonomy Control Software. (1999)
5. Peterson, J., Hudak, P., Elliott, C.: Lambda in motion: Controlling robots with Haskell. In: First International Workshop on Practical Aspects of Declarative Languages (PADL). (1999)
6. Hager, G., Peterson, J.: Frob: A transformational approach to the design of robot software. In: Proceedings of the 9th Int. Symp. of Robotics Research, ISRR'99. (1999) Snowbird, Utah, USA, October 9–12.
7. Goguen, J.: Semiotic morphisms, representations and blending for interface design. In: Proceedings of the AMAST Workshop on Algebraic Methods in Language Processing. (2003) Verona, Italy, August 25–27.
8. Makatchev, M., Tso, S.: Human-robot interface using agents communicating in an XML-based markup language. In: Procs. of the IEEE Int. Workshop on Robot and Human Interactive Communication. (2000) Osaka, Japan, September 27–29.
9. Nicolescu, M., Matarić, M.: Learning and interacting in human-robot domains. IEEE Transactions on Systems, Man, and Cybernetics, Part A: Systems and Humans **31** (2001) 419–430
10. Chandler, D.: Semiotics, The basics. Routledge (2003)
11. Malcolm, G., Goguen, J.: Signs and representations: semiotics for user interface design. In: Proceedings Workshop in Computing, Springer (1998) 163–172 Liverpool, UK.
12. Neumüller, M.: Applying computer semiotics to hypertext theory and the world wide web. In Reich, S., Anderson, K., (eds.): Proceedings of the 6th Int. Workshop and the 6th Int. Workshop on Open Hypertext Systems and Structural Computing. Lecture Notes in Computer Science, Springer-Verlag (2000) 57–65
13. Codognet, P.: The semiotics of the web. In: Leonardo. **35**(1). The MIT Press (2002)
14. Meystel, A., Albus, J.: Intelligent Systems: Architecture, Design, and Control. Wiley Series on Intelligent Systems. J. Wiley and Sons (2002)
15. Smirnov, G.: Introduction to the Theory of Differential Inclusions. Volume 41 of Graduate Studies in Mathematics. American Mathematical Society (2002)
16. Sequeira, J., Ribeiro, M.: Hybrid control of semi-autonomous robots. In: Proceedings of the 2004 IEEE/RSJ Int. Conf. on Intelligent Robots and Systems. (2004) Sendai, Japan.

PART III

Signal Processing, Systems Modeling and Control

Multimodelling Steps for Free-Surface Hydraulic System Control

Eric Duviella, Philippe Charbonnaud and Pascale Chiron

Laboratoire Génie de Production, EA 1905 69042 Heidelberg, Germany
Ecole Nationale d'Ingénieurs de Tarbes
47, avenue d'Azereix, BP 1629
65016 Tarbes Cedex, France
{Eric.Duviella,Philippe.Charbonnaud,Pascale.Chiron}@enit.fr

Abstract. The chapter presents multimodelling steps of free-surface hydraulic system for the design of control strategies. This method makes it possible to represent, simply and accurately, the non-linear hydraulic system dynamics with large operating conditions. The multimodelling steps are performed in order to lead to the determination of a finite number of models. The models are selected online by the minimization of a quadratic criterion. It is an interesting alternative to the use of Saint Venant partial differential equations because it allows the design, the tuning and the validation of control strategies.The evaluation of the proposed method is carried out by simulation within the framework of a canal with trapezoidal profile showing its effectiveness.

Keywords. Multimodelling, operating modes, online selection, hydrographic network.

1 Introduction

Hydrographic networks are systems geographically distributed conveying water quantities gravitationally. They are composed of open-surface hydraulic systems (canals, rivers, etc.) which are used to satisfy the demands related to human activities (ecological discharge, agricultural and industrial needs, drinkable water, etc.). According to the recognized importance of water resource, the efficient management of these systems is essential today. It requires the proposal, the design and the tuning of control strategies through simulation, before their implementation on real systems. The free-surface hydraulic system dynamics is characterized by nonlinearities and important transfer delays. Although the Saint Venant Partial Differential Equations (PDE) accurately represent hydraulic system dynamics [2] [10], their resolution involves numerical approaches according to discretization schemes which are rather complex to implement, especially for designing and tuning a control strategy. The PDE simplification and linearization around an operating point lead to simplify models of the hydraulic system dynamics [7]. In the literature, most authors have proposed control strategies based on the PDE linearization [9]. However, the accuracy of these models is only acceptable on restricted interval around the operating point, and their use on large operating conditions requires robust controller design, as proposed in [8].

The representation of the nonlinear systems with variable transfer delays involves the identification problem of a model with variable parameters or multiple models. In the literature, multimodelling approaches for the predictive control are described in [13] and in [12]. These methods are based on switching technics amongst several simple models. In the case of nonlinear systems with variable transfer delays, an algorithm for the estimation of the models most representative of the system dynamics is proposed in [14]. In these approaches, the number of models and their operating range are known a priori. The nonparametric modelling of the nonlinear system dynamics can also be carried out by Gaussian approaches [5]. The identification and control of open-channel systems using Linear Parameter Varying (LPV) models is proposed in [1] [15]. The hydraulic system dynamics is modelled by a first order differential equation with time delay. The parameters of the LPV model have been identified using a parameter estimation algorithm. These approaches require numerous adapted data.

In this chapter, the multimodelling steps which lead to the determination of a finite number of models, are proposed to design and tune control strategies for hydraulic systems subject to large operating conditions. In Sect. 2, the modelling of free surface hydraulic systems is detailed. The number of models and their validity boundaries are defined using the multimodelling steps described in Sect. 3. An online model selection criterion is presented in Sect. 4. In Sect. 5, the method is used to identify the dynamics of a canal with trapezoidal profile and the multimodelling steps are compared to the Saint Venant PDE showing its effectiveness.

2 Modelling of Free-Surface Hydraulic Systems

Open-channel systems are characterized by large sized, nonlinear dynamics and important transfer delays. They are generally broken down into several reaches which are located between two measurement points or two gates denoted herein G_i and G_{i+1} (see Fig. 1).

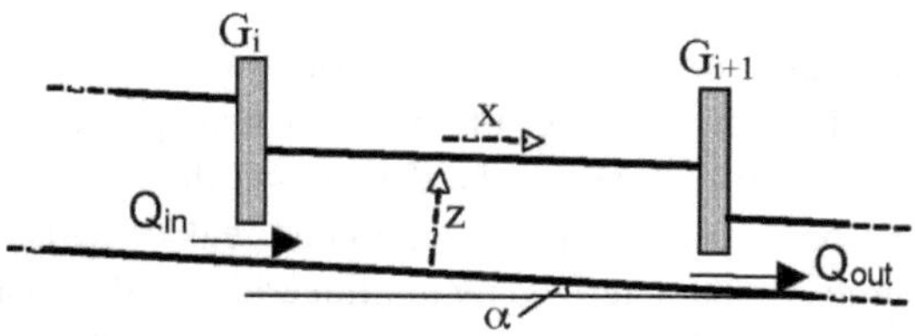

Fig. 1 Canal reach.

The diffusive wave equation [2], expressed by relation (1) can be used to represent the canal reach dynamics accurately.

$$\frac{\partial Q(x,t)}{\partial t} + C(Q,z,x)\frac{\partial Q(x,t)}{\partial x} - D(Q,z,x)\frac{\partial^2 Q(x,t)}{\partial x^2} = 0, \tag{1}$$

where $Q(x,t)$ is the reach flow discharge [m^3/s], $C(Q,z,x)$ the celerity coefficient [m/s] and $D(Q,z,x)$ the diffusion coefficient [m^2/s] expressed by:

$$\begin{cases} C(Q,z,x) = \dfrac{1}{L^2 \frac{\partial J}{\partial Q}} \left[\dfrac{\partial L}{\partial x} - \dfrac{\partial (LJ)}{\partial z} \right], \\ D(Q,z,x) = \dfrac{1}{L \frac{\partial J}{\partial Q}}, \end{cases} \tag{2}$$

where L is the water surface width and J is the friction slope. Several empirical formulas can be used to express J [6]. Generally, the Manning-Strickler formula (3) is used. When the flow depth is normal, J is considered equal to the canal slope α :

$$J = \frac{Q^2 n_M^2 P^{\frac{4}{3}}}{S^{\frac{10}{3}}}, \tag{3}$$

where n_M is the Manning coefficient associated with the hydraulic system considered (river, channel) and with his bed type (n_M is between 0.02 and 0.01 for a concrete canal). The Manning coefficient determination can be carried out from physical knowledge of the hydraulic system or by identification [11].

The diffusive wave equation (1) can be linearized around an operating discharge Q_e [7], and the identified celerity and diffusion parameters are denoted C_e and D_e.

$$\frac{dq(x,t)}{dt} + C_e \frac{dq(x,t)}{dx} - D_e \frac{d^2 q(x,t)}{dx^2} = 0, \tag{4}$$

where $Q = Q_e + q$. The discharge variation q from the reference discharge Q_e is flowed out with a mean speed of constant celerity C_e and is diffused with a constant diffusion D_e.

The linearization of the diffusive wave equation leads to a finite order transfer function:

$$F(s) = \frac{e^{-\tau s}}{1 + a_1 s + a_2 s^2}, \tag{5}$$

where the transfer function parameters a_1, a_2 and τ are calculated by the moment matching method as described in [4]:

$$\begin{cases} a_1 = \left(\dfrac{6XD_e^2}{C_e^5} + \sqrt{\dfrac{4X^2D_e^3}{C_e^9}\left(\dfrac{9D_e}{C_e} - 2X\right)} \right)^{\frac{1}{3}} \\ \quad + \left(\dfrac{6XD_e^2}{C_e^5} - \sqrt{\dfrac{4X^2D_e^3}{C_e^9}\left(\dfrac{9D_e}{C_e} - 2X\right)} \right)^{\frac{1}{3}}, \\ a_2 = \dfrac{2XD_e}{C_e^3}\left(1 - \dfrac{3D_e}{a_1 C_e^2}\right), \\ \tau = \dfrac{X}{C_e} - a_1. \end{cases}$$

The model order is determined according to the adimensional coefficient C_M [10]:

$$C_M = \frac{2C_e X}{9D_e}, \tag{6}$$

where X is the reach length. The canal reach dynamic is modelled by:

- a second order plus delay transfer function (5) when $C_M > 1$,
- a first order plus delay transfer function when $\frac{4}{9} < C_M \leq 1$, with $a_2 = 0$,
- a first order transfer function when $C_M \leq \frac{4}{9}$, with $a_2 = 0$ and $\tau = 0$.

The linearization of the diffusive wave model leads to the identification of free-surface hydraulic system dynamics. The model validity decreases as the operating point of system moves away from the identification point. Accurate representation of open-surface hydraulic system dynamic with large operating conditions requires a multimodelling method, which involves to determine the number of necessary models, i.e. Operating Modes (OM), and their validity boundaries.

3 Multimodelling Steps of Free-Surface Hydraulic Systems

The process model M of the hydraulic system is decomposed into a finite class of linear models $M = \{M_1, M_2, \ldots, M_n\}$, where the ith linear model of the hydraulic system is denoted M_i, and $n = \text{card}(M)$. The celerity coefficient values are used in order to determine the number n of OM.

The open-surface hydraulic system dynamics with large operating conditions are represented by the following state space relations:

$$\begin{cases} \dot{x} = \mathbf{A}_i x(t) + \mathbf{B}_i u(t-\tau), \\ y = \mathbf{C}_i x(t), \end{cases} \tag{7}$$

where u and y are respectively the input and output variables, x the state. Each of the matrices $\mathbf{A}_i$, $\mathbf{B}_i$ and $\mathbf{C}_i$ is computed for the model M_i associated to the ith OM. They are expressed, according to the transfer function parameters (5), by the relations:

$$\mathbf{A}_i = \begin{bmatrix} -\frac{a_{1i}}{a_{2i}} & 1 \\ -\frac{1}{a_{2i}} & 0 \end{bmatrix}, \mathbf{B}_i = \begin{bmatrix} 0 \\ \frac{1}{a_{2i}} \end{bmatrix} \text{ and } \mathbf{C}_i = \begin{bmatrix} 1 & 0 \end{bmatrix}.$$

The celerity coefficient can be considered as the most representative parameter of the open-channel system dynamics. Therefore, a model is considered as available as soon as the error on the celerity coefficient is inferior to a fixed percentage Π_c. The validity boundaries are defined for each OM. The value of parameter Π_c is chosen according to the system dynamics. The multimodelling steps are described by an algorithm (see Table 1), where the operating modes are determined starting with the medium celerity C_{med}. This one is computed with the parameters C_{min} and C_{max} corresponding, respectively, to the minimum and maximum discharges of the system Q_{min} and Q_{max}.

This algorithm leads to the determination of the celerity coefficient C_{id_r} used to identify the rth linear model and the OM validity boundaries C_{inf_r} and C_{sup_r}. According to C_{id_r}, the water elevation z_{id_r} of each rth OM is determined, with one millimeter precision, by the digital resolution of the relation (8) with Newton method.

$$C_{id} = \frac{\sqrt{J}S^{\frac{5}{3}}}{nP^{\frac{2}{3}}L^2}\left[-\frac{1}{2}\frac{\partial L}{\partial z} - \frac{L}{3P}\left(2\frac{\partial P}{\partial z} - 5\frac{P}{S}\frac{\partial S}{\partial z}\right)\right], \tag{8}$$

Table 1 Multimodelling algorithm.

Input : $C_{\max}, C_{\min}, \Pi_C$
Output : $C_{id_r}, C_{\sup_r}, C_{\inf_r}$,

$$C_{med} = \frac{C_{\max} + C_{\min}}{2},$$

$r = 1$,

For i : $\left[\frac{\ln \frac{C_{\min}}{C_{med}}}{\ln \frac{(1+\Pi_C)}{(1-\Pi_C)}} \right]$ to $\left\lfloor \frac{\ln \frac{C_{\max}}{C_{med}}}{\ln \frac{(1+\Pi_C)}{(1-\Pi_C)}} \right\rfloor$,

$$C_{id_r} = \left(\frac{1+\Pi_C}{1-\Pi_C} \right)^i C_{med},$$

$$C_{\sup_r} = (1+\Pi_C) \left(\frac{1+\Pi_C}{1-\Pi_C} \right)^i C_{med},$$

$$C_{\inf_r} = (1-\Pi_C) \left(\frac{1+\Pi_C}{1-\Pi_C} \right)^i C_{med},$$

$r++$,
EndFor.

Where L, P and S parameters are expressed, according to the geometrical characteristics, interms of the water elevation z_{id}. The computation of a water elevation z_{id_r} allows for the determination of the diffusion coefficient D_{id_r} (2). Finally, the matrices A_i, B_i and C_i are computed using C_{id_r} and D_{id_r} values.

The multimodelling method is used to identify the free-surface hydraulic system dynamics with several OM. The discharge boundary conditions $Q_{\inf_i}$ and $Q_{\sup_i}$ are computed according to the relation:

$$Q = \frac{\sqrt{J} S^{\frac{5}{3}}}{n_M P^{\frac{2}{3}}}, \tag{9}$$

where P and S parameters depend on the water elevation boundary conditions $z_{\inf_r}$ and $z_{\sup_r}$.

The detailed algorithm can be used for various hydraulic system profiles, i.e. rectangular, trapezoidal and circular profiles. The multimodelling steps were applied to a dam gallery with a circular profile to carry out its predictive control [3] . In order to simulate and implement control strategies for hydraulic systems with large operating conditions, it is necessary to propose an online model selection criterion.

4 Online Selection Criterion of Operating Mode

An analytical expression of the hydraulic system output y can be approached as:

$$y \simeq \hat{y} = \sum_{j=1}^{n} \delta_m^j . y_j, \tag{10}$$

where n is the number of OM, m denotes the actual OM, y_j is the response of the jth model M_j, and δ_m^j is equal to 1, if $m = j$ and to 0, otherwise. The main problem of OM detection lies in the real-time estimation of m at the boundary between two OM, i.e., the current behavior of the physical process. The selection criterion has to figure out the current OM value. For that, J_j, $0 < j \leq n$, is defined for each OM and computed at each sample period kT_s.

$$J_j(k) = \frac{1}{N-1} \sum_{i=0}^{N-1} \varepsilon_{j,i}(k), \tag{11}$$

where N is the size of a sliding window, $\varepsilon_{j,i}(k)$ is the jth modeling error and:

$$\varepsilon_{j,i}(k) = (y(k-i) - y_j(k-i))^2. \tag{12}$$

The multi-model output recursive square error criterion $\mathbf{J}(k) = [J_1(k)\ J_2(k) \dots J_j(k) \dots]$ is computed with the recursive formula:

$$J_j(k) = J_j(k-1) + \frac{1}{N-1}(\varepsilon_{j,0}(k) - \varepsilon_{j,0}(k-N)). \tag{13}$$

At each sampling period, a minimization of the criterion given by (11) is carried out to determine the right model. The correspondant OM number is identified and given by:

$$d(k) = \underset{1 \leq j \leq n}{\arg\min} J_j(k)\}, \tag{14}$$

where d is the decision vector.

At the starting time, it is assumed that $d(0) = 1$. The decision time is defined by:

$$t_d(k) = \{kT_s,\ d(k) \neq d(k-1)\}, \tag{15}$$

The multimodelling steps which lead to determine the different operating modes, associated to online selection criterion, make it possible to represent the hydraulic systems dynamics on the whole of operating range. In the following section, the multimodelling steps and online selection criterion are illustrated within the framework of a canal with trapezoidal profile. The multimodelling approach is then compared to a method based on Saint Venant equations.

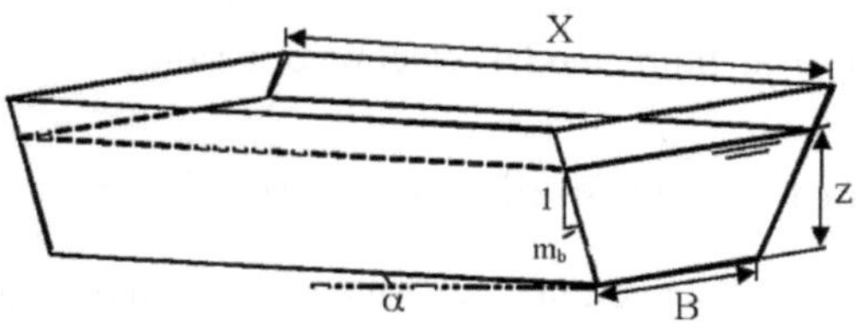

Fig. 2 Canal reach with trapezoidal profile.

Table 2 Canal reach downstream boundary conditions.

Q [m^3/s]	0.5	1.4	2.6	4.2	6.2	8.7	10
z [m]	0.3	0.5	0.7	0.9	1.1	1.3	1.4

5 Application to a Canal Reach with Trapezoidal Profile

The application of the multimodelling steps was carried out on a reach of the Neste canal with a trapezoidal profile, located in the French southwestern region (see Figure 2). Its geometrical characteristics are given below:
- bottom width $B = 2.85$ m,
- average fruit of the banks $m_b = 0.99$,
- profile length $X = 1732$ m,
- Manning coefficient $n_M = 0.02$,
- reach slope $\alpha = 0.13\%$,
- minimal discharge $Q_{\min} = 1$ m^3/s,
- maximal discharge $Q_{\max} = 10$ m^3/s.

This canal reach was both modelled by the Saint Venant equations, and by the proposed multimodelling step. The resolution of Saint Venant equations is carried out using the downstream boundary conditions (see Table 2) with the software SIC[1]. This one allows the dynamic simulation of rivers and canals based on the Preissmann scheme. Among the resolution algorithms proposed, the Newton algorithm which gives the best performances in spite of a longer simulation time, was chosen. The time and space steps must be tuned such as the Courant number (16) is equal to one in order to avoid the instability periods during simulation.

$$Cr = \frac{dt}{dx}(V + C), \tag{16}$$

where V is the mean velocity of the flow expressed by $V = \frac{Q}{S}$.
The multimodelling algorithm was applied (see Table 1) with tolerated error Π_c of 10% leads to the identification of three operating modes. In the case of a hydraulic system with a trapezoidal profile, celerity C_e and diffusion D_e are expressed by:

$$\begin{aligned} C_e &= \frac{Q_e}{L^2}\left[-m_b + \frac{L}{3}\left(\frac{2B}{Pz} + \frac{5L}{S} - \frac{2}{z}\right)\right], \\ D_e &= \frac{Q_e}{2LJ}, \end{aligned} \tag{17}$$

with:
- $L = B + 2m_b z$,
- $S = (B + m_b z)z$,
- $P = B + 2z\sqrt{1 + m_b{}^2}$.

[1] SIC user's guide and theorical concepts. CEMAGREF, Montpellier, 1992. http://canari.montpellier.cemagref.fr/

Table 3 Identified discharge Q_{id_r}, operating range Ω_r, and parameters z_{id_r}, C_{id_r} and D_{id_r} of each OM.

Q_{id_r} [m^3/s]	Ω_r [m^3/s]	z_{id_r}	C_{id_r}	D_{id_r}
1.5	[1 ; 2.2[	0.469	1.3	148
3.4	[2.2 ; 5.3[	0.775	1.6	300
8.7	[5.3 ; 10]	1.320	1.9	617

Table 4 Identified discharge Q_{id_r}, operating range Ω_r, and parameters a_{1i}, a_{2i} and τ_i of each OM.

Q_{id_r} [m^3/s]	Ω_r [m^3/s]	a_{1i}	a_{2i}	τ_i
1.5	[1 ; 2.2[	742	154520	608
3.4	[2.2 ; 5.3[	736	135470	369
8.7	[5.3 ; 10]	678	77690	226

The three operating modes, the correspondant identified discharge Q_{id_r}, operating range Ω_r, and parameters z_{id_r}, C_{id_r} and D_{id_r} are given in Table 3. Identified parameters a_{1i}, a_{2i} and τ_i are displayed in Table 4.

In order to show the operating mode identifiaction steps, parameters z_{id_r}, C_{id_r} and D_{id_r} are represented in Fig. 3, and parameters a_1, a_2 and τ are displayed in Fig. 4 according to the discharge Q. The values of each parameter were calculated beforehand for each discharge of the operating range, i.e. $[1, 5]$, with a step of 1 m^3/s. These values are represented by points in Figs. 3 and 5.

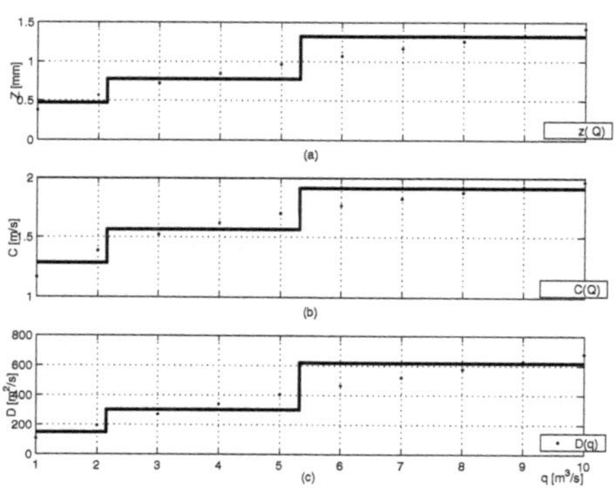

Fig. 3 Parameters Z, C and D according to the identified OM.

The evaluation of the multimodelling approach is carried out by comparison with the Saint Venant PDE. Figure 5 shows the responses to steps around the discharges used for the parameters identification of the three OM. The setpoints are in bold continuous line, the outputs resulting from SIC are also in bold continuous line, those resulting

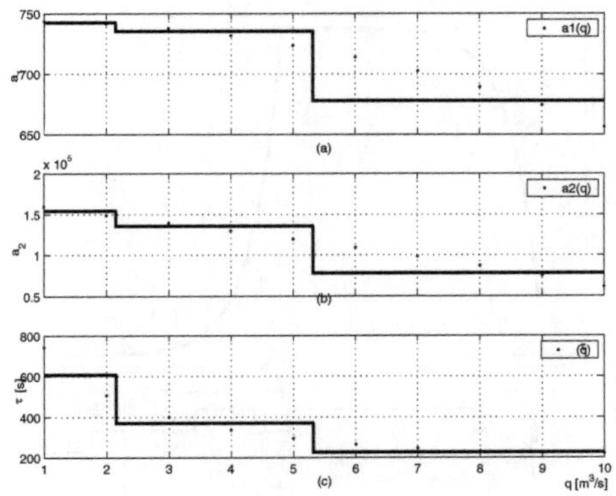

Fig. 4 Parameters a_1, a_2 and τ according to the identified OM.

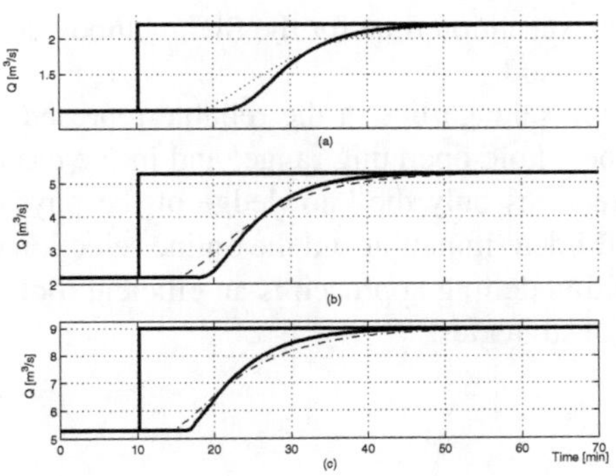

Fig. 5 Step responses arround (**a**) 1.5 m^3/s, (**b**) 3.5 m^3/s and (**c**) 8.5 m^3/s, resulting from SIC (bold continuous line), from M_1 (dot line), M_2 (dashed line) and M_3 (dash-dot line).

from the first model M_1 are in dot line, those resulting from the second M_2 are in dashed line, and finally those resulting from the third M_3 are in dash-dot line.

Figure 5a shows the M_1 and Saint Venant step responses of 1.2 m^3/s starting from a discharge of 1 m^3/s, Figure 5b, the step response of 3.1 m^3/s starting from a discharge of 2.2 m^3/s, and Figure 5c, the step response of 3.7 m^3/s starting from a discharge of 5.3 m^3/s. According to the simulation results, the modelled dynamics are close to those from SIC for each OM. Then, the comparison between the two modelling approaches is carried out by simulation on the whole operating range of the canal reach. The setpoint input is represented in bold continuous line in Figure 6a. It corresponds to setpoints with discharge amplitude from 1 to 9 m^3/s. The outputs resulting from the two approaches are very similar on the totality of the canal reach operating range. Some discrepancies between these outputs appear around 8 m^3/s when the discharge increase. The maximum output error between these two approaches is reached for 7.5 m^3/s and corresponds to an error percentage of 4.2%. These differences are sufficiently weak to conclude on the effectiveness of the multimodelling approach. Moreover, for

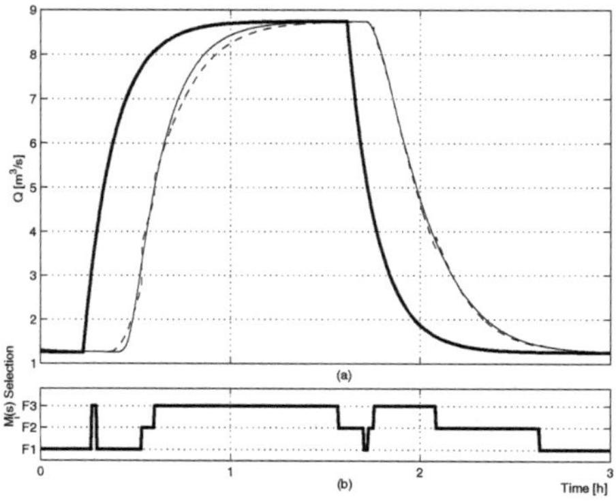

Fig. 6 (**a**) Tracking signal (bold continuous line), SIC response (continuous line) and multimodelling response (dashed line) and (**b**) β the selected model.

this simulation case, the excecution time for the SIC method was ten times longer than for the multimodelling method.

The multimodelling advantage lies in the reliable representation of the hydraulic systems dynamics on the whole operating range, and in the easiness of its implementation. This approach requires only the knowledge of the physical parameters of the hydraulic system, the OM determination and the online selection criterion implementation. Moreover, the multimodelling approach is an efficient tool for the design and the tuning of reactive control strategies.

6 Conclusions

Multimodelling steps of hydraulic systems subject to large operating conditions were proposed. They make it possible to identify a finite number of operating modes reproducing the hydraulic system dynamics, accurately. This multimodelling method based on an acceptable percentage of error on the celerity coefficient allows for the determination of the number of models and their corresponding validity boundaries. The online operating mode selection is based on the minimization of a quadratic criterion.

The multimodelling steps and the online operating modes selection criterion have been applied within the framework of a canal reach with a trapezoidal profile. The canal dynamics were identified by a multimodel and by a discretization scheme of the Saint Venant equations. The responses of theses models were compared, showing the multimodelling approach effectiveness.

References

1. Bolea, Y., Puig, V., Blesa, J., Gomez, M., and Rodellar, J. (2004). An LPV model for canal control. In *10th IEEE International Conférence on Methods and Models in Automation and Robotics, MMAR*, 30 August–2 September, Miedzyzdroje, Poland, 1:659–664.
2. Chow, V. T., Maidment, D. R., and Mays, L. W. (1988). *Applied Hydrology*. McGraw-Hill, New York, Paris.

3. Duviella, E., Chiron, P., and Charbonnaud, P. (2006). Multimodélisation des systèmes hydrauliques à surface libre. *Conférence Francophone de Modélisation et Simulation, MOSIM'06*, Rabat (Maroc), 3–5 avril.
4. Georges, D. and Litrico, X. (eds) (2002). *Automatique Pour la Gestion Des Ressources En Eau.* Hermès Science Publications, Lavoisier, 11 rue Lavoisier, 75008 Paris.
5. Gregorcic, G. and Lightbody, G. (2002). Gaussian processes for modelling of dynamic nonlinear systems. In *Proceedings of the Irish Signals and Systems Conference, ISSC'02*, Cork, Ireland, 25–26 Juin 2002.
6. Kovacs, Y. (1988). *Modèles de Simulation D'écoulement Transitoire En Reseau D'assainissement.* PhD thesis, ENPC - CERGRENE.
7. Litrico, X. and Georges, D. (1999a). Robust continuous-time and discrete-time flow control of a dam-river system. (i) modelling. *Applied Mathematical Modelling*, 23:809–827.
8. Litrico, X. and Georges, D. (1999b). Robust continuous-time and discrete-time flow control of a dam-river system. (II) controller design. *Applied Mathematical Modelling*, 23:829–846.
9. Malaterre, P., Rogers, D., and Schuurmans, J. (1998). Classification of canal control algorithms. *Journal of Irrigation and Drainage Engineering*, 124(1):3–10.
10. Malaterre, P.-O. and Baume, J.-P. (1998). Modeling and regulation of irrigation canals: Existing applications and ongoing researches. *IEEE International Conference on Systems, Man, and Cybernetics*, 4:3850–3855.
11. Ooi, S. K., Krutzen, M., and Weyer, E. (2003). On physical and data driven modelling of irrigation channels. *Control Engineering Practice*, 13:461–471.
12. Özkan, L. and Kothare, M. V. (2005). Stability analysis of a multi-model predictive control algorithm with application to control of chemical reactors. *Journal of Process Control, In Press, Corrected Proof.*
13. Palma, F. D. and Magni, L. (2004). A multi-model structure for model predictive control. *Annual Reviews in Control*, 1, 28:47–52.
14. Petridis, V. and Kehagias, A. (1998). A multi-model algorithm for parameters estimation of time-varying nonlinear systems. *Automatica*, Elsevier Science Ltd, 34(4):469–475.
15. Puig, V., Quevedo, J., Escobet, T., Charbonnaud, P., and Duviella, E. (2005). Identification and control of an open-flow canal using LPV models. *CDC-ECC'05, 44th IEEE Conference on Decision and Control and European Control Conference*, Seville, Spain, 12–15 December 2005, pp. 1893–1898.

Model-based Reconstruction of Distributed Phenomena using Discretized Representations of Partial Differential Equations

Felix Sawo, Kathrin Roberts and Uwe D. Hanebeck

Intelligent Sensor-Actuator-Systems Laboratory
Institute of Computer Science and Engineering
Universität Karlsruhe (TH)
Karlsruhe, Germany
{sawo,roberts}@ira.uka.de, uwe.hanebeck@ieee.org

Abstract. This article addresses the model-based reconstruction and prediction of distributed phenomena characterized by partial differential equations, which are monitored by sensor networks. The novelty of the proposed reconstruction method is the systematic approach and the integrated treatment of uncertainties, which occur in the physical model and arise naturally from noisy measurements. By this means it is possible not only to reconstruct the entire phenomenon, even at non-measurement points, but also to reconstruct the complete density function of the state characterizing the distributed phenomenon. In the first step, the partial differential equation, i.e., *distributed*-parameter system, is spatially and temporally decomposed leading to a *finite*-dimensional state space form. In the next step, the state of the resulting *lumped*-parameter system, which provides an approximation of the solution of the underlying partial differential equation, is dynamically estimated under consideration of uncertainties. By using the estimation results, several additional tasks can be achieved by the sensor network, e.g. optimal sensor placement, optimal scheduling, model improvement, and system identification. The performance of the proposed model-based reconstruction method is demonstrated by means of simulations.

Keywords. Stochastic systems, Baysian estimation, model-based reconstruction, distributed phenomenon, environmental monitoring, sensor-actuator-network.

1 Introduction

Recent developments in wireless sensor network technology and miniaturization of sensor nodes make it possible to exploit sensor networks [14] [18] for monitoring the natural environment [22] [4]. Such sensor networks can be used in industrial, medical, urban, and many other environments. Furthermore, applying sensor networks can provide new data for environmental science, such as climate models, as well as growth and reproduction of coral reefs indicated by physical environmental factors, e.g. temperature, water movement, salt concentration, and pollutant concentration. Additional examples are reconstruction of fluid-flow in a thermal reactor [6], and reconstruction of the surface motion of a beating heart in minimally invasive surgery [16].

In practical implementations the sensor nodes are normally densely deployed either inside the phenomenon or very close to it. For such scenarios the number of sensor nodes should be as low as possible because of economical and energy reasons; the same holds for the measurement rates. In general, it can be stated the lower the measurement rate of the sensor nodes the higher their durability. Thus, a trade-off between energy costs and accuracy has to be found. In order to get meaningful and accurate information not only at the sensor nodes itself but also between these nodes, the *model-based* reconstruction of the distributed phenomenon is of major significance. By the consideration of additional background information in form of a physical model, the approach proposed in this chapter achieves accurate reconstruction results even for a low number of sensor nodes and even between the individual sensor nodes.

In general, physical phenomena can be classified into *distributed*-parameter systems and *lumped*-parameter systems. The key characteristic of a lumped-parameter system is that the state, which uniquely describes the system behavior, depends only on time, e.g. bird swarms or swarms of robots. Such systems can be conveniently described by a set of *ordinary* differential equations. On the other hand, the key characteristic of distributed-parameter systems is that the state not only depends on time but also on the location, e.g. irrotational fluid flow, heat conduction, and wave propagation [13] [19]. The behavior of distributed-parameter systems can be described by a set of *partial* differential equations.

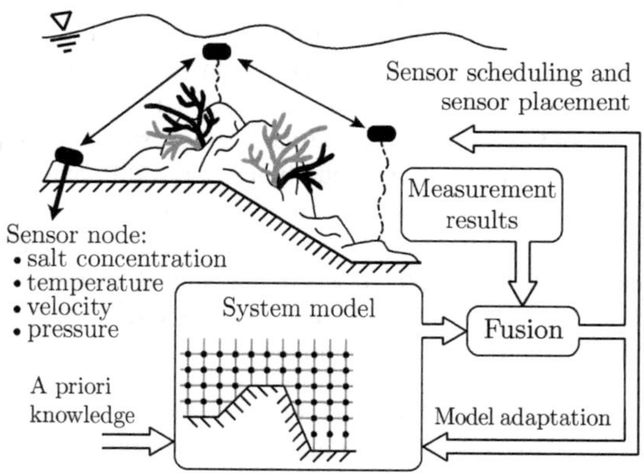

Fig. 1 Visionary scenario for the reconstruction and the observation of a distributed phenomenon by means of a sensor network: observation of a coral reef.

For the prediction of the state characterizing uniquely the distributed phenomenon we present a novel systematic approach, which allows the integrated treatment of uncertainties both occuring in the physical model and arising from noisy measurements. Basically, data from the sensor network in form of measurements as well as data from a physical model are used to derive the best possible estimate for the state of a distributed phenomenon; even at non-measurement points. The estimation results on the other hand can be used for finding the optimal placement and measurement time sequences for the sensor nodes similar to [3], and furthermore can be used for improving the parameter values of the used physical model, or for system identification to name just a few applications, (see Fig. 1).

Usually, a model-based reconstruction approach based on a distributed-parameter system description is quite complicated. The reason is for a Bayesian estimation method usually a lumped-parameter system description is used. To cope with this problem, the system description has to be converted from a distributed-parameter description to a lumped-parameter description. This means the partial differential equations characterizing the distributed phenomena has to be solved. Such equations can be solved for example by the finite-element method or the finite-difference method [2]. The unstructured nature of those methods allows a detailed geometrical description of complex geometries. Additionally, it is possible to solve even nonlinear partial differential equations. However, many parameters describing uniquely the state of the distributed phenomenon are necessary to achieve appropriate convergence.

On the other hand, there are modal analysis methods using a set of *global* expansion functions. These methods just need a few parameters for characterizing a smooth solution of the partial differential equation [19]. The drawback, however, is that for modal analysis global expansion functions can be found only for problems with simple boundary conditions.

Combining these two general approaches, leads to the so-called finite-spectral method [5] [7] [11] [12] [15]. Basically, this method approximates the solution within each element with a set of orthogonal polynomials, such as Legendre polynomials, or Chebyshev polynomials. Thus, it combines the geometrical flexibility of conventional finite-element methods with the exponential convergence rates associated with the modal analysis.

The novelty of this research work is the model-based observation of distributed phenomena by a sensor network under consideration of uncertainties both occuring in the physical model, i.e., system model, and arising from noisy measurements. Here, we extend and generalize our previous research work [19] in such a way that both the system model and the measurement model are derived by the finite-spectral method. Using this method, it turns out that even nonlinear phenomena with complex boundary conditions can be reconstructed and predicted in a systematic manner.

The remainder of this chapter is organized as follows. In Sect. 2, a rigorous formulation of the problem of the reconstruction of distributed phenomena characterized by partial differential equations is given. In Sects. 3 and 4, a spatial and temporal decomposition of the partial differential equation is performed. This allows the approximation of the partial differential equation (distributed-parameter system) by means of a finite-dimensional system in state space form (lumped-parameter system). Finally, in Sect. 5 the centralized estimator is derived and its performance is demonstrated by means of simulations: As an example, the temperature distribution in a heat conductor is reconstructed by means of a sensor network. The results presented in this chapter were prepublished in the *Proceedings of the 3rd International Conference on Informatics in Control, Automation and Robotics*, [20]. However explanations of the proposed model-based reconstruction method are presented in a considerably extended form in this chapter.

2 Problem Formulation

The main goal is to design a dynamic system model for the purpose of estimating the state of a distributed phenomenon monitored by a sensor network. A large number of distributed phenomena, such as irrotational fluid flow, heat conduction, and wave propagation [19], can be described by means of a set of linear partial differential equations.

In this article, for simplicity and brevity only a one-dimensional linear partial differential equation is used, although similar expressions can be found for the multi-dimensional nonlinear case. The one-dimensional linear partial differential equation is given by

$$\mathbb{L}(f) = \frac{\partial f(z,t)}{\partial t} - c\,\frac{\partial^2 f(z,t)}{\partial z^2} - s(z,t) = 0 \ , \tag{1}$$

where $f(z,t)$ denotes the solution of the partial differential equation, e.g. temperature at a certain location z and at a certain time t, $s(z,t)$ represents the source term, and c is a positive constant. Considering the solution in a domain $\Omega = \{z \,|\, 0 \leq z \leq L\}$, we assume following boundary conditions

$$f(z=L,t) = g_D \ , \qquad \frac{\partial f(z=0,t)}{\partial z} = g_N \ , \tag{2}$$

where g_D is referred to as a *Dirichlet* boundary condition and g_N, specifying a condition on the derivative, is the so-called *Neumann* boundary condition.

The aforementioned partial differential equation (1) describes the distributed phenomenon in an *infinite*-dimensional state space. However, in order to estimate and reconstruct the state of a distributed phenomenon by means of a Bayesian estimation approach, the partial differential equation has to be characterized by a *finite*-dimensional state space form. Thus, the partial differential equation (distributed-parameter system) has to be approximated by means of a finite-dimensional system in state space form, according to

$$\underline{x}_{k+1} = \mathbf{A}_k\,\underline{x}_k + \mathbf{B}_k\,\underline{u}_k + \underline{w}_k \ , \tag{3}$$

where $\underline{x}_k$ contains the individual states characterizing the time evolution of the distributed phenomenon and the matrix $\mathbf{A}_k$ maps the current state vector at time step k to the next state vector at time step $k+1$. The noise vector $\underline{w}_k$ contains both driving input $\Delta\underline{u}_k$ noises and subsumed endogenous uncertainties $\underline{d}_k$, e.g. modeling errors, according to

$$\underline{w}_k = \begin{bmatrix}\mathbf{I}\;\mathbf{I}\end{bmatrix}\begin{bmatrix}\mathbf{B}\Delta\underline{u}_k \\ \underline{d}_k\end{bmatrix} \ , \tag{4}$$

where $\mathbf{I}$ is the identity matrix. At this point it is worthwhile to mention that linear partial differential equations can always be approximated by a linear system model according to (3). However, the approximation of nonlinear partial differential equations leads to a nonlinear system model.

Furthermore, it is assumed that the measurements $\hat{\underline{y}}_k$ are related linearly to the state vector $\underline{x}_k$, according to

$$\hat{\underline{y}}_k = \mathbf{H}_k\,\underline{x}_k + \underline{v}_k \ , \tag{5}$$

where $\mathbf{H}_k$ denotes the measurement gain matrix and $\underline{v}_k$ represents the measurement uncertainties.

Once the system equation and the measurement equation are derived, the state vector $\underline{x}_k$ characterizing the distributed phenomenon can be estimated by means of the Kalman filter for the linear case [1], or nonlinear estimation procedures [9] for the nonlinear case.

3 Spatial Decomposition (ODE-System)

In this section, we present the spatial decomposition allowing the conversion of the partial differential equation (1) into a set of ordinary differential equations (ODE-System). It is well-known that the finite-difference, the finite-element, and the finite-spectral method may be used with the same numerical methodology described in the next section, namely the *Galerkin* formulation [2] [11].

3.1 Galerkin Formulation

The first basic ingredient for the conversion of the partial differential equation into a set of ordinary differential equations is the decomposition of the solution domain $\Omega = \{z \,|\, 0 \leq z \leq L\}$ into N_{el} subdomains Ω^e, which are the so-called *finite elements*. These subdomains are defined by the element boundary-nodes, z_i, for $i = 0, 1, \ldots, N_{dof} - 1$, where $z_0 = 0$ and $z_{N_{dof}-1} = L$. The elemental decomposition of the solution domain Ω into several subdomains Ω^e is visualized by means of an example in Fig. 2a.

For the second basic ingredient, it is assumed that the solution $f(z,t)$ in the solution domain Ω can be represented by a piecewise approximation $\tilde{f}(z,t)$ according to

$$\tilde{f}(z,t) = \sum_{i=0}^{N_{dof}-1} \Psi_i(z)\, \alpha_i(t) \ , \tag{6}$$

where $\Psi_i(z)$ are analytic functions called the *shape functions*, e.g. see Fig. 2b, and N_{dof} denotes the aforementioned degree of freedom of the resulting system description.

It is important to note that the individual shape functions $\Psi_i(z)$ are defined in the entire solution domain Ω. Furthermore, because of numerical advantages the shape functions Ψ_i should be constructed in such way that the so-called *cardinal interpolation property* is satisfied

$$\Psi_i(z_j) = \delta_{ij} \ , \tag{7}$$

where z_j represents the element end-nodes and δ_{ij} is Kronecker's delta,

$$\delta_{ij} = \begin{cases} 1 & i = j \\ 0 & i \neq j \end{cases} \tag{8}$$

In other words, the ith shape function Ψ_i takes the value of unity at the ith node, and remains zero at the other nodes. This essential property of the shape function can be easily recognized in Fig. 2b. Furthermore, because of the cardinal interpolation property, the coefficients $\alpha_i(t)$ have a physical interpretation in that they represent the approximated solution $\tilde{f}(z_j, t)$ at the element end-nodes z_j.

The essence of all spatial decomposition methods such as the finite-difference, the finite-element, and the finite-spectral method lies in the choice of the shape functions $\Psi_i(z)$. This function is to be specified in more detail later.

Due to the fact that the approximated solution $\tilde{f}(z,t)$ cannot satisfy the partial differential equation (1) everywhere in the region of interest, a *residual* R_Ω according to

$$\mathbb{L}\left(\tilde{f}(z,t)\right) = R_\Omega\left(\tilde{f}(z,t)\right) \neq 0 \tag{9}$$

remains. A common method known as the *Galerkin* formulation is based on finding a way to make this residual small in some sense; achieved by minimizing an appropriate number of *weighted integrals*. The most popular choice for the weighting functions is the *shape function* itself leading to following weighted integrals

$$\int_\Omega \Psi_i(z)\mathbb{L}\left(f(z,t)\right) dz = 0 \ , \tag{10}$$

with $i = 0, 1, \ldots, N_{dof} - 1$. Replacing both the solution function $f(z,t)$ and the input function $s(z,t)$ by the finite expansion (6), denoted by $\tilde{f}(z,t)$ and $\tilde{s}(z,t)$, respectively, leads to following weighted integrals over the solution domain Ω

$$\int_\Omega \Psi_i(z)\left[\frac{\partial \tilde{f}(z,t)}{\partial t} - c\,\frac{\partial^2 \tilde{f}(z,t)}{\partial z^2} - \tilde{s}(z,t)\right] dz = 0 \tag{11}$$

with $i = 0, 1, \ldots, N_{dof} - 1$. It can be stated that the minimization of this integral automatically leads to the best numerical approximation of the solution $f(z,t)$ of the partial differential equation [2].

Expression (11) can now be reformulated by using the rules of product differentiation, according to

$$\begin{aligned}\int_\Omega \Psi_i(z)\frac{\partial \tilde{f}(z,t)}{\partial t}dz = \int_\Omega \Psi_i(z)\tilde{s}(z,t)dz + \\ c\int_\Omega \left[\frac{\partial}{\partial z}\left(\Psi_i(z)\frac{\partial \tilde{f}(z,t)}{\partial z}\right) - \frac{d\Psi_i(z)}{dz}\frac{\partial \tilde{f}(z,t)}{\partial z}\right] dz\end{aligned} \tag{12}$$

Applying the fundamental theorem of calculus for the evaluation of the last term on the right-hand side leads to

$$\begin{aligned}\int_\Omega \Psi_i(z)\frac{\partial \tilde{f}(z,t)}{\partial t}dz = \int_\Omega \Psi_i(z)\tilde{s}(z,t)dz \\ - c\int_\Omega \frac{d\Psi_i(z)}{dz}\frac{\partial \tilde{f}(z,t)}{\partial z}dz + c\left[\Psi_i(z)\frac{\partial \tilde{f}(z,t)}{\partial z}\right]_{\partial\Omega_l}^{\partial\Omega_r}\end{aligned} \tag{13}$$

where the last term on the right-hand side contains the boundary conditions, i.e., $\partial\Omega_r$ and $\partial\Omega_l$ denote the right-hand and left-hand boundary, respectively. This weighted

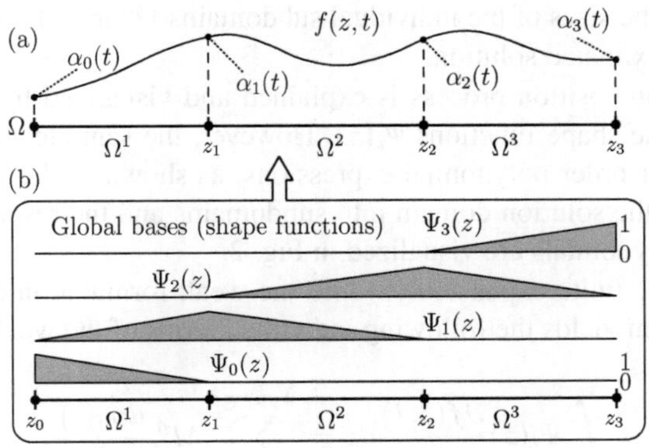

Fig. 2 The solution $f(z,t)$ of the partial differential equation is approximated by $\tilde{f}(z,t)$ depending on the shape functions $\Psi_j(z)$ and their weighting coefficients α_j(t). **(a)** Elemental decomposition of the solution domain Ω into several subdomains Ω^e. **(b)** Shape functions $\Psi_i(z)$ for the linear case.

residual statement is known as the *weak formulation* of the partial differential equation (1).

As it is shown in the next sections, by specifying the shape functions $\Psi_i(z)$ the weighted residual statement (13) can be reduced to a system of ordinary differential equations in terms of $\alpha_i(t)$, i.e., the semi-discrete version of problem (1),

$$\mathbf{M}_G\,\dot{\underline{x}}(t) = \mathbf{M}_G\,\underline{u}(t) - c\,\mathbf{D}_G\,\underline{x}(t) + \underline{b}^*(t) \tag{14}$$

In this system, $\mathbf{M}_G$ is called the global *mass* matrix and $\mathbf{D}_G$ is the global *diffusion* matrix, and $\underline{b}^*$ represents the boundary conditions. The vector $\underline{x}(t)$ and $\dot{\underline{x}}(t)$ are the state vectors of the unknown weighting coefficients $\alpha_i(z)$ and their derivatives

$$\underline{x}(t) = \left[\alpha_0(t), \alpha_1(t), \cdots, \alpha_{N_{dof}-1}(t)\right]^T \tag{15}$$

It is important to note that the entries of the matrices $\mathbf{M}_G$ and $\mathbf{D}_G$, and the state vector $\underline{x}(t)$ merely depend upon the choice of the shape functions $\Psi_i(z)$, as shown in the following sections.

The next sections are devoted to appropriate definitions of the shape functions $\Psi_i(z)$. First, the h-type expansion functions decomposing the solution domain into small subdomains are considered; $\Psi_i(z)$ are usually linear functions or simple polynomials. Then, based on this decomposition, it is possible to introduce additional supporting nodes within the subdomains; the so-called nodal p-type expansion functions using higher-order orthogonal polynomials. Finally, the shape functions $\Psi_i(z)$ can be defined by a set of orthogonal polynomials of different order or modes, called modal p-type expansion functions. For each type it is shown that the global mass matrix $\mathbf{M}_G$ and the global diffusion matrix $\mathbf{D}_G$ can be calculated in a similar manner.

3.2 Elemental Decomposition (*h*-type extension)

In this section, the decomposition of the solution domain Ω into smaller subdomains Ω^e is demonstrated in more detail. This process, which is referred to as the *h-type extension*, basically reduces the sizes of the individual subdomains Ω^e in order to achieve convergence of the approximated solution.

Here, the decomposition process is explained and visualized by means of simple polynomials for the shape functions $\Psi_i(z)$. However, the same techniques can be exploited with higher-order polynomial expressions, as shown in the next sections. The decomposition of the solution domain into subdomains and the respective shape functions for linear polynomials are visualized in Fig. 2.

Substituting the finite expansion (6) into the weak formulation (13) of the partial differential equation yields the following individual terms of the weak formulation

$$\int_\Omega \Psi_i(z) \frac{\partial \tilde{f}(z,t)}{\partial t} dz = \sum_{j=0}^{N_{dof}-1} M_{ij}^g \frac{d\alpha_j(t)}{dt} \tag{16}$$

$$\int_\Omega \frac{d\Psi_i(z)}{dz} \frac{\partial \tilde{f}(z,t)}{\partial z} dz = \sum_{j=0}^{N_{dof}-1} D_{ij}^g \alpha_j(t) \tag{17}$$

$$\int_\Omega \Psi_i(z) \tilde{s}(z,t) dz = \sum_{j=0}^{N_{dof}-1} M_{ij}^g \hat{s}_j(t) \tag{18}$$

The individual entries M_{ij}^g and D_{ij}^g, defined by

$$\begin{aligned} M_{ij}^g &= \int_\Omega \Psi_i(z) \Psi_j(z) dz \ , \\ D_{ij}^g &= \int_\Omega \frac{d\Psi_i(z)}{dz} \frac{d\Psi_j(z)}{dz} dz \ , \end{aligned} \tag{19}$$

can be assembled to the global mass matrix $\mathbf{M}_G$ and the global diffusion matrix $\mathbf{D}_G$, respectively. At this point it can be easily recognized that this always leads to the ordinary differential equation given in (14).

Furthermore, it is noted that in the case of piecewise affine shape functions $\Psi_i(z)$ this leads to the finite-difference formula, which can be regarded as a special case of the finite-element method [2].

3.3 Nodal Polynomial Expansion

This section describes how the weak formulation (13) is decomposed in space using higher order polynomials; also called *p-type expansion*. Assuming a fixed mesh, these methods achieve a higher accuracy of the solution by increasing the polynomial order *inside* the element.

Since a consideration of the expansion in terms of global modes $\Psi_i(z)$ is uneconomical and numerically intractable, especially when using a large number of elements, it is reasonable to introduce a standard element Ω_{st}, such that

$$\Omega_{st} = \{\xi \,|\, -1 \le \xi \le 1\} \ . \tag{20}$$

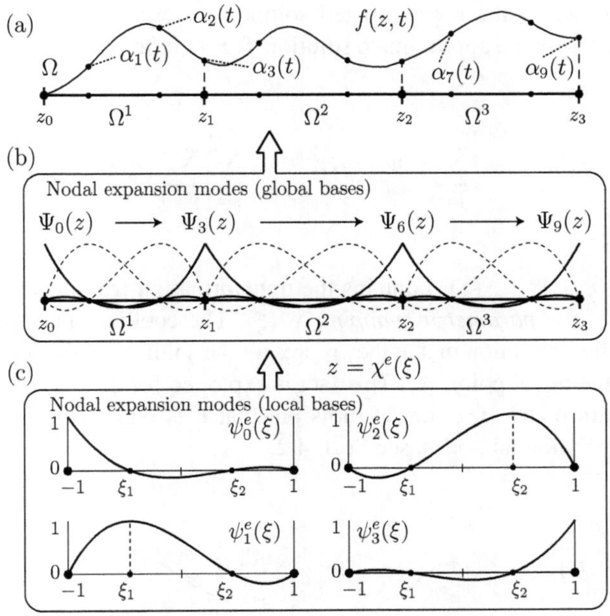

Fig. 3 Approximation of the solution $f(z,t)$ by means of a nodal polynomial expansion, e.g. Legendre polynomials for $m = 4$ refinement. **(a)** Elemental decomposition of the solution domain Ω into several subdomains Ω^e. **(b)** Shape functions $\Psi_i(z)$ for a nodal polynomial expansion visualized in global bases, and **(c)** in local bases (standard element).

This standard element Ω_{st} can be mapped to any elemental domain Ω^e using the isoparametric transformation $\chi^e(\xi)$, which expresses the global coordinate z in terms of the local coordinate ξ, depicted in Fig. 3b–c.

A class of nodal p-type expansions, which also have become known as *spectral elements*, are based on Legendre polynomials. The polynomials are defined in the standard domain Ω_{st} according to

$$L_m(\xi) = \frac{1}{2^m\, m!}\, \frac{d^m(\xi^2-1)^m}{d\xi^m} \ , \tag{21}$$

where m denotes the degree of the used polynomial. Based on the Legendre polynomials the *spectral elements* are given by

$$\psi_p^e(\xi) = \frac{(1-\xi^2)L_m^{'}(\xi)}{m(m+1)L_m(\xi_p)(\xi_p-\xi)} \ , \tag{22}$$

where L_m is the Legendre polynomial of degree m, $L_m^{'}$ denotes the differentiation with respect to the argument, and ξ_p is the p-th Gauss-Lobatto-Legendre quadrature point defined by the corresponding root of $(1-\xi^2)\, L_m^{'}(\xi) = 0$.

The choice of these quadrature point plays an important role in the stability of the approximation. The spectral elements $\psi_p^e(\xi)$ are shown in the standard domain Ω^e in Fig. 3 for $m = 4$, [7].

Finally, considering the approximated solution in terms of the spectral elements $\psi_p^e(\xi)$, we can express the approximate solution $\tilde{f}(z,t)$ in terms of $\psi_p^e(\xi)$ according to

$$\tilde{f}(z,t) = \sum_{j=0}^{N_{dof}-1} \Psi_j(z)\,\alpha_j(t) = \sum_{e=1}^{N_{el}} \sum_{p=0}^{m} \psi_p^e(\xi)\,\alpha_p^e(t) \tag{23}$$

where $\psi_p^e(\xi) = \psi_p([\chi^e]^{-1}(z))$ contains the transformation from local bases to global bases in terms of the *parametric mapping* $\chi^e(\xi)$. The coefficients $\alpha_j(t)$ in this form have a physical interpretation in that they represent the solution of the partial differential equation (1) at the nodal points x_j. This fact is exploited for the derivation of a simple measurement gain matrix $\mathbf{H}_k$, i.e., it turns out that it is reasonable to put the sensor nodes onto the polynomial nodes, see Sect. 4.2.

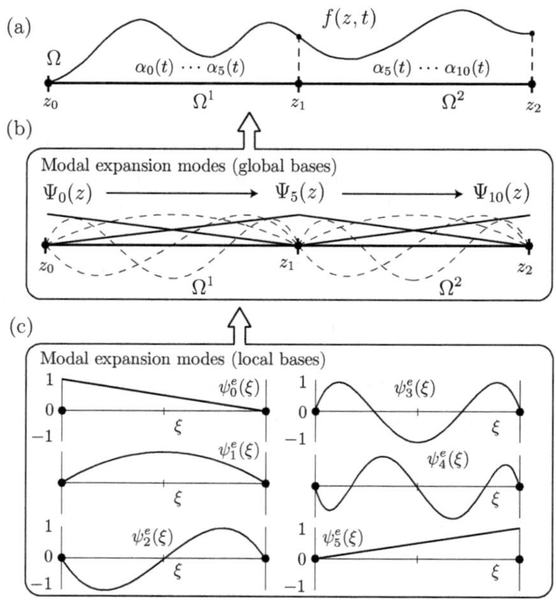

Fig. 4 Approximation of the solution $f(z,t)$ by means of a modal polynomial expansion, e.g. Legendre polynomials for $m = 5$ refinement. **(a)** Elemental decomposition of the solution domain Ω into several subdomains Ω^e. **(b)** Shape functions $\Psi_i(z)$ for a modal polynomial expansion visualized in global bases, and **(c)** in local bases (standard element).

Upon substituting the nodal expansion (23) into the weak formulation of the partial differential equation (13) the entries of the local mass matrices M_{ij}^e and local diffusion matrices D_{ij}^e can be derived as

$$M_{ij}^e = \int_{-1}^{1} \psi_i^e(\xi)\psi_j^e(\xi)d\xi \, ,$$

$$D_{ij}^e = \int_{-1}^{1} \frac{d\psi_i^e(\xi)}{d\xi} \frac{d\psi_j^e(\xi)}{d\xi} d\xi \, . \tag{24}$$

Considering the boundary conditions at the elemental nodes, e.g. z_1 and z_2 in Fig. 3, the local matrices M_{ij}^e and D_{ij}^e can be easily assembled to the global mass matrix $\mathbf{M}_G$ and global diffusion matrix $\mathbf{D}_G$. This is an automatic procedure known as *global assembly*.

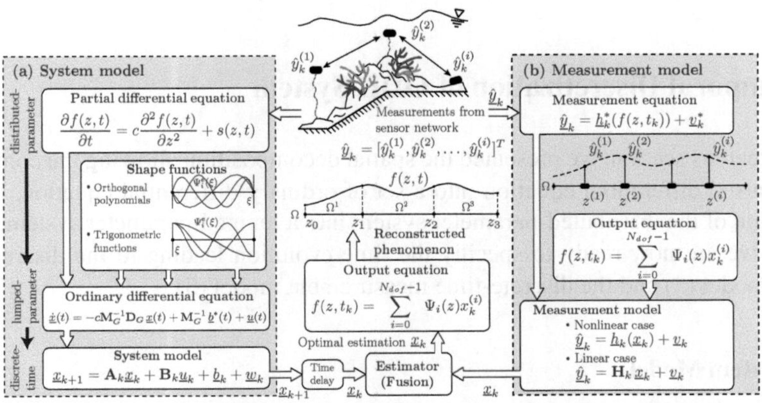

Fig. 5 Overview and components of the procedure for model-based reconstruction of distributed phenomena. **(a)** System model: conversion of the distributed-parameter system into a lumped-parameter discrete-time system, **(b)** Measurement model: relating specific measurements $\hat{\underline{y}}_k$ to the underlying distributed phenomenon. The estimated state vector $\underline{x}_k$ from the system model and the measurement model is used to obtain optimal estimation result, and thus the reconstruction of the distributed phenomenon.

3.4 Modal Polynomial Expansions

For the construction of p-type expansions it is often favorable to select a set of orthogonal functions (polynomials), such as Legendre polynomials, Chebyshev polynomials or sine functions. The most commonly used orthogonal polynomials in computational fluid dynamics, which offer some advantages compare to the other polynomials, are based upon the Legendre polynomials. The p-type modal expansion modes in the standard element Ω_{st} are defined as

$$\psi_p^e(\xi) = \begin{cases} \frac{1-\xi}{2} & p = 0 \\ (1-\xi^2)L_{p-1}^{'}(\xi) & 0 < p < m \\ \frac{1-\xi}{2} & p = m \end{cases} \, , \tag{25}$$

where the lowest expansion modes $\psi_0(\xi)$ and $\psi_p(\xi)$ are the same as in the linear finite element expansion. These modes are denoted as boundary modes since they are the only modes which are nonzero at the ends of the interval. In Fig. 4 the p-type modal expansion based upon the Legendre polynomials is depicted for $m = 5$, [11].

Thus, the approximate solution $\tilde{f}(z,t)$ in terms of a modal polynomial expansion can be expressed in the same manner as the nodal polynomial expansion (23); the function $\psi_p^e(\xi)$ is just replaced by p-type modal expansion modes (25) and the coefficients $\alpha_j(t)$ denote the weighting coefficients for the individiual *modes*. The local mass matrices M_{ij}^e and local diffusion matrices D_{ij}^e can be derived similar to the nodal polynomial expansion (24). Considering the boundary conditions at the end of each element, e.g. z_1 in Fig. 4, the local matrices M_{ij}^e and D_{ij}^e can be easily assembled to the global mass matrix $\mathbf{M}_G$ and global diffusion matrix $\mathbf{D}_G$.

4 Temporal Discretization of ODE-System

In the previous section, we presented the spatial decomposition allowing the conversion of the partial differential equation into a set of ordinary differential equation, i.e., the conversion of the distributed-parameter system into a lumped-parameter system. In this section, we are now ready to specifiy the time evolution leading to the discrete-time system model (3) and the discrete-time measurement model (5).

4.1 System Model

To circumvent the restriction on the time step Δt, it is reasonable to integrate the set of ordinary differential equations (14) by means of *implicit* methods, such as the Crank-Nicolson discretization. Basically, this discretization method selects a time step Δt, evaluates the differential equation at time $t + \frac{1}{2}\Delta t$, and finally approximates the time derivative on the left-hand side of (14) with a centered finite difference and the rest of the terms with averages, leading to

$$\mathbf{M}_G \frac{\underline{x}_{k+1} - \underline{x}_k}{\Delta t} = \mathbf{M}_G \underline{u}_k - \frac{c}{2}\mathbf{D}_G \left[\underline{x}_{k+1} + \underline{x}_k\right] + \underline{b}_k^*$$

Rearranging, the Crank-Nicolson discretization of the differential equation produces a linear system of equations for the state vector $\underline{x}_{k+1}$ containing the unknown weighting factors α_i at the $k+1$ time step,

$$\left(\mathbf{M}_G + \frac{1}{2}c\Delta t\mathbf{D}_G\right)\underline{x}_{k+1} = \left(\mathbf{M}_G - \frac{1}{2}c\Delta t\mathbf{D}_G\right)\underline{x}_k$$
$$+\Delta t\mathbf{M}_G\underline{u}_k + \Delta t\underline{b}_k^* \ .$$

It is important to note that this linear system is *unconditionally* stable for any time step Δt. Using the following abbreviations

$$\mathbf{A}_k = \left(\mathbf{M}_G + \frac{1}{2}c\Delta t\mathbf{D}_G\right)^{-1}\left(\mathbf{M}_G - \frac{1}{2}c\Delta t\mathbf{D}_G\right),$$

$$\mathbf{B}_k = \Delta t \left(\mathbf{M}_G + \frac{1}{2} c \Delta t \mathbf{D}_G \right)^{-1} \mathbf{M}_G \ ,$$

$$\underline{b}_k = \Delta t \left(\mathbf{M}_G + \frac{1}{2} c \Delta t \mathbf{D}_G \right)^{-1} \underline{b}_k^* \ ,$$

and adding noise terms and modelling error terms the linear system can be stated in the well-known compact state space form

$$\underline{x}_{k+1} = \mathbf{A}_k \, \underline{x}_k + \mathbf{B}_k \, \underline{u}_k + \underline{b}_k + \underline{w}_k \ , \tag{26}$$

where the matrices $\mathbf{A}_k$ and $\mathbf{B}_k$ are determined using the global mass matrix $\mathbf{M}_G$ and the global diffusion matrix $\mathbf{D}_G$. The vector $\underline{b}_k$ containing the boundary conditions is determined using the matrices $\mathbf{M}_G$, $\mathbf{D}_G$, and the boundary vector $\underline{b}_k^*$. Fig. 5a visualizes the conversion of a distributed phenomenon characterized by means of a partial differential equation from the distributed-parameter form into a lumped-parameter form, and finally into a discrete-time finite-dimensional state space form.

The resulting physical model in finite-dimensional state space form (26) could be used for the simulation of the underlying distributed phenomenon by propagating the finite-dimensional state vector $\underline{x}_k$ over time. Then, the desired result $\tilde{f}(z,t)$ of the underlying distributed phenomenon could be directly derived by using the equation for the approximated solution (6).

However, for a model-based reconstruction method by means of a sensor network the aim is not just the simulation of the distributed phenomenon, rather the reconstruction of that phenomenon by means of discrete-time measurements from the sensor network. Unfortunately, for most applications the measurements have to be regarded as uncertain values containing the uncertainties of the actual sensor node. In order to take these uncertainties of both the measurements and the system model into account, it is common to use appropriate parameterized density functions for the description of the *estimated* state vector $\underline{x}_k$. This will be specified in more detail in Sect. 5. Before it is essential to derive a description mapping the specific measurements to the solution of the observed phenomenon; the so-called measurement model.

4.2 Measurement Model

For model-based reconstruction not only a system model of the distributed phenomenon is necessary but also a measurement model, which maps the specific measurements $\underline{\hat{y}}_k$ obtained from the sensor network to the solution $f(z,t)$ of the observed phenomenon. This section is devoted to the derivation of the discrete-time measurement model.

By means of the spatial decomposition, introduced in Sect. 3, it is possible to derive a relation between the individual measurements $\underline{\hat{y}}_k$ and the entire observed phenomenon. The sensor nodes, which are densely deployed inside the distributed phenomenon, are used to observe that phenomenon and to improve its estimated state $\underline{x}_k$. The measurement model consists of two parts: the measurement equation and the output equation.

The measurement equation relates the actual measurements $\hat{y}_k^{(i)}$ at location $z^{(i)}$ and at time t_k to the distributed phenomenon $f(z, t_k)$, according to

$$\underline{\hat{y}}_k = \underline{h}_k^*(f(z, t_k)) + \underline{v}_k^* \ , \tag{27}$$

where the vector $\underline{v}_k^*$ contains the uncertainties arising from the actual sensor node. Even for simple measurement principles the mapping $\underline{h}_k^*(.)$ consists of non-linear functions. In such cases non-linear approaches for the estimation algorithm have to be applied.

The output equation, on the other hand, relates the punctiform measurements of the distributed phenomenon $f(z^{(i)}, t_k)$ directly to the finite-dimensional state vector $\underline{x}_k$, according to

$$f(z^{(i)}, t_k) = \sum_{j=0}^{N_{dof}-1} \Psi_j(z^{(i)}) x_k^{(j)} \ , \tag{28}$$

which is identical to the representation of the approximated solution $\tilde{f}(z,t)$ of the partial differential equation introduced in Sect. 3.1.

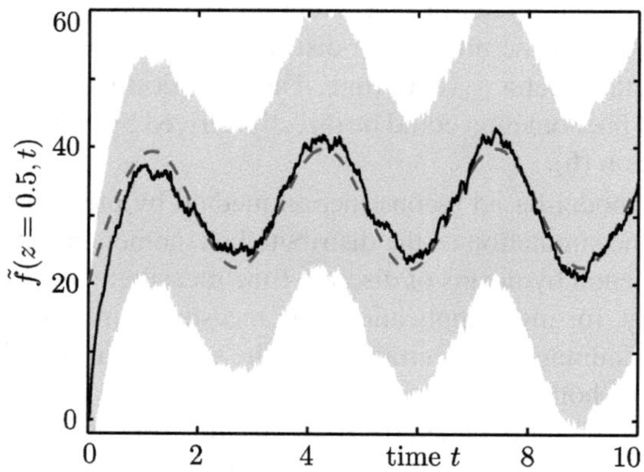

Fig. 6 Prediction: estimated solution $\tilde{f}(z,t)$ (black), 3σ-Bounds (gray shaded), and analytic solution $f(z,t)$ (black dotted).

Substitution of the output equation (28) into the measurement equation (27) leads to the complete measurement model, which provides the mapping of the individual discrete-time measurements $\underline{\hat{y}}_k$ obtained from the sensor network to the finite-dimensional state vector $\underline{x}_k$ characterizing the state of the distributed phenomenon. For the nonlinear case the measurement model is given by

$$\underline{\hat{y}}_k = \underline{h}_k(\underline{x}_k) + \underline{v}_k \ ,$$

and for the linear case the measurement model is given by

$$\underline{\hat{y}}_k = \mathbf{H}_k \underline{x}_k + \underline{v}_k \ ,$$

where $\underline{v}_k$ are the measurement uncertainties arising from both the sensor node and the modeling errors. Fig. 5b shows the complete measurement model relating specific measurements $\underline{\hat{y}}$ to the underlying distributed phenomenon in terms of the state vector $\underline{x}_k$.

In this article, for simplicity and brevity only the linear case is considered. That means, the measurement of the i-th sensor node at the location $z^{(i)}$ and at time step k, i.e., $t = t_k$, is related to the state vector $\underline{x}_k$ according to

$$\begin{aligned}\hat{y}_i(z^{(i)}, t = t_k) &= \sum_{j=0}^{N_{dof}-1} \Psi_j(z^{(i)})\, \alpha_j(t = t_k) \\ &= \underline{\Psi}^T(z^{(i)})\, \underline{x}_k .\end{aligned}$$

Assuming a sensor network consisting of L sensor nodes, the measurement gain matrix $\mathbf{H}_k$ is set up by the shape function $\underline{\Psi}^T(z^{(i)})$ of order N, according to

$$\underline{\hat{y}}_k = \underbrace{\begin{pmatrix} \Psi_1(z^{(1)}) & \cdots & \Psi_N(z^{(1)}) \\ \vdots & \ddots & \vdots \\ \Psi_1(z^{(L)}) & \cdots & \Psi_N(z^{(L)}) \end{pmatrix}}_{\mathbf{H}_k} \underline{x}_k + \underline{v}_k \ , \tag{29}$$

where $\underline{v}_k$ are the measurement uncertainties. The number of rows L of the matrix $\mathbf{H}_k$ depends on the number of used sensors, i.e., number of measurement points, and the number of columns N depends on the desired number of modes.

However, in the case of nodal expansion and assuming that the sensors are located at the polynomial nodes, i.e., $z^{(i)} = z_i$, the weighting coefficients can be measured directly, i.e., $\alpha_j(t) = \hat{y}_j$. In other words, the diagonal entries of the measurement gain matrix $\mathbf{H}_k$ are either 1 or 0, depending on the location of the sensor nodes,

$$\underline{\hat{y}}_k = \underbrace{\begin{pmatrix} 1 & \cdots & 0 \\ \vdots & \ddots & \vdots \\ 0 & \cdots & 1 \end{pmatrix}}_{\mathbf{H}_k} \underline{x}_k + \underline{v}_k \ . \tag{30}$$

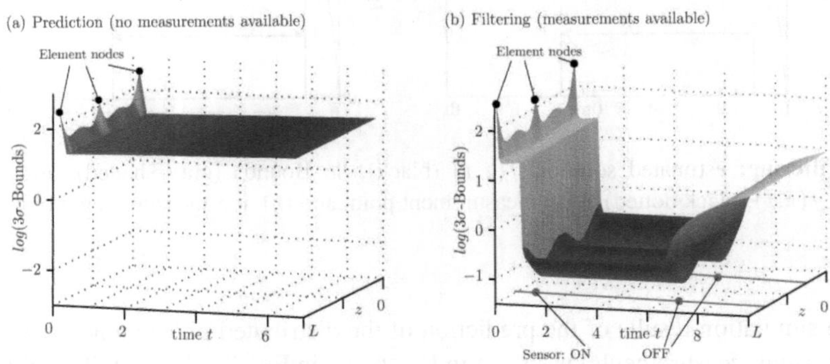

Fig. 7 Logarithm of the 3σ-Bounds as a measure for the uncertainty of the estimated solution $\tilde{f}(z,t)$. **(a)** Uncertainty for the prediction (no measurements available), **(b)** Uncertainty for filtering (measurements available). Due to the measurements the uncertainty can be decreased rapidly, even between the sensor nodes.

5 Centralized Estimation Approach

In this section, the prediction step and the filter step for a centralized estimation approach are derived. Due to the fact that both the system equation (26) and the measurement equation (5) are linear, it is sufficient to use the linear Kalman filter [10] [17] to obtain the best possible estimate for the system state characterizing the distributed phenomena.

5.1 Prediction Step

The purpose of the prediction step is to propagate the current state estimate $\hat{\underline{x}}_k^e$ through the system equation (26) to the next time step. Thus, the mean of the predicted state vector $\hat{\underline{x}}_{k+1}^p$ at time step $k+1$ is given by

$$\hat{\underline{x}}_{k+1}^p = \mathbf{A}_k \, \hat{\underline{x}}_k^e + \mathbf{B}_k \, \hat{\underline{u}}_k + \hat{\underline{b}}_k \ .$$

Assuming that the input vector $\underline{u}_k$ and the state vector $\underline{x}_k$ are uncorrelated the covariance matrix is obviously given by

$$\mathbf{C}_{k+1}^p = \mathbf{A}_k \, \mathbf{C}_k^e \, \mathbf{A}_k^T + \mathbf{B}_k \, \mathbf{C}_k^u \mathbf{B}_k^T + \mathbf{C}_k^d \ ,$$

where $\mathbf{C}_k^e$ is the covariance matrix of the state estimate $\underline{x}_k^e$, $\mathbf{C}_k^u$ is the covariance matrix of the input noise, and $\mathbf{C}_k^d$ is the covariance matrix of subsumed endogenous uncertainties, e.g. modeling errors.

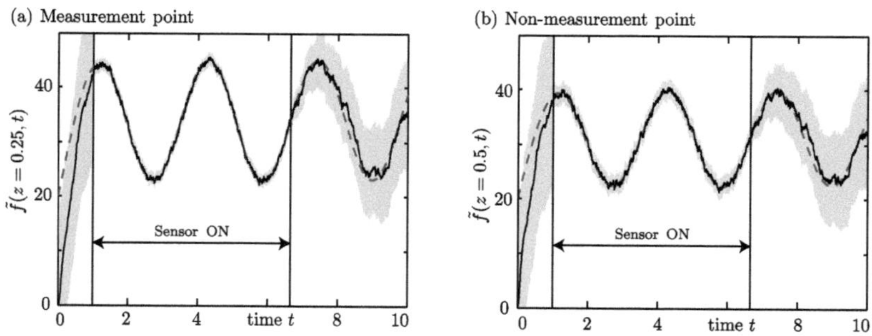

Fig. 8 Filtering: estimated solution $\tilde{f}(z,t)$ (black), 3σ-Bounds (gray shaded), and analytic solution $f(z,t)$ (black dotted) at **(a)** measurement point and **(b)** non-measurement point.

The simulation results of the prediction of the distributed phenomenon (estimation without using measurements) is shown in Fig. 6 and in Fig. 7a. The details of the simulation are explained in more detail in Sect. 6. Here, it can be easily recognized that the certainty of the estimated solution $\tilde{f}(z,t)$ cannot be increased, i.e., only the propagation of the estimated state vector $\underline{x}_k$ seems not to be sufficient.

5.2 Filter Step

For the purpose of reducing the estimation uncertainty, measurements are used that are related to the state via the measurement equation (5). Given the predicted state $\hat{\underline{x}}_k^p$ with covariance $\mathbf{C}_k^p$ and a vector observation $\hat{\underline{y}}_k$ with covariance $\mathbf{C}_k^v$, the estimated state vector $\hat{\underline{x}}_k^e$ is derived by

$$\begin{aligned}\hat{\underline{x}}_k^e =&\hat{\underline{x}}_k^p + \mathbf{C}_k^p \mathbf{H}_k^T \left(\mathbf{C}_k^v + \mathbf{H}_k \mathbf{C}_k^p \mathbf{H}_k^T\right)^{-1} \\ &\cdot \left(\hat{\underline{y}}_k - \mathbf{H}_k \hat{\underline{x}}_k^p\right)\end{aligned}$$

In the uncorrelated case the covariance matrix of the estimate is given by

$$\mathbf{C}_k^e = \mathbf{C}_k^p - \mathbf{C}_k^p \mathbf{H}_k^T \left(\mathbf{C}_k^v + \mathbf{H}_k \mathbf{C}_k^p \mathbf{H}_k^T\right)^{-1} \mathbf{H}_k \mathbf{C}_k^p \ .$$

The simulation results of the filtering (estimation using measurements) is shown in Fig. 7b and in Fig. 8. The details of the simulation are explained in more detail in Sect. 6. Here, it can be easily recognized that the certainty of the estimated solution $\tilde{f}(z,t)$ can be significantly increased by using additional information in terms of measurements from the sensor network.

6 Simulation Results

In this section we demonstrate the performance of the proposed estimation method by means of simulations. The goal is the reconstruction of the temperature distribution in a

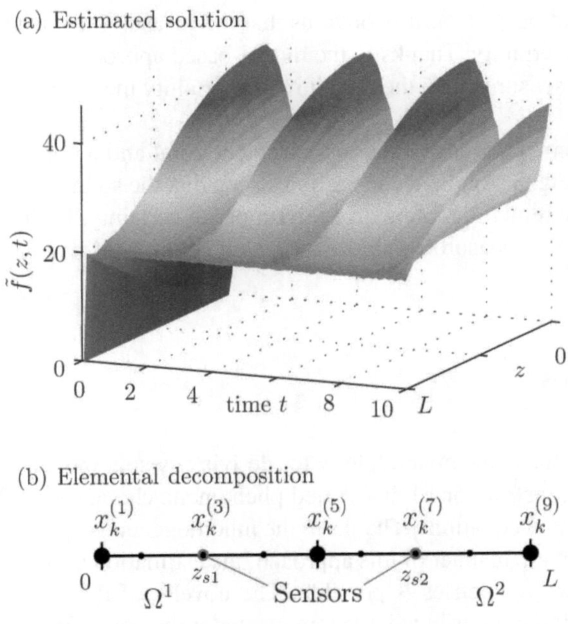

Fig. 9 (a) Estimated solution $\tilde{f}(z,t)$. **(b)** Elemental decomposition into two elements Ω_1 and Ω_2 with a polynomial expansion of order $m = 5$ and location of sensor nodes.

heat rod using both a physical model and measurements obtained by a sensor network. It is important to note that a novelty of our approach is to consider the uncertainties arising from noisy measurements and occuring in the physical model (state vector and input vector).

The evolution of the temperature is modelled by the well-known one-dimensional heat equation (1) of a bar with the length $L = 1\,\mathrm{m}$. The noisy input function is given by $s(z,t) = 30\,\sin(2\,t) + 30 + \Delta u_k$, where Δu_k denotes the input uncertainties. Applying a nodal expansion (23) for the approximated solution $\tilde{f}(z,t) = \underline{\Psi}^T(z)\,\underline{\alpha}(t)$, the partial differential equation (1) can be spatially and temporally decomposed leading to the *finite*-dimensional state space form (26). The state vector $\underline{x}_k$ can be derived by temporal discretization of the weighting factors $\underline{\alpha}(t)$ of the approximated solution, as shown in Sect. 4. Here, we assume a nodal expansion based on a Legendre polynomial of the order $m = 5$. The spatial decomposition of the heat rod into two elements Ω_1 and Ω_2 is visualized in Fig. 9b.

Furthermore, it is assumed that for simplicity the sensor network consists of two sensor nodes located at $z_{s1} = 0.25$ and $z_{s2} = 0.75$, shown in Fig. 9b. In the case of a nodal expansion the general relation between the measurement vector $\hat{\underline{y}}_k$ and the state vector $\underline{x}_k$ described by (29) can be simplified to (30). For this simulation example that means the state variables $x_k^{(3)}$ and $x_k^{(7)}$ can be measured directly, as it is visualized in Fig. 9b.

The estimated solution $\tilde{f}(z,t)$ is visualized in Fig. 9a. The logarithm of the 3σ-bound of the estimated solution $\tilde{f}(z,t)$, which is a measure for its uncertainty, is depicted in Fig. 7a for prediction which means the propagation of the estimated solution $\tilde{f}(z,t)$ without using measurements from the sensor network and in Fig. 7b for the filtering which means measurements from the sensor network are exploited to improve the estimated solution $\tilde{f}(z,t)$. It is obvious that the measurements are used for rapidly decreasing the uncertainty. Thanks to the model-based approach they can be decreased even *between* the sensor nodes; the remaining uncertainty merely arises from the model uncertainties.

The estimation results for both a measurement point and a non-measurement point are visualized in Fig. 8. Again, it can easily be seen that the solution $\tilde{f}(z,t)$ is estimated even at non-measurement points with an appropriate certainty. Furthermore, it is clear that by means of the measurements the estimated solution $\tilde{f}(z,t)$ can be significantly improved.

7 Conclusions

This chapter introduced the methodology for deriving system models and measurement models for the reconstruction of distributed phenomena characterized by means of linear partial differential equations. Thanks to the inhomogeneous approximation capabilities and the systematic manner of this approach, the estimation of nonlinear phenomena even with complex geometries is possible. The novelty of this chapter is the model-based reconstruction of distributed phenomena under the consideration of uncertainties both occuring in the physical model and arising from noisy measurements.

It is believed that applying such methods in sensor network applications for the reconstruction provides novel prospects to optimal sensor node placement, optimal measurement time sequences, and improvement of the used physical model. The performance of the proposed model-based approach for the reconstruction of distributed pheonmena was demonstrated by means of simulation results for the one-dimensional partial differential equation.

For the one-dimensional linear partial differential equation, the decomposition may seem unnecessarily involved. However, the same principles can easily be applied to the decomposition in multiple dimensions and even for nonlinear partial differential equation. This is left for future research work.

As it was mentioned in the introduction, the estimation results can be exploited for several additional tasks, such as optimal sensor placement, model improvement, and system identification. Especially the problem of system identification would be essential for sensor networks performing self-organization in a completely unknown sourrounding. By using such methods for the identification of an appropriate physical model of the sourrounding it would be possible for the sensor nodes to identify, observe, and reconstruct unknown distributed phenomena. This is also part of future research work.

Furthermore, especially for large sensor networks it is essential to find a decentralized reconstruction approach. It is believed that the decomposition method introduced in this chapter is well-suited for such an estimation approach. The application of decentralized data fusion methods such as introduced in [8] [21] for a decentralized reconstruction approach is left for future research work.

Acknowledgements

This work was partially supported by the German Research Foundation (DFG) within the Research Training Group GRK 1194 "Self-organizing Sensor-Actuator-Networks".

References

1. Anderson, B. D. O. and Moore, J. B. (1979). *Optimal Filtering*. Prentice-Hall.
2. Baker, A. J. (1983). *Finite Element Computational Fluid Mechanics*. Taylor and Francis.
3. Brunn, D. and Hanebeck, U. D. (2005). A model-based framework for optimal measurements in machine tool calibration. In *IEEE International Conference on Robotics and Automation (ICRA 2005)*, Barcelona, Spanien.
4. Culler, D. E. and Mulder, H. (2004). Smart sensors to network the world. In *Scientific American*, volume 6.
5. Fagherazzi, S., Furbish, D. J., Rasetarinera, P., and Hussaini, M. Y. (2004). Application of the discontinuous spectral Galerkin method to groundwater flow. In *Advances in Water Resources*, volume 27, pp. 129–140.
6. Faulds, A. L. and King, B. B. (2000). Sensor location in feedback control of partial differential equation systems. In *Proceedings of the 2000 IEEE International Conference on Control Applications (CCA 2000)*, pp. 536–541, Anchorage, AK.
7. Fournier, A., Bunge, H.-P., Hollerbach, R., and Vilotte, J.-P. (2004). Application of the spectral-element method to the axisymmetric Navier-Stokes equation. In *Geophysical Journal International*, number 156, pp. 682–700.

8. Hanebeck, U. D. and Briechle, K. (2001). New results for stochastic prediction and filtering with unknown correlations. In *Proceedings of the IEEE Conference on Multisensor Fusion and Integration for Intelligent Systems (MFI 2001)*, pp. 147–152, Baden-Baden, Germany.
9. Hanebeck, U. D., Briechle, K., and Rauh, A. (2003). Progressive Bayes: a new framework for nonlinear state estimation. In *Proceedings of SPIE, AeroSense Symposium*, pp. 256–267, Orlando, Florida.
10. Kalman, R. E. (1960). A new approach to linear filtering and prediction problems. In *Transactions of the ASME Journal of Basic Engineering*, issue 82, pp. 34–45.
11. Karniadakis, G. E. and Sherwin, S. (2005). *Spectral/HP Element Methods for Computational Fluid Dynamics*. Oxford University Press.
12. Komatitsch, D., Ritsema, J., and Tromp, J. (2002). The spectral-element method, Beowulf computing, and global seismology. In *Science*, volume 298, pp. 1737–1742.
13. Komatitsch, D., Vilotte, J.-P., Vai, R., Castillo-Covarrubias, J. M., and Sanchez-Sesma, F. J. (1999). The spectral element method for elastic wave equations – application to 2-D and 3-D seismic problems. In *International Journal for Numerical Methods in Engineering*, volume 45, pp. 1139–1164.
14. Kumar, T., Zhao, F., and Shepherd, D. (2002). Collaborative signal and information processing in microsensor networks. In *IEEE Signal Processing Magazine*, volume 19, pp. 13–14.
15. Levin, J. G., Iskandarani, M., and Haidvogel, D. B. (2000). A nonconforming spectral element ocean model. In *International Journal for Numerical Methods in Fluids*, volume 34, pp. 495–525.
16. Ortmaier, T., Groeger, M., Boehm, D. H., Falk, V., and Hirzinger, G. (2005). Motion estimation in beating heart surgery. In *IEEE Transactions on Biomedical Engineering*, volume 52, issue 10, pp.1729–1740.
17. Papoulis, A. (1991). *Probability, Random Variables and Stochastic Processes*. McGraw-Hill.
18. Rao, B., Durrant-Whyte, H., and Sheen, J. (1993). A fully decentralized multisensor system for tracking and surveillance. In *International Journal of Robotics Research*, volume 12, pp. 20–44.
19. Roberts, K. and Hanebeck, U. D. (2005). Prediction and reconstruction of distributed dynamic phenomena characterized by linear partial differential equations. In *The 8th International Conference on Information Fusion (Fusion 2005)*, Philadelphia, USA.
20. Sawo, F., Brunn, D., and Hanebeck, U. D. (2006a). Parameterized joint densities with gaussian mixture marginals and their potential use in nonlinear robust estimation. In *IEEE International Conference on Control Applications (CCA 2006)*, Munich, Germany.
21. Sawo, F., Roberts, K., and Hanebeck, U. D. (2006b). Bayesian estimation of distributed phenomena using discretized representations of partial differential equations. In *Proceedings of the 3rd International Conference on Informatics in Control, Automation and Robotics (ICINCO 2006)*, pp. 16–23, Setubal, Portugal.
22. Thuraisingham, B. (2004). Secure sensor information management and mining. In *IEEE Signal Processing Magazine*, volume 5, pp. 14–19.

GA-based Approach to Pitch Recognition of Musical Consonance

Masanori Natsui, Shunichi Kubo and Yoshiaki Tadokoro

Department of Information and Computer Sciences, Toyohashi University of Technology
1-1 Hibarigaoka, Tempaku-cho, Toyohashi-shi
Aichi 441-8580, Japan
{natsui, skubo, tadokoro}@signal.ics.tut.ac.jp

Abstract. This chapter presents a novel method for the pitch recognition of the musical consonance (i.e., unison or octave) using genetic algorithm (GA). GA is a kind of optimization techniques based on natural selection and genetics. In our method, the pitch recognition is performed by the following two-step procedure: (i) search space reduction using comb filter estimation, and (ii) evolutionary parameter estimation of tone parameters such as notes and volumes by minimizing error between a target waveform and a synthesized waveform using sound templates with estimated parameters. The potential capability of the system is demonstrated through the pitch estimation of randomly-generated consonances. Experimental results show that the system can successfully estimate chords with more than 84% success rate for two-note consonances, and more than 71% success rate for three-note consonances.

Keywords. Pitch recognition, genetic algorithm, musical consonance, template matching.

1 Introduction

Automatic music transcription is important for many applications including music archival, music retrieval, supports of music composition/arrangement, and also significant problems in machine perception, [11] [8] [9] [10] [7] [5]. The study of automatic musical transcription can be classified into some categories, and that of the pitch detection is the most important task and many studies have been done. Most of old studies are for monophony, and based on the spectrum analysis using the fast Fourier transform (FFT). On the other hand, the novel technologies such as neural network, fuzzy logic, and hidden Marcov model have also been proposed in the recent studies, [6].

For the pitch estimation of polyphonic sounds, we have proposed a unique method based on comb filters ($H(z) = 1 - z^{-N}$), [14] [12] [13]. The comb filter can eliminate a fundamental frequency and its harmonic components of a sound by simple subtraction. So far, we have presented that cascade or parallel connections of the comb filters enable the polyphonic pitch estimation and can be effective for the realization of the automatic music transcription system.

A difficult problem in the polyphonic pitch estimation is that some frequency components of one note may be overlapped with harmonics of other notes. In fact, composers

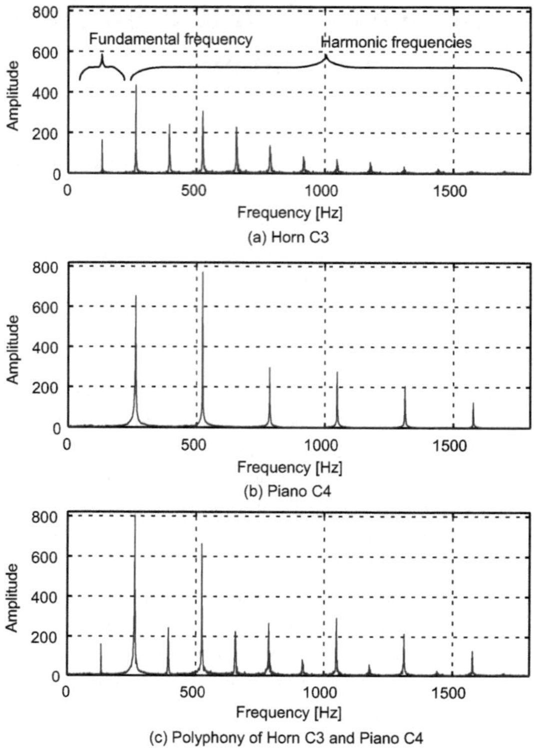

Fig. 1 Spectra of music instruments: (**a**) Horn C3, (**b**) Piano C4, and (**c**) polyphony of the two tones.

often use chords containing notes that have a simple ratio between their fundamental frequencies, such as 1:1 (perfect unison), 2:1 (perfect octave), or 3:2 (perfect fifth), since these codes called *consonances* typically produce sounds which are pleasing to the human ear. If one note having a fundamental frequency of f Hz and another note having that of $2f$ Hz are produced at the same time, then every harmonic of the upper note will be overlapped to the even harmonics of the lower note (Fig. 1). To infer the presence of the upper note, we have to use some other information which is obtained by a technique except traditional methods such as spectrum analysis.

From this viewpoint, we propose a unique method of the pitch estimation based on *genetic algorithm* (GA). GA is an optimization algorithm based on a model of evolution in life. In this chapter, we demonstrate the possibility of the GA-based pitch estimation method through the experimental pitch estimation of musical consonance. The key ideas presented here are: (i) time-domain template matching based on GA, and (ii) search space reduction using the pitch estimation result based on the comb filters.

This chapter is organized as follows: Section 2 presents the basic concept of the pitch estimation system using GA. Section 3 shows an overview of the proposed pitch estimation system. Section 4 demonstrates the experimental result of the pitch estimation. Section 5 is the conclusion and future prospects.

2 Genetic Algorithm

Genetic algorithm (GA) can be regarded as a unique variation of evolutionary computation techniques, [1] [4] [2]. In general, evolutionary methods mimic the process of natural evolution, the driving process for emergence of complex structures well-adapted to the given environment. The better an individual performs under the conditions the greater is the chance for the individual to live for a longer while and generate offspring. As a result, the individuals are transformed to the suitable forms on the designer's defined constraint.

Figure 2 shows the overall procedure of GA. At first, GA generates embryonic individuals randomly to form the initial population $P(0)$. In the traditional GA, each individual is represented by a fixed-length bit string. The next step is to evaluate a fitness function at all individuals in $P(t)$. A value for fitness is assigned to each individual depending on how close it actually is to solving the problem. After the evaluation, the system selects a set of individuals having higher fitness values to perform evolutionary operations: *crossover* and *mutation*. The probability of selecting an individual for crossover and mutation depends on its fitness value. The offsprings generated by these evolutionary operations form the populations $C(t)$ and $M(t)$, where $C(t)$ and $M(t)$ are obtained by crossover and mutation operations, respectively. The individuals for the next generation $P(t+1)$ are selected from the current population $C(t) \cup M(t) \cup P(t)$. The crossover recombines two individuals into two new bit strings. The mutation operation, on the other hand, flips the values of chosen bits to their complements. There are many ways how to do crossover and mutation as shown in Fig. 3. For example, Fig. 3a shows an example of crossover operation called *uniform crossover*, which generates offsprings by selecting each genes randomely from the corresponding genes of the parents.

```
program Genetic Algorithm;
 begin
  t := 0;
  {t: Number of generations.}
  initialize(P(t));
  {P(t): Population.}
  evaluate(P(t));
  while t ≤ Max. num. of gen. do
   begin
    C(t) := crossover(P(t));
    M(t) := mutation(P(t));
    evaluate(C(t) ∪ M(t));
    P(t+1) := select(C(t) ∪ M(t) ∪ P(t));
    t := t + 1;
   end
 end.
```

Fig. 2 Typical flow of GA.

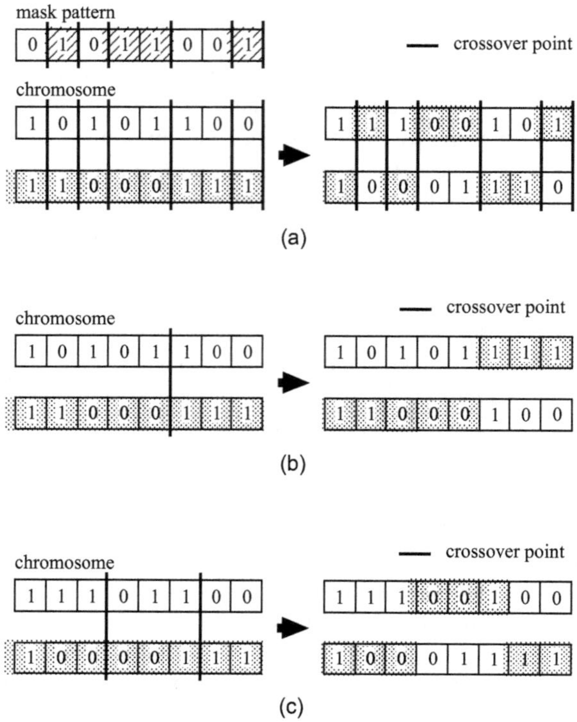

Fig. 3 Various crossover operations: (**a**) uniform crossover, (**b**) one-point crossover, and (**c**) two-point crossover.

3 Pitch Estimation System using GA

Figure 4 shows the overview of the proposed pitch estimation system. The estimation process of the system can be roughly divided into two phases.

Table 1 Instruments and pitches stored in the sound template.

Instrument	Pitch
Alt Saxophone (AS)	D3 - A5
Clarinet (CL)	D3 - B5
Horn (HR)	C3 - F5
Trumpet (TR)	E3 - B5
Viora (VL)	C3 - B5
Violin (VN)	G3 - B5
Piano (PF)	C3 - B5

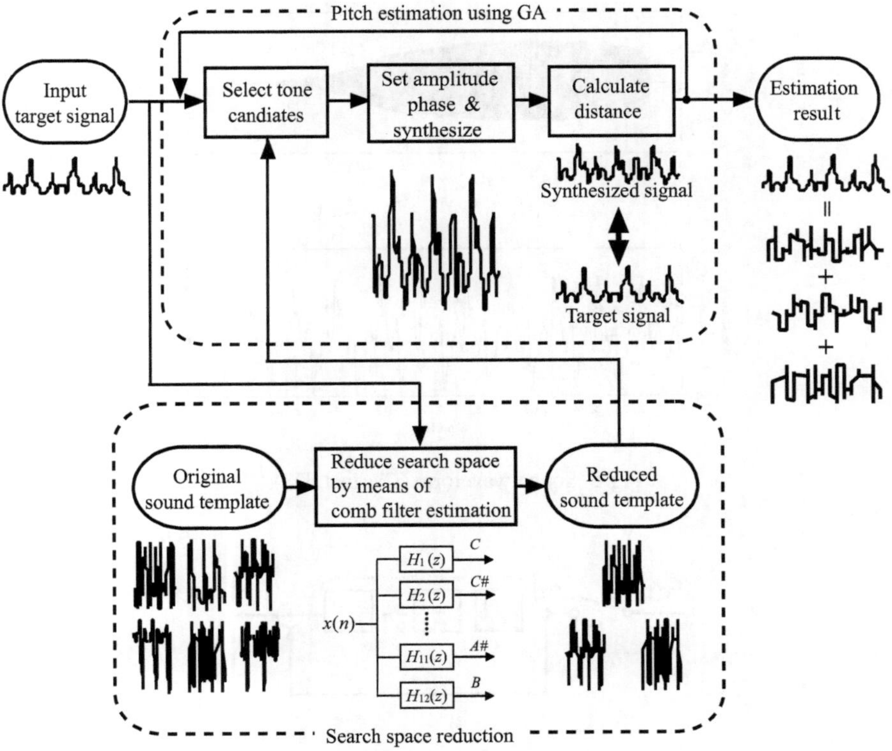

Fig. 4 Overview of the proposed system.

At first, we make a database which contains waveforms of sound templates. Then the system determines waveforms for the estimation by using the result of comb-filter-based pitch estimation.

After that, the system applies GA to search the optimal parameters such as amplitudes and phases, which minimize the squared error between a target waveform and a generated one. A more detailed description is provided in the following.

3.1 Sound Template

We make a database which contains a waveform of each instrument and tone shown in Table 1. In this experiment, the RWC music database, [3], is used as the original data. We use a waveform appeared after 500 ms from the beginning of a sound data as shown in Fig. 5, which is considered to have the level of the steady state amplitude (*sustain*), as a sound template. The sampling rate is 44.1 kHz, and the length of each template is 54 ms (2,385 points). Note that the maximum amplitude of each template is normalized to 1.0.

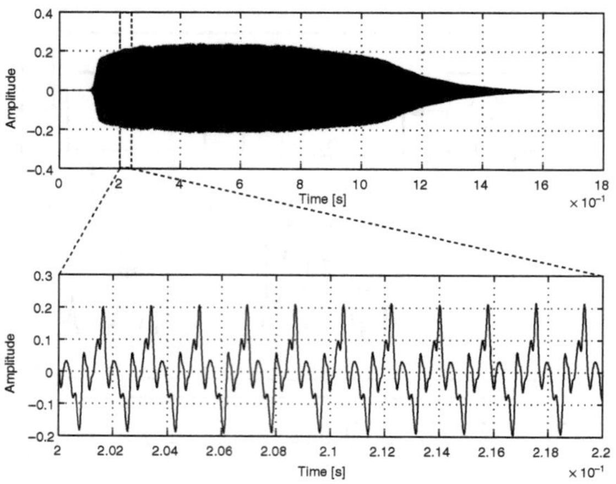

Fig. 5 Sound waveform (Clarinet, D3).

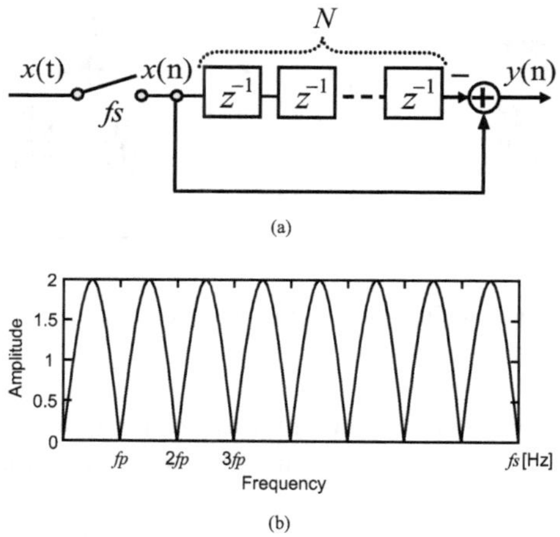

Fig. 6 Notch-type comb filter: (**a**) block diagram, and (**b**) frequency characteristic.

3.2 Search Space Reduction Using Comb Filter Estimation

The transfer function of the notch-type comb filter for a tone p is written by

$$H_p(z) = 1 - z^{-N_p}, N_p = [f_s/f_p],$$

where f_s is a sampling frequency, and f_p is a fundamental frequency for the tone p. The block diagram and its frequency characteristic are shown in Fig. 6a and 6b, respectively. The spectrum of a single note from a musical instrument usually has a set of peaks at harmonic ratios. That is, if the fundamental frequency is f_p, there are peaks at f_p, and also at $2f_p$, $3f_p$, $4f_p$, etc. Consequently, the operation of $H_p(z)$ eliminates all frequency components of the target tone at once.

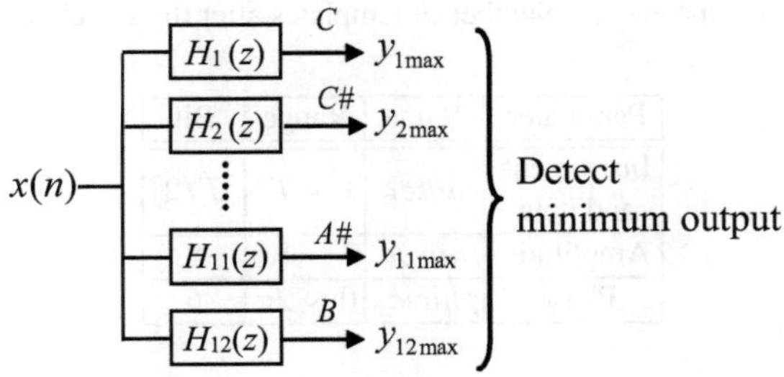

Fig. 7 Parallel-connected comb filters for pitch estimation.

As in the case of usual methods based on spectrum analysis, the comb filtering cannot separate the overlapped frequency components. For example, let a polyphony be composed of C3, E3, G3, and C4. We can estimate only the presence of C3, E3, and G3 by using a parallel-connected comb filters (Fig. 7), while the note C4 cannot be detected since every harmonic of the note C4 is overlapped to the even harmonics of the note C3. However, if we know the input sound is a chord of four notes, we can also estimate that a fundamental frequency of another tone is equal to the one of the known three tones. That is, the search space of sound templates can be reduced to the one which contains only the following tones:

Harmonics of C3: C4, G4, C5, E5,
Harmonics of E3: E4, B4, E5, G#5,
Harmonics of G3: G4, D4, G5, B5.

It achieves a reduction in the size of the search space and an improvement in the search efficiency.

3.3 Parameter Optimization Using GA

Figure 8 shows an individual representation in the system, which corresponds to a unique polyphonic sound. Here K is the number of tones included in a target polyphony, and amp_k, $note_k$, and $phase_k$ are parameters for kth notes, which have bit lengths and ranges of values shown in Table 2.

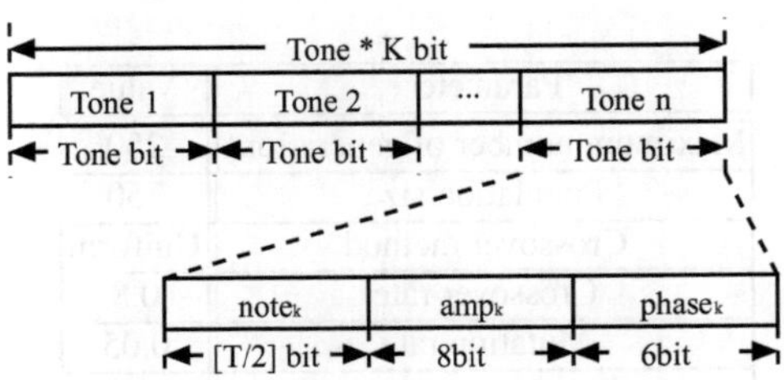

Fig. 8 Individual representation.

Table 2 Target parameters (T: Number of templates after the search space reduction).

Paramater	Name	Range	Bit
Instrument and note	$note_k$	$1 \sim T^*$	$\lceil T/2 \rceil$
Amplitude	amp_k	$0 \sim 1.0$	8
Phase	$phase_k$	$0 \sim 2\pi$	6

At the beginning of a new evolutionary run, the system creates a set of randomly generated bit strings with above data format to form the initial population. A generated bit string is interpreted as its corresponding waveform defined as follows:

$$x'(n) = \sum_{k=1}^{K} \sum_{n=1}^{N} amp_k \cdot x_{note_k}(n + phase_k \frac{f_s}{f_k \cdot 2\pi}),$$

where $x_{note_k}(n)$ is a waveform of template $note_k$, N is a window size, f_s is a sampling frequency, and f_k is a fundamental frequency of a template $note_k$. Through an evolutionary run, the system searches the optimal waveform which minimizes the error as follows:

$$error = \sum_{n=1}^{N} \{x(n) - x'(n)\}^2,$$

where $x(n)$ is a target waveform.

4 Experiments

We demonstrate the potential capability of the GA-based pitch estimation of musical consonance. Note that the input sounds used in this experiment are polyphony which contain at least one consonance whose fundamental frequency ratio can be simply represented by $1 : n$ (n: integer). Table 3 summarizes the system parameters in this experiment. A set of evolutionary runs were carried out on a Linux PC (CPU: Intel Xeon 2.8 GHz dual, RAM: 2GByte).

Table 3 Main parameter values for GA.

Parameter	Value
Maximum number of generations	250
Population size	750
Crossover method	Uniform
Crossover rate	0.8
Mutation rate	0.05
Mutation method	Uniform

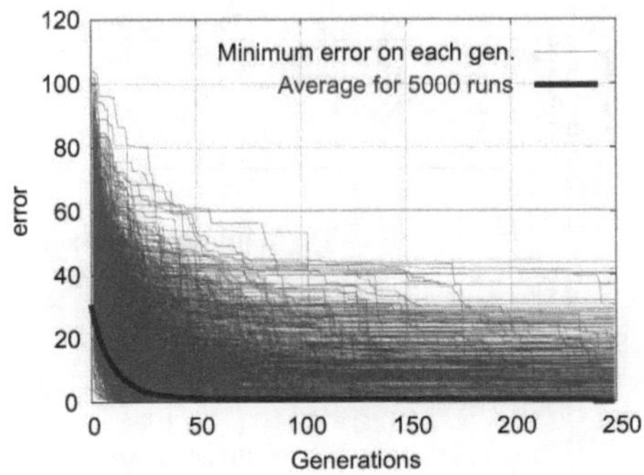

Fig. 9 Transition of the sum of squared errors between the target waveform and the synthesized one.

Figure 9 shows a fitness transition of the error between a generated waveform and a target one. Target sounds considered here are polyphony of randomly-selected three tones. Note that the input waveform includes at least one consonance. The horizontal axis indicates the error between a generated waveform having the best fitness value and a target waveform, and the vertical axis indicates the number of generations. We can see the staircase improvements of the best individual fitness on average.

Figures 10 and 11 depict an example snapshot of the population and a waveform of the best fitness individual on each generation. In Fig. 10, the horizontal axes indicate types of music instruments and pitches of tones, and the vertical axis indicates the

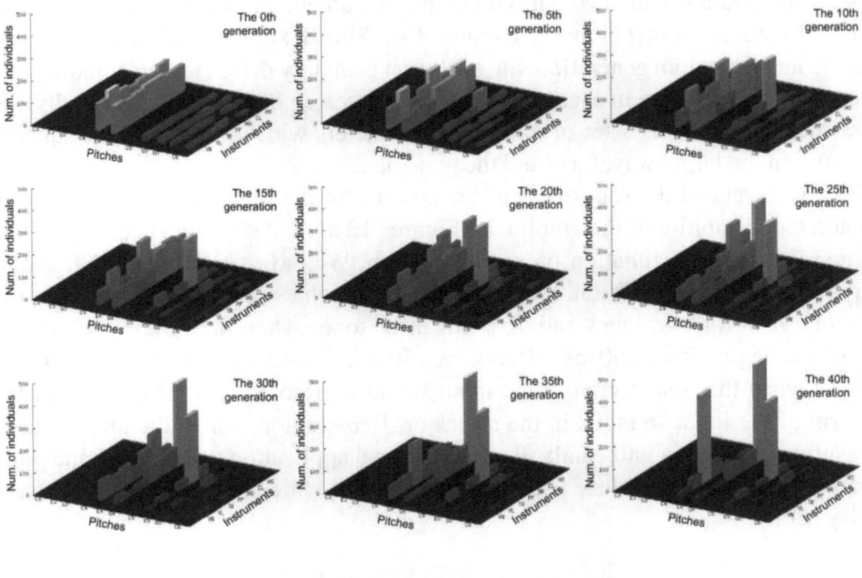

Fig. 10 Population transition.

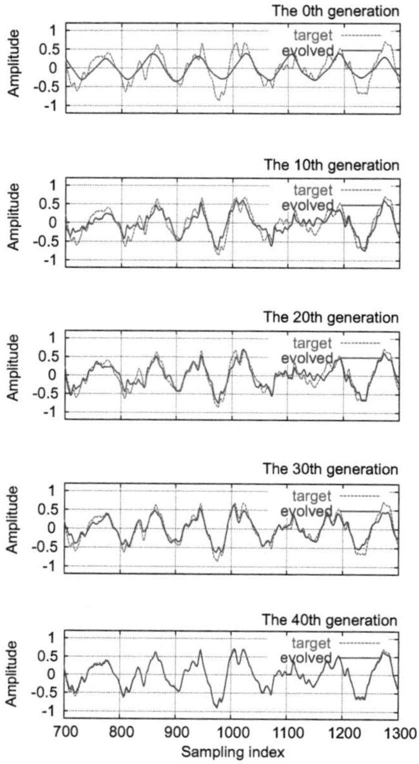

Fig. 11 Waveform transition.

number of individuals which contains a certain instrument and pitch. The input sound considered here is a polyphony of Trumpet C4, Alto Saxophone E4, and Horn C5. Given the initial random generation, the evolution is mainly driven towards finding one correct tone, then the system eventually shifts to the search of another tone. Finally, the system successfully finds a set of tones and parameters which minimizes the squared error between the target waveform and the generated one.

Next, we evaluate the robustness of the system for the increase of the number of tones and the instability of the amplitude. Figures 12 and 13 show the accuracy of the pitch and instrument estimation on each condition. We have performed 5,000 evolutionary runs for every condition. We can obtain the pitch estimation accuracy of more than about 90% for two tones and 70% for three tones when the time region of the target sound begins from 250 ms, 500 ms, or 750 ms. On the other hand, the accuracy decreases when the time region of the input sound is from 125 ms. This is because a waveform of the input sound is in the *attack* or *decay* region, where the amplitude of the waveform changes significantly. For the practical application to the automatic pitch estimation, we should introduce additional information which improves the robustness of the system.

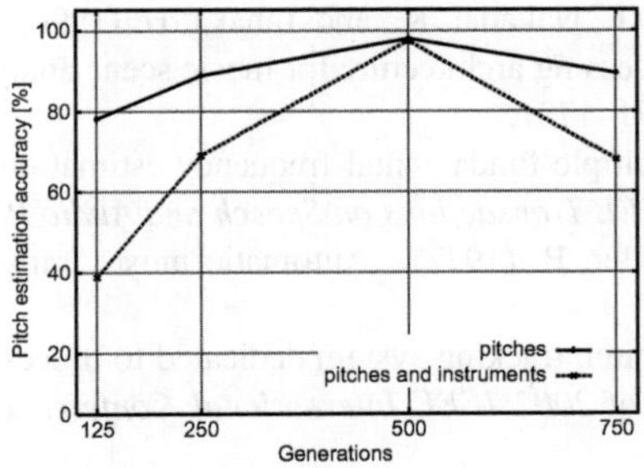

Fig. 12 Accuracy of pitch/instrument estimation for two notes.

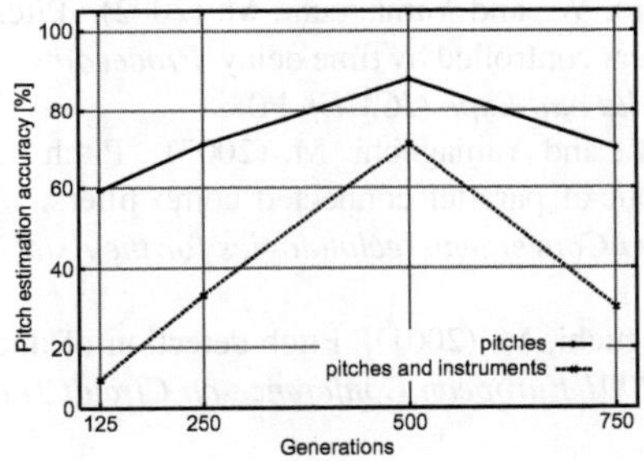

Fig. 13 Accuracy of pitch/instrument estimation for three notes.

5 Conclusions

In this chapter, we have presented a possibility of the GA-based pitch estimation method through the experimental pitch estimation of the musical consonance. An experimental pitch estimation system has a capability of analyzing the pitches of consonances, which have not been realized by the conventional methods.

References

1. Back, T., Hammel, U., Schwefel, P.H. (1997). *Evolutionary Computation: Comments on the History and Current State. IEEE Transactions Evolutionary Computation*, 1(1):3 – 13.
2. Goldberg, D. (1989). *Genetic Algorithms in Search, Optimization and Machine Learning*. Addison-Wasley Publishing Company.
3. Goto, M. (2004). Development of the RWC music database. *Proceedings of the 18th International Congress on Acoustics (ICA 2004)*, I–553–556.
4. Holland, J. H. (1975). *Adaptation in Natural and Artificial Systems*. University of Michigan Press.

5. Kashino, K., Kinoshita, T., Nakadai, K., and Tanaka, H. (1996). Chord recognition mechanisms in the optima processing architecture for music scene analysis. *Transactions IEICE of Japan*, J79-D-II(11):1771–1781.
6. Klapuri, A. (2003). Multiple fundamental frequency estimation based on harmonicity and spectral smoothness. *IEEE Transactions on Speech and Audio Processing*, 11(6):804–816.
7. Piszczalski, M. and Galler, B. (1977). Automatic music transcription. *Computer Music Journal*, 1(4):24–31.
8. Pollastri, E. (2002). A pitch tracking system dedicated to process singing voice for musical retrieval. *Proceedings of 2002 IEEE International Conference on Multimedia and Expo (ICME)*.
9. Roads, C. (1985). Research in music and artifical intelligence. *ACM Computing Surveys*, 17(2):163–190.
10. Roads, C. (1996). *The Computer Music Tutorial.* MIT Press.
11. Sterian, A. and Wakefield, G. (2000). Music transcription systems: from sound to symbol. *Proceedings of AAAI-2000 Workshop on Artifical Intelligence and Music.*
12. Tadokoro, Y., Matsumoto, W., and Yamaguchi, M. (2002). Pitch detection musical sounds using adaptive comb filters controlled by time delay. *Proceedings of 2002 IEEE International Conference on Multimedia and Expo (ICME)*, P03.
13. Tadokoro, Y., Morita, T., and Yamaguchi, M. (2003). Pitch detection of musical sounds noticing minimum output of parallel connected comb filters. *Proceedings of 2003 IEEE Region 10 Conference on Convergent Technologies for the Asia-Pacific (TENCON)*, tencon-072.
14. Tadokoro, Y. and Yamaguchi, M. (2001). Pitch detection of duet song using double comb filters. *Proceedings of 2001 European Conference on Circuit Theory and Design (ECCTD)*, I:57–60.

Controlling the Lorenz System with Delay

Yechiel J. Crispin

Department of Aerospace Engineering, Embry-Riddle University
Daytona Beach, FL 32114, USA
crispinj@erau.edu

Abstract. A generalized method for adaptive control, synchronization of chaos and parameter identification in systems governed by ordinary differential equations and delay-differential equations is developed. The method is based on the Lagrangian approach to fluid dynamics. The synchronization error, defined as a norm of the difference between the state variables of two similar and coupled systems, is treated as a scalar fluid property advected by a fluid particle in the vector field of the controlled response system. As this error property is minimized, the two coupled systems synchronize and the time variable parameters of the driving system are identified. The method is applicable to the field of secure communications when the variable parameters of the driver system carry encrypted messages. The synchronization method is demonstrated on two Lorenz systems with variable parameters. We then apply the method to the synchronization of hyperchaos in two modified Lorenz systems with a time delay in one the state variables.

Keywords. Adaptive control, chaos, hyperchaos, parameter estimation, signal processing, Lagrangian fluid dynamics, chaotic advection, nonlinear dynamics, nonlinear systems and modeling.

1 Introduction

In this chapter we develop a generalized method for controlling chaos in dynamical systems and for synchronizing two coupled chaotic or hyperchaotic systems. The systems are described by ordinary differential equations with initial conditions or delay-differential equations with initial functions. The independent parameters appearing in the equations can be constant or variable. The synchronization error , defined as a norm of the difference between the state variables of the two chaotic systems, is viewed as a fluid property advected by a marker particle moving along a trajectory in the vector field of the response system. The controlled parameters of the response system are varied continuously such as to minimize the synchronization error while the two chaotic systems evolve in time. For the initial value problem, we derive a system of differential equations governing the evolution of the controlled parameters required for synchronization. For the case of the initial function problem, we derive a system of delay-differential equations governing the controlled parameters for synchronizing the response system. In both cases, the fluid dynamical approach mentioned above is used to develop the equations for the controlled parameters.

The method is demonstrated by studying several examples of chaotic and hyperchaotic systems. The synchronization method is demonstrated on two Lorenz systems with variable parameters. We then apply the method to the synchronization of hyperchaos in two modified Lorenz systems with a time delay in one of the state variables.

The possibility of synchronizing two chaotic physical systems has attracted considerable attention in recent years [2]. A major motivation is the potential of applying the synchronization methods to the field of secure communications by chaotic masking or scrambling of messages [11] [25]. Other important applications are in the field of nonstationary time series analysis and system identification [21] [5]. So far, most of the attention has been directed towards the study of stationary chaotic systems, that is systems with constant parameters. To date, the possibility of using systems with nonstationary parameters for secure communications has received less attention, although such systems are good candidates for secure communications, especially when the nonstationary parameters rather than the state variables are used to hide the secret messages [8].

Several methods for the control of chaos in dynamical systems have been proposed in recent years [3]. The conjecture of chaos control by means of perturbations of an accessible parameter is based on an inherent property of chaotic systems, namely, their sensitivity to small perturbations in the parameters [22] [4] [20].

Model reference adaptive control methods have been suggested for chaos suppression and synchronization [1]. For example, a model reference method for the adaptive control of chaos in dynamical systems with periodic forcing has been proposed by Crispin and Ferrari [10]. Adaptive control and synchronization of chaos in discrete time systems has been studied by Crispin [9]. Another method for controlling chaos in dynamical systems consists of using parametric forcing and adaptive control, see for example [6]. A more recent method based on an analogy from fluid dynamics has been described by Crispin [5]. Other applications include parameter estimation [22], synchronization of chaotic systems with variable parameters [8] and the control of chaos in fluids [7].

Many physical, physiological and biological systems display time delay in their dynamics. Nonlinear dynamical systems with time delay can have periodic orbits or very complex dynamics depending upon the range of values of the time delay and the independent parameters of the system. This complex behavior has attracted a lot of interest in the study of time delayed systems from the mathematical point of view [14] as well as from the physical and physiological points of view, see for example [16] [17] [18]. The use of time delay in feedback control systems has also been proposed, see for example [12] [11] [13]. Control of chaos in systems with time delay has also been studied in [17].

2 The Fluid Dynamical Approach

Consider two similar dynamical systems described by ordinary differential equations. We define two dynamical systems as similar if the right hand sides of the equations are represented by the same function $f(x(t), p(t))$, except that the independent parameters $p(t)$ or $q(t)$ can be represented either by different or the same functions of time. We first present the method for the initial value problem and then we consider the case when one of the state variables has a time delay.

$$d\boldsymbol{x}/dt = \boldsymbol{f}(\boldsymbol{x}(t), \boldsymbol{p}(t)) \tag{1}$$

$$dy/dt = f(y(t), q(t)) \tag{2}$$

Here t is time, $x \in \mathrm{R}^n$ are the state variables of the driver and $y \in \mathrm{R}^n$ are the state variables of the response system. The parameters $p(t) \in \mathrm{R}^k$ and $q(t) \in \mathrm{R}^k$ are independent time variable parameters of the respective systems and f:R^nx $\mathrm{R}^k \mapsto \mathrm{R}^n$ is a nonlinear vector function of the state variables. It is assumed that the initial values of the state variables $x(0) = \mathrm{x}_0$ and $y(0) = \mathrm{y}_0$ for $t = 0$ are not necessarily the same. Similarly the initial values of the parameters $p(0) = p_0$ and $q(0) = q_0$ of the driver and response systems are different. Since chaotic systems are sensitive to initial conditions, the driver and response systems will not synchronize, unless the response system is controlled and forced to synchronize with the driver system using some kind of coupling, such as a transmitted scalar signal. For instance, a single scalar signal $s(t)$, which is a function of the state $x(t)$, can be transmitted by the driver and used to enslave the response system [24] [23].

$$s(t) = h(x(t)) \tag{3}$$

As stated above, the purpose of this chapter is to propose a generalized method of control, stabilization, synchronization and parameter identification of chaotic systems in the more general case where the parameters $p(t)$ of the driver system vary as a function of time. In the context of secure communications, this means that it would be possible to encode a message in one of the parameters of the driver system rather then in a state variable, as has been proposed so far. Once a variable parameter is identified by the response system using the proposed generalized method, the encoded message can be recovered. The method allows more flexibility in masking information in chaos. The message can be encoded in a state variable or in a time variable parameter. The useful information can also be split into two messages, where one message is modulated by a state variable and a second message modulated by a parameter. Synchronization of the state variables of an eavesdropping response system with the state variables of the driver will be difficult because of the sensitivity to small variations in the parameters of the system, in addition to the sensitivity to initial conditions and the divergence of nearby trajectories in chaotic systems. Also, synchronization can be achieved even when the parameters $p(t)$ and $q(t)$ are initially substantially different. This is accomplished by controlling the response system y such that the parameters $p(t)$ of the driver system x are eventually identified, that is,

$$\lim_{t\to\infty} |p(t) - q(t)| = 0 \tag{4}$$

In order to achieve synchronization, the dynamics of the response system parameters $q(t)$ need to be determined. In other words, the differential equations governing the evolution of the response parameters $q(t)$ need to be derived for dynamical systems of the form of (1), (2).

The proposed fluid dynamical approach is based on the Lagrangian description of fluid motion. It follows the motion of a fluid particle as it moves in the velocity vector field $w(x, p)$ created by a fluid flow [15] [19]. According to this approach, the equations of motion of two marker particles advected in the fluid flows described by the vector fields $w(x, p) = f(x, p)$ and $w(y, q) = f(y, q)$, are given by (1), (2) where the right hand sides are to be interpreted as the local velocity vectors $w(x, p)$ and $w(y, q)$ at any given point $x \in \mathrm{R}^n$ and $y \in \mathrm{R}^n$ of the two flow fields, respectively, i.e.,

$$d\boldsymbol{x}/dt = \boldsymbol{w}(\boldsymbol{x}, \boldsymbol{p}) = \boldsymbol{f}(\boldsymbol{x}, \boldsymbol{p}) \tag{5}$$

$$d\boldsymbol{y}/dt = \boldsymbol{w}(\boldsymbol{y}, \boldsymbol{q}) = \boldsymbol{f}(\boldsymbol{y}, \boldsymbol{q})$$

In fluid dynamics, the vector fields $\boldsymbol{w}(\boldsymbol{x}, \boldsymbol{p}), \boldsymbol{w}(\boldsymbol{y}, \boldsymbol{q}) \in \mathrm{R}^n$ and the state variables $\boldsymbol{x}, \boldsymbol{y} \in \mathrm{R}^n$ have a dimension $n \leq 3$. Here the analogy is extended to vector fields of higher dimensions. Consider the time variation of a scalar property $J(\boldsymbol{x}, \boldsymbol{y})$ of the flow along a trajectory of the response system as it evolves in the state space $\boldsymbol{y} \in \mathrm{R}^n$. The rate of change of this scalar property is due to two contributions: a local contribution due to its time variation, plus a contribution which is due to the rate of change of the property as it is advected along the trajectory in the state space. The total rate of change is then given by the substantial derivative DJ/Dt of the scalar property J, which is the derivative following the flow:

$$DJ/Dt = \partial J/\partial t + \boldsymbol{f}(\boldsymbol{y}(t), \boldsymbol{q})\nabla J \tag{6}$$

$$\nabla = (\partial/\partial y_1, \partial/\partial y_2, ..., \partial/\partial y_n)^t$$

Here the product $\boldsymbol{f}(\boldsymbol{y}, \boldsymbol{q})\nabla J$ is a scalar or dot product. Consider a scalar property J based on the Euclidean distance between the state vectors $\boldsymbol{x}$ and $\boldsymbol{y}$:

$$J = \frac{1}{2}|\boldsymbol{y} - \boldsymbol{x}|^2 = \frac{1}{2}\sum_{i=1}^{n}(y_i - x_i)^2 = \frac{1}{2}\sum_{i=1}^{n} e_i^2 \tag{7}$$

where $e_i = y_i - x_i$ are components of the error vector function $\boldsymbol{e} = \boldsymbol{y} - \boldsymbol{x}$. Using (7), the local component of the derivative of J with respect to time is given by:

$$\partial J/\partial t = \sum_{i=1}^{n} e_i \, de_i/dt \tag{8}$$

whereas the components of the gradient ∇J are given by

$$\partial J/\partial y_i = y_i - x_i = e_i \tag{9}$$

Using (5), (6), (8) and (9), the substantial derivative of J is written as:

$$DJ/Dt = \sum_{i=1}^{n} e_i[de_i/dt + f_i(\boldsymbol{y}(t), \boldsymbol{q})] \tag{10}$$

and since $\boldsymbol{e} = \boldsymbol{y} - \boldsymbol{x}$ and

$$d\boldsymbol{e}/dt = d\boldsymbol{y}/dt - d\boldsymbol{x}/dt = \boldsymbol{f}(\boldsymbol{y}, \boldsymbol{q}) - \boldsymbol{f}(\boldsymbol{x}, \boldsymbol{p})$$

Equation (10) reduces to:

$$DJ/Dt = \sum_{i=1}^{n} e_i[2f_i(\boldsymbol{y}(t), \boldsymbol{q}) - f_i(\boldsymbol{x}(t), \boldsymbol{p})] \tag{11}$$

Equation (11) defines the rate of change of the positive scalar property J in terms of the state variables $\boldsymbol{x}$ and $\boldsymbol{y}$ of the driver and response systems, the independent driver parameters $\boldsymbol{p}$ and the controllable parameters $\boldsymbol{q}$ of the response system. The question now is how should the control vector $\boldsymbol{q}$ be varied such as to continuously minimize J? As the driver and response systems evolve, the substantial derivative should be continuously decreased in order to achieve control and synchronization, as can be seen from (11), where perfect synchronization is reached when $\boldsymbol{e} = \boldsymbol{y} - \boldsymbol{x} = \boldsymbol{0}$, and from (7), J reaches the minimum $J = 0$. A possible control law is to vary the control vector $\boldsymbol{q}$ such as to decrease the substantial derivative DJ/Dt, that is, to continuously change the control $\boldsymbol{q}$ in a direction opposite to the gradient of DJ/Dt with respect to the control $\boldsymbol{q}$:

$$d\boldsymbol{q}/dt = -G^{'}\,\nabla_{\boldsymbol{q}}(DJ/Dt) =$$

$$= -G^{'}\,\nabla_{\boldsymbol{q}}\{\sum_{i=1}^{n} e_i[2f_i(\boldsymbol{y}(t),\boldsymbol{q}) - f_i(\boldsymbol{x}(t),\boldsymbol{p})]\} \quad (12)$$

where

$$\nabla_{\boldsymbol{q}} = (\partial/\partial q_1, \partial/\partial q_2, ..., \partial/\partial q_k)^t$$

and $G^{'}$ is a kxk matrix of control gains. Since $f_i(\boldsymbol{x},\boldsymbol{p})$ does not depend on the control $\boldsymbol{q}$, it follows that

$$\nabla_{\boldsymbol{q}} f_i(\boldsymbol{x}(t),\boldsymbol{p}) = 0,$$

and (12) becomes:

$$d\boldsymbol{q}/dt = -G\,\nabla_{\boldsymbol{q}}\,[\sum_{i=1}^{n} e_i f_i(\boldsymbol{y}(t),\boldsymbol{q})] \quad (13)$$

where $G = 2G^{'}$.

3 Synchronizing the Lorenz System

We now apply the method to a dynamical system without delay, the chaotic Lorenz system. Consider the case of synchronization between two Lorenz systems with variable parameters, where the driver system is given by:

$$dx_1/dt = \sigma(x_2 - x_1)$$

$$dx_2/dt = p_1(t)x_1 - p_2(t)x_2 - p_3(t)x_1x_3 \quad (14)$$

$$dx_3/dt = x_1x_2 - bx_3$$

and the transmitted coupling signal for synchronization is chosen as the single variable:

$$s(t) = h(\boldsymbol{x}(t)) = x_2(t) \quad (15)$$

The response system is defined by the following Lorenz system with variable parameters. It is driven by the transmitted signal $s(t) = x_2(t)$.

$$dy_1/dt = \sigma(s(t) - y_1)$$

$$dy_2/dt = q_1(t)y_1 - q_2(t)y_2 - q_3(t)y_1y_3 \tag{16}$$

$$dy_3/dt = y_1s(t) - by_3$$

The next step is to derive a system of differential equations governing the evolution of the controlled parameters $q_1(t)$, $q_2(t)$ and $q_3(t)$. As the response system evolves and synchronizes with the driver system, the parameters $q_1(t)$, $q_2(t)$ and $q_3(t)$ will follow the original parameters $p_1(t)$, $p_2(t)$ and $p_3(t)$ of the driver system. According to (13) of the previous section, the first step is to develop the term $\sum_{i=1}^{3} e_i f_i(\boldsymbol{y}, \boldsymbol{q})$ for the response system (16). Using the right hand sides of (16), we have:

$$\sum_{i=1}^{3} e_i f_i(\boldsymbol{y}, \boldsymbol{q}) = (y_1 - x_1)[\sigma(s(t) - y_1)]+$$

$$+(y_2 - x_2)[q_1(t)y_1 - q_2(t)y_2 - q_3(t)y_1y_3]+$$

$$+(y_3 - x_3)(y_1s(t) - by_3) \tag{17}$$

The gradient with respect to the parameters $\boldsymbol{q}$ is given by:

$$\nabla_{\boldsymbol{q}} [\sum_{i=1}^{3} e_i f_i(\boldsymbol{y}, \boldsymbol{q})] =$$

$$= [(y_2 - x_2)y_1\,,\, -(y_2 - x_2)y_2\,,\, -(y_2 - x_2)y_1y_3]^T \tag{18}$$

The differential equations governing the evolution of the controlled parameters $\boldsymbol{q}$ are given by:

$$dq_1/dt = -G_{11}(y_2 - s(t))y_1 + G_{12}\,(y_2 - s(t))y_2+$$

$$+G_{13}\,(y_2 - s(t))y_1y_3$$

$$dq_2/dt = -G_{21}(y_2 - s(t))y_1 + G_{22}\,(y_2 - s(t))y_2+$$

$$+G_{23}\,(y_2 - s(t))y_1y_3 \tag{19}$$

$$dq_3/dt = -G_{31}(y_2 - s(t))y_1 + G_{32}\,(y_2 - s(t))y_2+$$

$$+G_{33}\,(y_2 - s(t))y_1y_3$$

For example, consider the case where only one parameter, say $p_1(t)$ is to be identified, whereas the other two parameters $p_2(t)$ and $p_3(t)$ are known constants. If the synchronized systems are used for secure communication, the parameter $p_1(t)$ can be used to carry a hidden message, which can be identified by the response system. In this case, (17), (18), (19) reduce to:

$$\sum_{i=1}^{3} e_i f_i(\boldsymbol{y}, \boldsymbol{q}) = (y_1 - x_1)[\sigma(s(t) - y_1)]+$$

$$+(y_2 - x_2)[q_1(t)y_1 - p_2(t)y_2 - p_3(t)y_1y_3]+$$

$$+(y_3 - x_3)(y_1y_2 - by_3) \tag{20}$$

$$\nabla_{\boldsymbol{q}}[\sum_{i=1}^{3} e_i f_i(\boldsymbol{y}, \boldsymbol{q})] = [(y_2 - x_2)y_1\,,\,0\,,\,0]^T \tag{21}$$

$$dq_1/dt = -G_{11}(y_2 - s(t))y_1 \tag{22}$$

$$dq_2/dt = 0$$

$$dq_3/dt = 0$$

We now show results of a computer simulation with the following values of the parameters:

$$\sigma = 10 \;\; b = 8/3 \tag{23a}$$

$$p_1(t) = r_0 + \delta r\; sin\,\omega t \tag{23b}$$

$$\delta r = 2 \;\; \omega = 2\pi/10 \;\; r_0 = 28 \tag{23c}$$

$$q_2 = p_2 = 1 \;\; q_3 = p_3 = 1 \tag{23d}$$

together with the initial condition:

$$q_1(0) = r_0 = 28 \tag{23e}$$

The results of this example are given in Figs. 1, 2, 3. Figure 1 shows the chaotic state variables of the driver system. The chaotic attractor is shown in Fig. 2. Similar results are obtained for the response system as it synchronizes with the driver system. Figure 3 displays the synchronized state variables $y_1 = x_1$, $y_2 = x_2$ and $y_3 = x_3$, all three eventually converging to straight lines as shown in the figure. The identified signal $q_1(t) - r$ converges to the driver signal $q_1(t) - r = \delta r\; sin\,\omega t$ and is also shown in the figure.

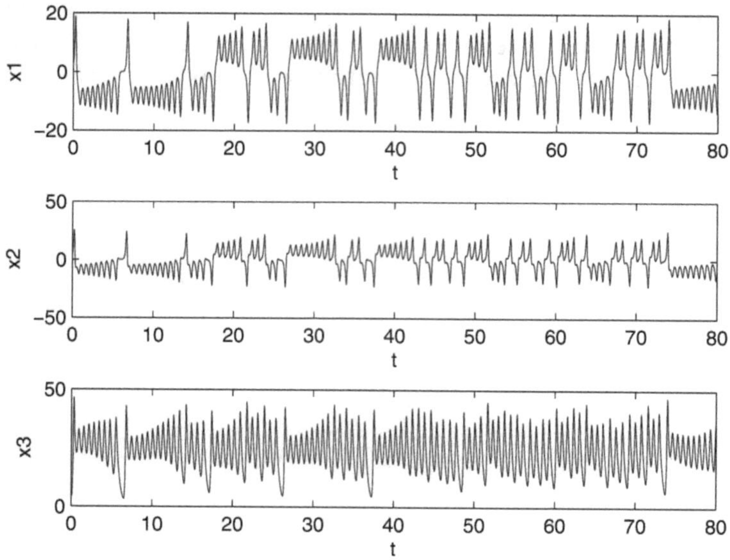

Fig. 1 The chaotic state variables of the driver Lorenz system with variable parameter.

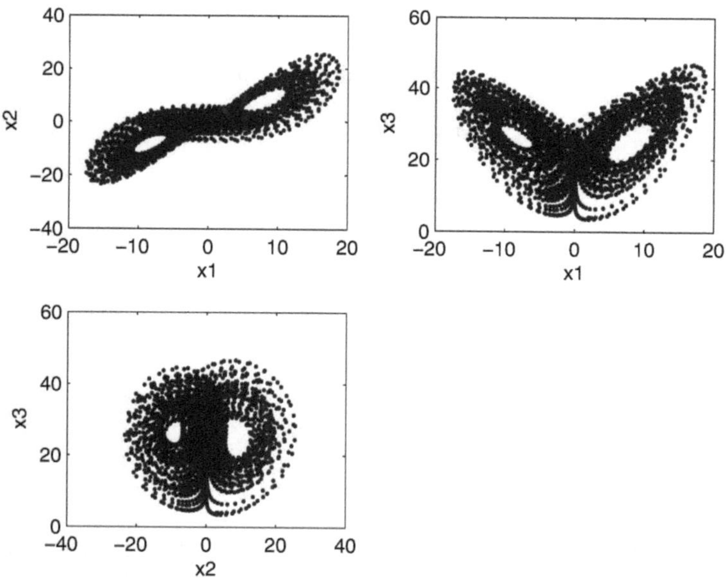

Fig. 2 The chaotic attractor of the driver Lorenz system with variable parameter. The response system has a similar attractor.

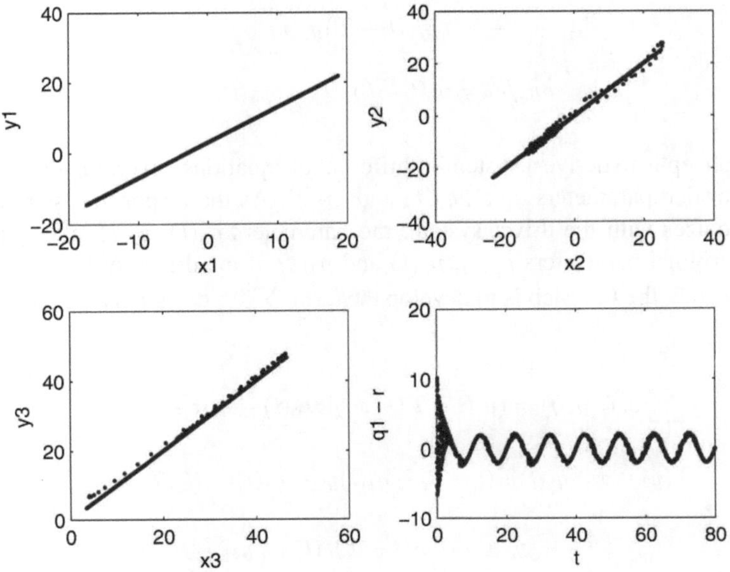

Fig. 3 Synchronization between the driver and response Lorenz systems with variable parameters and the parameter q_1(t) of the response system as it identifies the driver signal p_1(t).

4 Synchronizing Lorenz System with Delay

In this section we apply the method to a hyperchaotic dynamical system, the Lorenz system with a time delay in one of the state variables. Consider the case of synchronization between two Lorenz systems with time delay. We treat the case where the state variable $x_1(t-T)$ is delayed by a time delay T. Here the driver system is given by:

$$dx_1/dt = \sigma(x_2(t) - x_1(t-T))$$

$$dx_2/dt = p_1(t)x_1(t-T) - p_2(t)x_2(t) - \\ -p_3(t)x_1(t-T)x_3(t)$$

$$dx_3/dt = x_1(t-T)x_2(t) - bx_3(t) \tag{24}$$

Suppose the transmitted signal for synchronization is chosen as the single variable:

$$s(t) = h(\boldsymbol{x}(t)) = x_2(t) \tag{25}$$

The response system is defined by the following nonstationary and delayed Lorenz system, where the transmitted scalar signal $s(t) = x_2(t)$ is used to drive the system.

$$dy_1/dt = \sigma(s(t) - y_1(t-T))$$

$$dy_2/dt = q_1(t)y_1(t-T) - q_2(t)y_2(t) - \\ -q_3(t)y_1(t-T)y_3(t)$$

$$dy_3/dt = y_1(t-T)s(t) - by_3(t) \tag{26}$$

The next step is to derive a system of differential equations governing the evolution of the controlled parameters $q_1(t)$, $q_2(t)$ and $q_3(t)$. As the response system evolves and synchronizes with the driver system, the parameters $q_1(t)$, $q_2(t)$ and $q_3(t)$ will follow the original parameters $p_1(t)$, $p_2(t)$ and $p_3(t)$ of the driver system. According to (13) of Sect. 2, the first step is to develop the term $\sum_{i=1}^{3} e_i f_i(\boldsymbol{y},\boldsymbol{q})$ for the response system (26).

$$\sum_{i=1}^{3} e_i f_i(\boldsymbol{y},\boldsymbol{q}) = (y_1(t-T) - x_1)[\sigma(s(t) - y_1(t-T))] + \\ +(y_2 - x_2)[q_1(t)y_1(t-T) - q_2(t)y_2 - q_3(t)y_1(t-T)y_3] + \\ +(y_3 - x_3)[y_1(t-T)s(t) - by_3] \tag{27}$$

The gradient with respect to the parameters $\boldsymbol{q}$ is given by:

$$\nabla_{\boldsymbol{q}} [\sum_{i=1}^{3} e_i f_i(\boldsymbol{y},\boldsymbol{q})] = \\ = [(y_2 - x_2)y_1(t-T)\,,\, -(y_2 - x_2)y_2\,,\, -(y_2 - x_2)y_1(t-T)y_3]^t \tag{28}$$

The differential equations governing the evolution of the controlled parameters $\boldsymbol{q}$ are given by:

$$dq_1/dt = -G_{11}(y_2 - s(t))y_1(t-T) + G_{12}\,(y_2 - s(t))y_2 + \\ +G_{13}\,(y_2 - s(t))y_1(t-T)y_3$$

$$dq_2/dt = -G_{21}(y_2 - s(t))y_1(t-T) + G_{22}\,(y_2 - s(t))y_2 + \\ +G_{23}\,(y_2 - s(t))y_1(t-T)y_3$$

$$dq_3/dt = -G_{31}(y_2 - s(t))y_1(t-T) + G_{32}\,(y_2 - s(t))y_2 + \\ +G_{33}\,(y_2 - s(t))y_1(t-T)y_3 \tag{29}$$

For example, consider the case where only one parameter, say $p_1(t)$ is to be identified, whereas the other two parameters $p_2(t)$ and $p_3(t)$ are known constants. If the synchronized systems are used for secure communication, the parameter $p_1(t)$ can carry a hidden message, which can be identified by the response system. In this case, (27), (28), (29) reduce to:

$$\sum_{i=1}^{3} e_i f_i(\boldsymbol{y}, \boldsymbol{q}) = (y_1 - x_1)[\sigma(s(t) - y_1)] +$$

$$+(y_2 - x_2)[q_1(t)y_1 - p_2 y_2 - p_3 y_1 y_3] + (y_3 - x_3)(y_1 y_2 - b y_3) \quad (30)$$

$$\nabla_{\boldsymbol{q}} [\sum_{i=1}^{3} e_i f_i(\boldsymbol{y}, \boldsymbol{q})] = [(y_2 - x_2) y_1(t - T), 0\ , 0]^T \quad (31)$$

$$dq_1/dt = -G_{11}(y_2 - s(t)) y_1(t - T) \quad (32)$$

$$dq_2/dt = 0$$

$$dq_3/dt = 0$$

Here we show results of a computer simulation with the following values of the parameters:

$$\sigma = 10\ b = 8/3\ T = 0.1\ G11 = 10 \quad (33a)$$

$$p_1(t) = r_0 + \delta r\ sin\,\omega t \quad (33b)$$

$$\delta r = 2\ \omega = 2\pi/10\ r_0 = 28 \quad (33c)$$

$$q_2 = p_2 = 1\ \ q_3 = p_3 = 1 \quad (33d)$$

together with the initial condition:

$$q_1(0) = r_0 = 28 \quad (33e)$$

The results of this example are given in Figs. 4, 5, 6. Figure 4 shows the hyperchaotic state variables of the driver system. A comparison with Fig. 1 above, it can be seen that the state variables in Fig. 4 display a more complex type of chaos, because of the time delay. Comparing with the attractor of Fig. 2 above, it is apparent that the hyperchaotic attractor shown in Fig. 5 displays a more complex behavior. Similar results are obtained for the response system as it synchronizes with the driver system. Figure 6 displays the synchronized state variables $y_1 = x_1, y_2 = x_2$ and $y_3 = x_3$, all three eventually converging to straight lines as shown in the figure. The identified signal $q_1(t) - r$ converges to the driver signal $q_1(t) - r = \delta r\ sin\,\omega t$ and is also shown in the figure.

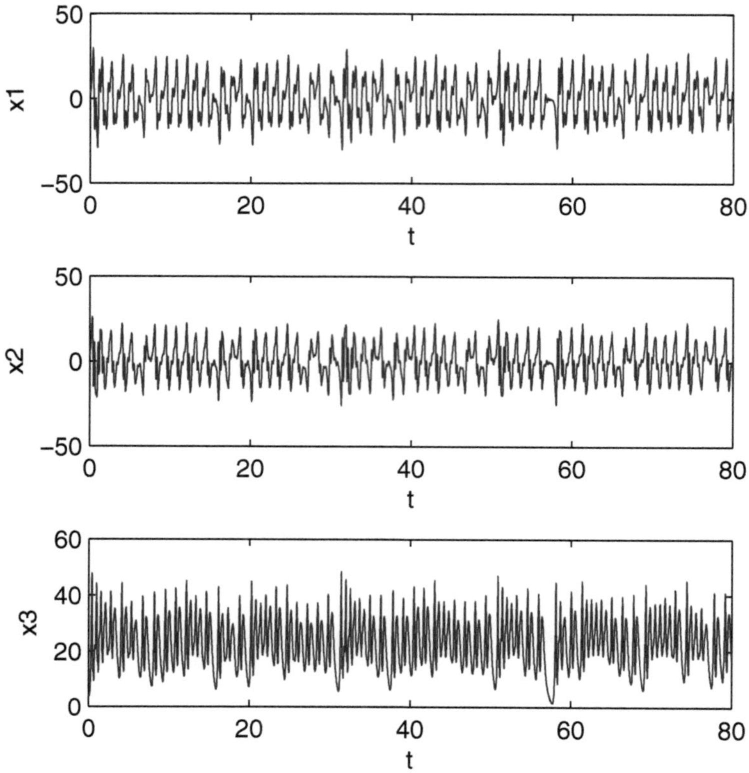

Fig. 4 The hyperchaotic state variables of the driver Lorenz system with delay and variable parameter.

5 Conclusions

A generalized method for adaptive control, synchronization of chaos and parameter identification in systems governed by ordinary differential equations and delay – differential equations has been presented. The method is based on the Lagrangian approach to fluid dynamics. The synchronization error is treated as a scalar fluid property advected in the vector field of the controlled response system. Upon minimizing this error, the two coupled systems synchronize and the time variable parameters of the driving system are identified. The method was used to synchronize two Lorenz systems with variable parameters. The method was also applied to the synchronization of hyperchaos in two modified Lorenz systems with a time delay in one the state variables. Some implications of using the method in the field of secure communications where the transmitted information is masked by chaos or hyperchaos have been discussed.

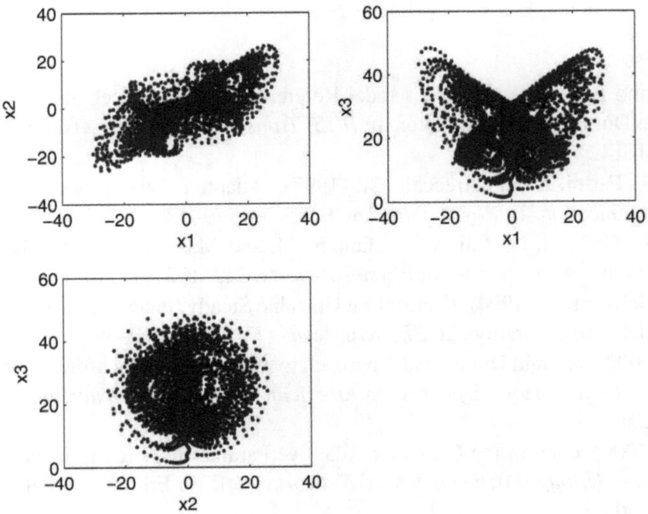

Fig. 5 The hyperchaotic attractor of the driver Lorenz system with delay and variable parameter. The response system has a similar attractor.

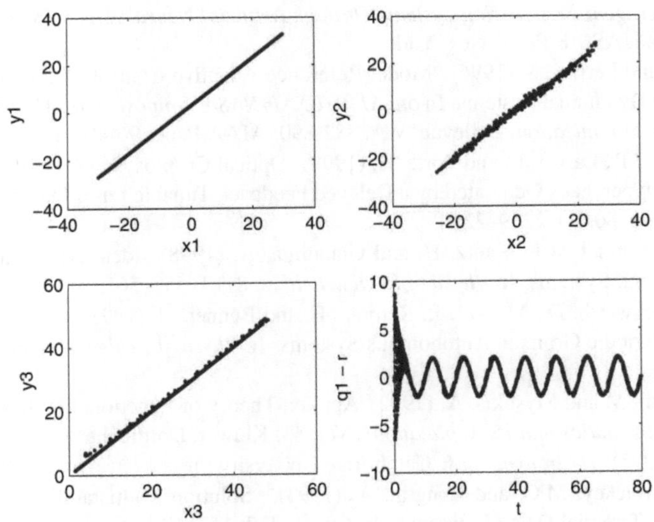

Fig. 6 Synchronization between the hyperchaotic driver and response Lorenz systems with delay and variable parameters and the parameter q_1(t) of the response system as it identifies the driver signal p_1(t).

References

1. Aguirre, L. and Billings, S. (1994). Model Reference Control of Regular and Chaotic Dynamics in the Duffing-Ueda Oscillator. In *IEEE Transactions on Circuits and Systems I*, 41, 7, 477–480. IEEE.
2. Boccaletti, S., Farini, A. and Arecchi, F.T. (1997). Adaptive Synchronization of Chaos for Secure Communication. In *Physical Review E*, 55, 5, 4979–4981.
3. Boccaletti, S., Grebogi, C., Lai, Y.C., Mancini, H. and Maza, D. (2000). The Control of Chaos: Theory and Applications. In *Physics Reports*, 329, 103–197, Elsevier.
4. Carr, T. and Schwartz, I. (1994). Controlling Unstable Steady States Using System Parameter Variation and Control Duration. In *Physical Review E*, 50, 5, 3410–3415.
5. Crispin, Y. (2002). A Fluid Dynamical Approach to the Control, Synchronization and Parameter Identification of Chaotic Systems. In *American Control Conference, ACC*, Anchorage, AK, 2245–2250.
6. Crispin, Y. (2000) Controlling Chaos by Adaptive Parametric Forcing. In *Intelligent Engineering Systems Through Artificial Neural Networks*, Vol. 10, Edited by Dagli et al., ASME Press, New York.
7. Crispin, Y. (1999) Control and Anticontrol of Chaos in Fluids. In *Intelligent Engineering Systems Through Artificial Neural Networks*, Vol. 9, Edited by Dagli et al., ASME Press, New York.
8. Crispin, Y. (1998). Adaptive Control and Synchronization of Chaotic Systems with Time Varying Parameters. In *Intelligent Engineering Systems Through Artificial Neural Networks*, Vol. 8, Edited by Dagli et al., ASME Press, New York.
9. Crispin, Y. (1997) Adaptive Control and Synchronization of Chaos in Discrete Time Systems. In *Intelligent Engineering Systems Through Artificial Neural Networks*, Vol. 7, Edited by Dagli et al., ASME Press, New York.
10. Crispin, Y. and Ferrari, S. (1996). Model Reference Adaptive Control of Chaos in Periodically Forced Dynamical Systems. In *6th AIAA/USAF/NASA Symposium on Multidisciplinary Analysis and Optimization*, Bellevue, WA, 882-890, AIAA Paper 96-4077.
11. Goedgebuer, J.P., Larger, L. and Porte, H. (1998). Optical Cryptosystem Based on Synchronization of Hyperchaos Generated by a Delayed Feedback Tunable Laser Diode. In *Physical Review Letters*, 80, 10, 2249–2252
12. Hegger, R., Bunner M.J., Kantz, H. and Giaquinta, A. (1998) Identifying and Modeling Delay Feedback Systems. In *Physical Review Letters*, 81, 3, 558–561.
13. Just, W., Reckwerth, D., Mockel, J., Reibold, E. and Benner, H. (1998) Delayed Feedback Control of Periodic Orbits in Autonomous Systems. In *Physical Review Letters*, 81, 3, 562–565.
14. Kolmanovskii, V. and Myshkis, A. (1992) Applied Theory of Functional Differential Equations. In *Mathematics and Its Applications*, Vol. 85, Kluwer, Dordrecht.
15. Lamb, H. (1995). *Hydrodynamics*. Cambridge University Press, New York, sixth edition.
16. Losson, J., Mackey, M.C. and Longtin, A. (1993). Solution Multistability in First-order Nonlinear Differential Delay Equations. In *Chaos*, 3, 2, 167–176.
17. Mansour, B. and Longtin, A. (1998a) Chaos Control in Multistable Delay Differential Equations and Their Singular Limit Maps. In *Physical Review E*, 58, 1, 410–422.
18. Mansour, B. and Longtin, A. (1998b) Power Spectra and Dynamical Invariants for Delay Differential and Difference Equations. In *Physica D* 113, 1, 1–25.
19. Milne-Thomson, L. (1968). *Theoretical Hydrodynamics*. MacMillan, New York, fifth edition.
20. Ott, E. and Spano, M. (1995). Controlling Chaos. In *Physics Today*, 48, 34.

21. Parlitz, U. (1996). Estimating Model Parameters From Time Series by Auto-Synchronization. In *Physical Review Letters*, 76, 8, 1232–1235.
22. Parlitz, U., Junge, L. and Kocarev, L. (1996). Synchronization Based Parameter Estimation From Time Series. In *Physical Review E*, 54, 6, 6253–6259.
23. Peng, J.H., Ding, E.J., Ding, M. and Yang, W. (1996). Synchronizing Hyperchaos with a Scalar Transmitted Signal. In *Physical Review Letters*, 76, 6, 904–907.
24. Tamasevicius, A. and Cenys, A. (1997). Synchronizing Hyperchaos with a Single Variable. In *Physical Review E*, 55, 1, 297–299.
25. Yang, T. (2004). A Survey of Chaotic Secure Communication Systems. In *International Journal of Computational Cognition*, 2, 2, 81–130.

Hardware-in-the-Loop Simulations for FPGA-based Digital Control Design

Carlos Paiz, Christopher Pohl and Mario Porrmann

Heinz Nixdorf Institute, University of Paderborn
Fürstenallee 11, 33102 Paderborn, Germany
{paiz,pohl,porrmann}@hni.upb.de

Abstract. A framework to perform hardware-in-the-loop (HIL) simulations in the designflow of digital controllers, based on Field Programmable Gate Array (FPGA) technology, is presented. The framework allows the interaction of digital controllers, implemented on our rapid prototyping system RAPTOR2000 with a Matlab/Simulink simulation running on a host computer. The underlying hardware and software designs supporting the interaction of the digital control and the simulation are presented. The designflow of FPGA-based digital controllers when using HIL is described and examples are given. Results from HIL simulations are presented, showing that the acceleration of the simulation increases with the complexity of the design when the number of I/Os stays constant. Furthermore, using the proposed HIL framework the clock accurate verification of the design can be achieved within the design phase.

Keywords. FPGA, Hardware-in-the-Loop, digital control, reconfigurable hardware.

1 Introduction

Reconfigurable hardware has been successfully used to implement a variety of applications, ranging from digital signal processing [1] to digital control [2], among others. The key feature of this technology is its ability to perform computations spatially (i.e., in hardware) to increase performance, while retaining much of the flexibility of a software solution (i.e., reconfigurability).

Applications of digital controllers using reconfigurable hardware have been reported since the early 90s. However, it is only until recently, that researchers have started to explore the potentials offered by this technology. This is because of the higher computational demands of digital control systems and the fast evolution of Field Programmable Gate Array (FPGA) technology.

Among the many kinds of existing reconfigurable architectures, we research FPGAs as implementation platform for control algorithms. Modern FPGAs are heterogeneous architectures, constituted by programmable functional blocks interconnected by a reconfigurable network and embedded application specific hardware, such as embedded processors, block RAM, or multipliers. This kind of architecture allows the implementation of complete Systems on Chip (SoC). More information on architectures and design methods for reconfigurable hardware can be found, e.g., in [3] [4].

FPGAs have been used to implement control algorithms in areas such as motor control [5] [6], power electronics [7] [8], industrial control [9] [10], sensor monitoring [11] [12], motion control [13] [14], among others. Some of the reported advantages of using FPGAs are acceleration, flexibility, reduced costs, and low energy consumption.

The increasing interest in using this technology is evident. However, the migration from traditional platforms, such as DSPs or microprocessors, to reconfigurable hardware is not a straightforward process, since a different background is required to use this technology. To ease the migration, high-abstraction-level hardware descriptions are used (e.g., DSP builder from Altera or System Generator from Xilinx), which enable non-experts in digital hardware design to easily implement algorithms using FPGAs.

For the design, implementation and testing of digital control systems, Hardware-in-the-Loop (HIL) simulations are becoming an essential tool. According to Isermann et al. [15], a HIL simulation is characterized by the operation of real components in connection with real-time simulated components. The simulated components are often the processes being controlled and/or sensors and actuators.

The utilization of HIL simulations has been extensively reported in literature. Terwiesch et al. [16] have presented a HIL simulation setup based on several commercially available boards for rail vehicle control system integration. Their approach is based on the acceleration of the simulated model by using several processors concurrently to achieve a real-time simulation based on the controller timing, which has a sampling period of 30 μs. Similar approaches have been described in [17] [18] [19] [20] and [21] where a simulated process is accelerated in order to couple it in real-time with the control system, which is running on the final implementation platform.

A different approach was presented in [22] [23] [24] and [25]. Real parts of the system being controlled were actually used in the simulation loop. This method was said to bring more accuracy to the simulation, making the design process shorter for that specific application. However, since the setup was application specific, it was not possible to use it for other designs.

In all previously mentioned publications software-based architectures (e.g., DSPs or microcontrollers) were used as the final implementation platform and therefore are not suitable for an FPGA-based control designflow. Different groups have been working on acceleration and debug environments for FPGA designs. In [26] a framework for the design of control algorithms for mechatronic systems, including HIL simulations, was presented. The design and implementation of linear, time-invariant (LTI) control systems on FPGA technology was described using a self-developed software called Computer-Aided Mechatronics Laboratory (CAMeL), as design environment.

There are commercially available FPGA-based prototyping boards which can perform HIL simulations. In [27] a HIL system, DIME from Nallatech, was presented. This board is connected to the host PC via the PCI bus. Their approach is not universal and requires the user to develop on the DIME board. Our approach is platform independent and can be adapted to any existing prototyping environment with a reasonably fast communication link.

Our current implementation is based on the modular rapid prototyping system RAPTOR2000 [28], which is connected to the host computer via the PCI bus, but it is universally adaptable to other platforms. Our framework enables the utilization of RAP-

TOR2000 for HIL simulation under Matlab/Simulink. Simulations are coupled with the prototyped controller by controlling its system clock. In that way, the digital controller can be accurately tested, without the need of accelerating the simulated process, yet the HIL simulation experiences a noticeable acceleration in comparison to a full software simulation.

In the following subsection various terms used in this chapter are introduced. Section two describes in detail the framework used for Hardware-in-the-Loop simulations, starting from the underlying hardware platform and supporting hardware blocks, followed by the software to integrate the digital design with a simulation running under Matlab/Simulink. The utilized tool flow is also described in this section. In section three the design flow of digital controllers that utilizes our Hardware-in-the-Loop framework is shown and three examples with the corresponding performance data are presented. Finally, in section four conclusions are drawn and future work is sketched.

1.1 Definitions

In the following paragraphs some of the terms used in this chapter are introduced.
Sample Rate: from the point of view of a digital hardware designer, this is the frequency at which a given input/output port is driven/updated.
Single-Rate Design: is a design where all elements share a common clock (all input/output ports share the same sample rate and there is no internal up/down sampling). This is the simplest clock scheme.
Multi-Rate Design: a design is multi-rate if it has signals running at different clock frequencies. A multi-rate design can either be driven by a single clock (in which case all other frequencies are derived from it) or by many clocks.
Periodic Design: a design is periodic if its latency is known and stays constant. Examples are digital filters or classical control algorithms (e.g., a PID controller).
Aperiodic Design: a design is aperiodic if its latency varies depending on the input data. This variation might happen because an adaptation to different inputs or operative regions. However, the processing time is bounded to a maximum (e.g., the latency is known to be never greater than the required sampling period). Designs that fall in this category are soft-processors and adaptive control schemes (e.g., multi-model based control).
DUT: this is the acronym for Design Under Test. It refers to the algorithm being developed. This is the part of the simulation, which is implemented in reconfigurable hardware. The DUT is implemented on an FPGA module of our RAPTOR2000 (see Sect. 2.1).
DUT Clock: the clock frequency used by the DUT. This clock frequency may be different (lower) from that of the hardware part of the HIL framework (Hardware Wrapper and Synchronizer, cf. Sect. 2.1).
System Clock: this is the maximum clock frequency (used by the Hardware Wrapper and the Synchronizer, see Sect. 2.1).

2 Simulation Framework

The Hardware-in-the-Loop simulation framework consists of hardware and software interfaces, which enable the interaction of the DUT with Matlab/Simulink. These interfaces are described in the following sections.

2.1 Hardware

In the following section our rapid prototyping system and the hardware designs, which support our HIL framework, are described.

RAPTOR2000. The underlying hardware platform is the RAPTOR2000 system, an FPGA-based rapid prototyping environment, which has been developed at our department. The board itself has been used in many research projects, e.g., [29] [30] [28] and therefore only the communication structure is described here.

The interface between the RAPTOR2000 FPGA module(s) on the one hand and the PCI bus on the other hand is done by a PLX9054 chip, which is a bus master compatible PCI bridge (see Fig. 1). This bridge translates the PCI protocol to a 32 bit local bus on the board and back. FPGA designs that have to access host data have to implement a local bus (LB) interface. The LB interface is very small (less than 1% slice usage on a Xilinx XC2V3000 FPGA) and easy to use. All communication between FPGA and host is processed by this interface, so its transfer rate is critical for the resulting simulation speed. Measurements show maximum rates of 25 MByte/s in PIO (Programmed Input Output, the processor manages the data transfer) mode and 95 MByte/s in DMA (Direct Memory access, the processor initiates the data transfer, but is not involved after the initialization) mode.

Clock Management. In this section we describe the clock management, which is a critical part of the HIL simulation. In order to generate a accurate simulation, it is necessary to precisely control the number of clock cycles during which the DUT must run.

All memory elements (registers, latches, block RAM, etc.) have two clock related inputs: the clock input and the clock enable input. In single-rate designs the clock enable input is usually set to one, which means that the register operation depends on the

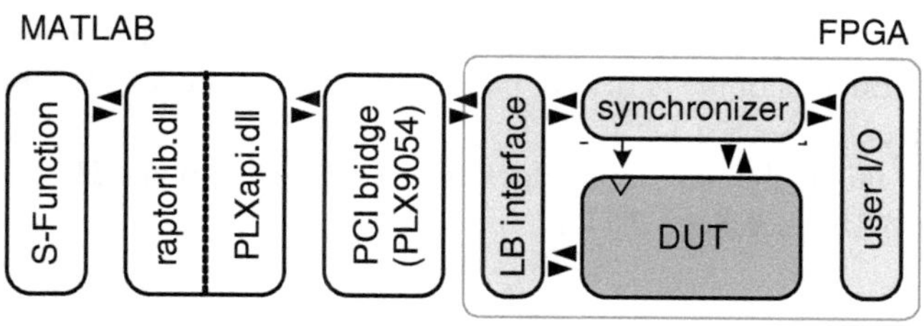

Fig. 1 Information flow of the Hardware-in-the-Loop simulation framework.

clock (usually the rising edge of the clock) only. In our case it is used to control the hardware simulation; while the system clock keeps running, the clock enable signal is set to zero and therefore all memory elements keep their current state. This is a better option than using clock-gating, since it avoids having delays in the clock path and the problems derived from them (e.g., additional delays in the clock tree). In effect, this is the same as halting the DUT-clock(s) and therefore allows cycle accurate simulation (like debug-stepping) of the behaviour of the circuit, while the timing behaviour is not correctly simulated. Both clock gating and clock enable generate implicit multi-cycle paths, which do not render the normal operation mode.

Synchronizer. In order to coordinate a simulation with a DUT, a Finite State Machine (FSM), the Synchronizer, has been implemented as a dedicated hardware block. The Synchronizer enables the DUT clocks (see Sect. 2.1) after a request from a Matlab/Simulink simulation. Therefore, if a design is to be simulated using our HIL framework, it needs to have a clock enable port associated with its clock. The states diagram of the Synchronizer is depicted in Fig. 2.

The Synchronizer controls a given DUT in two possible modes: periodic and aperiodic (see Sect. 1.1). Before explaining these operative modes, the ports of the Synchronizer are introduced. The current version of the Synchronizer has four input ports and four output ports to interact with the simulation and the DUT. When the simulation

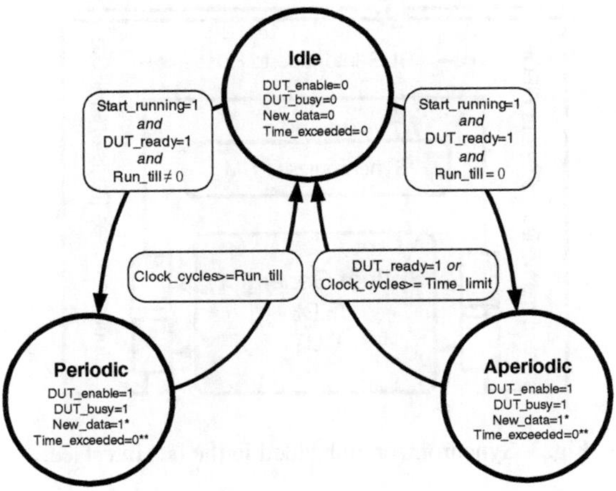

Fig. 2 Synchronizer state machine (*This signal is enabled during one System clock cycle, ** Not used in this mode, *** Set to one if a time overflow occurs).

sends a non-zero value to the *running_time* port, the Synchronizer enters the *periodic* mode. The DUT is then enabled during the requested time, keeping the hand-shaking signal *DUT_busy* enabled and ignoring the *DUT_ready* port. The design updates its output ports, which are read and stored by the Synchronizer and sent to the host computer, where the simulation reads and propagates them to the other blocks (e.g., a plant model) in the simulation.

If the simulation sends a zero to the *running_time* port, the Synchronizer enters the *aperiodic* mode and waits until the DUT sends a hand-shaking signal through the *DUT_ready* port to disable it. As in the periodic mode, the output port *DUT_busy* is set to *high*, in order to prevent that the simulation sends new values while the DUT is busy. Although this case might happen very rarely (cf. Sect. 2.1), this signal avoids to loose the synchronization between the DUT and the simulation.

The Synchronizer also detects whether there is a discrepancy between the given sampling period and the time required by the DUT to complete a cycle. This happens if its latency is greater than the sampled period reported by the simulation. In this case the DUT is disabled and a warning signal is sent to the simulation through the *time_-exceeded* port. The simulation can then react to this exception.

Hardware Wrapper. Both the Synchronizer and the DUT, are embedded in a hardware wrapper, as depicted in Fig. 3. The Wrapper provides specialized hardware for interfacing the Synchronizer and the DUT with the Matlab simulation running on the host computer through the PCI bus (see Sect. 2.1). In order to embed the DUT into the Wrapper, the bus interface is adapted to the input and output ports of the DUT. This process is done automatically as described in Sect. 2.2.

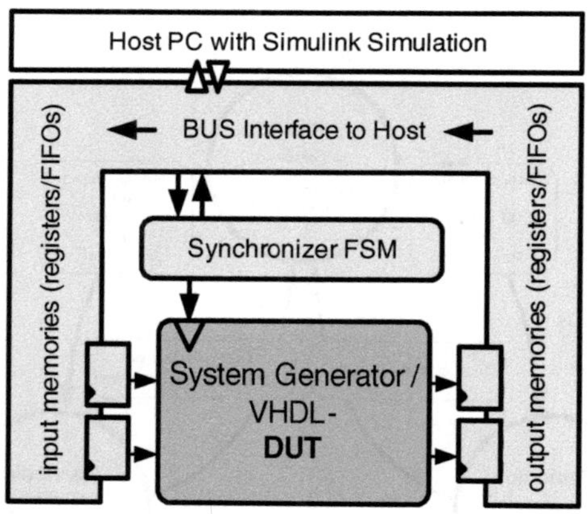

Fig. 3 Synchronizer embedded in the bus interface.

The Wrapper enables reading and writing data from and to the input/output ports from the simulation. There are two methods to realize these operations: using a registers bank and using a FIFO memory. In control applications, one often has feedback loops, which means, that the output(s) (or some function of it) has to be fed back to the input(s) *without delay*. This reduces the FIFOs-depth to one, which is, however, the slowest possible way of communication because only one input and output can be write/read per cycle; the amount of data to be transferred is rather small, so that the communication overhead is high. For multi-inputs multi-outputs (MIMO) systems, this process can be accelerated by utilizing DMA transfers, which has to be implemented in the future. A further increase of communication throughput can be achieved by implementing FIFOs instead of a register bank. This is, however, not used for control applications with a feedback loop and is therefore postponed to future work.

Hardware Performance. There are several pre- and post-processing steps needed to simulate one (or several) DUT cycles (cf. Sect. 2.1 and Sect. 2.3). Taking into account these actions a theoretical maximum for the simulation frequency is

$$F_{sim} = \frac{1}{T_{update} + T_{run} + T_{fetch}} \tag{1}$$

where T_{update} is the time required to update the memories at the input of the DUT, T_{run} is equivalent to the time needed by the DUT to produce a new output, and T_{fetch} is the time for retrieving the data from the output memories. If $T_{DUT} \approx T_{PCI}$ (period of the DUT clock and Period of the PCI clock correspondingly), for a filter running at $F_{DUT} = 50$ MHz

$$F_{sim} \leq \frac{R_T}{W_{Bus} * (1 + N_I + N_O)} \tag{2}$$

is a good approximation. Here, N_I and N_O are the numbers of input and output ports of the DUT, W_{Bus} is the width of the PCI bus (PCI: 32 bit), and R_T is the transfer rate, which can be achieved in the current mode (PIO or DMA). An example for an application with a lower frequency might be a controller for some mechanical system, where the control update rate is in the magnitude of kHz, so $T_{DUT} \gg T_{PCI}$. Here

$$F_{sim} \leq (T_{DUT} + \frac{W_{Bus}}{R_T} * (1 + N_I + N_O))^{-1} \tag{3}$$

is a reasonable approximation. Software issues, such as calculation of the test vectors (Simulink model), are not included in this consideration and will influence the results according to their complexity. On the software side, PIO and DMA transfers can be initiated via simple library functions, which have been integrated into an S-Function block (for details cf. Sect. 2.3).

In Fig. 4, the theoretical maximum for the simulation frequency (cf. (2)) is given. The points indicate real measurements made with our examples (see Sect. 3). The actual values are lower than the theoretical maximum, because a lot of calculations have to be performed in software. This is the software model on the one hand and the preprocessing and postprocessing of data for the hardware implementation on the other hand. For a detailed description of the necessary translation steps for the hardware, see Sect. 2.3.

2.2 Hardware Integration

The Hardware Wrapper described in the previous section stays the same between different hardware implementations, except for the embedded DUT and the corresponding bus interface. To simplify and accelerate the process of generating the wrapper and, there upon, the hardware, a JAVA based application (jvToolsLib) was developed, which is embedded in the toolflow as depicted in Fig. 5.

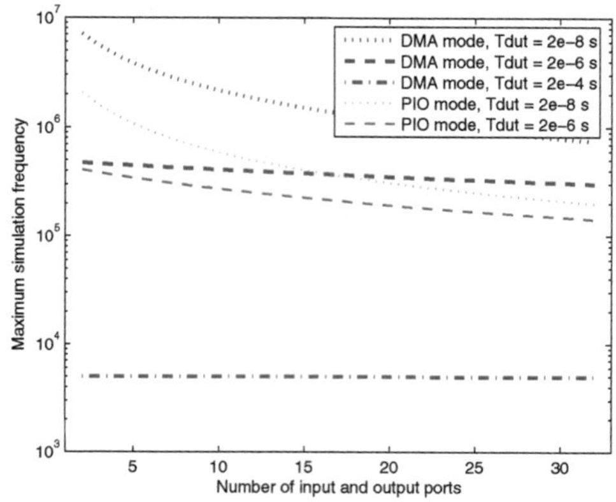

Fig. 4 Theoretical maximum for simulation performance.

jvToolsLib. The purpose of this library is to provide an easy-to-use API (Application Programming Interface) for programmers, who want to manipulate their designs in an object oriented manner rather than in VHDL code. Therefore, a graphical user interface (GUI) was developed, which enables users without VHDL experience to set up a complete HIL simulation. The setup process includes analyzing the user design, inserting it into a wrapper template (see Sect. 2.1), and customize memory elements and the bus controller. Additionally, the FPGA flow (synthesis, place and route, bitstream generation) can be started from the tool.

In the following a general overview of the functionality of the jvToolsLib is given, and its benefits for control designers without specific HDL skills are presented. The library provides a set of classes to represent most synthesizable vhdl constructs, and several methods to manipulate them. Furthermore, all classes implement a *constructFromVHDL(String vhdl)* method, which allows the construction of an object-tree from a VHDL Template file. In this way any VHDL file can be parsed, manipulated through the API and finally stored in a file. As an example, the steps performed during the generation of a design in our HIL framework will be described:

1. The user opens the VHDL file containing an entity description of his design, which is then parsed by jvToolsLib. An entity description is basically an interface definition of the design, which will be exported from the design tool while generating

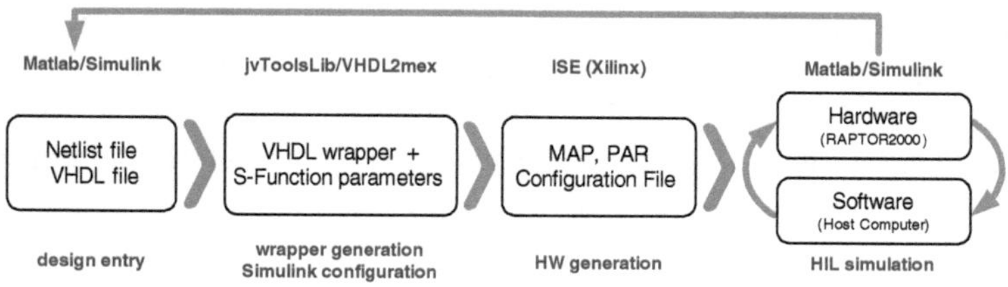

Fig. 5 Toolflow for HIL simulations.

the netlist (a textual representation of the synthesized design). In case of a VHDL design, the entity description is an integral part of the design.
2. A list of all entities defined in the file is presented to the user, who can then choose the appropriate entity defining his design. This design will be referred to as the DUT (Design Under Test) from here on.
3. The user can change certain properties of the DUT IO-interface (ports) such as sample rate and number format, which are important for the Matlab interface. These properties in conjunction with the hardware internals are used to create a Matlab file, containing the information to configure the simulation.
4. The HIL tool comes with a template file containing the static parts of the hardware wrapper. This file is parsed, too, and the DUT is integrated into it by adding the entity-object to the wrapper object tree. The memory elements (see Sect. 2.1) and the bus decoder are customized and connected to the DUT ports automatically.
5. In order to start the FPGA flow, the newly generated object tree is translated into a VHDL file, which can then be processed by Xilinx ISE or any other FPGA toolchain.

These simple steps complete the setup of the HIL simulation on the hardware side, and all HDL specific tasks are performed by the jvToolsLib and the jvHIL program without user interaction. The jvToolsLib is being developed as an open source project. It is going to be released in the near future.

Apart from these "template-style" tasks, the jvToolsLib can be used for other purposes, too, which will be presented in the future. One of the projects planned is the implementation of an automatic documentation tool like VHDLDoc (see http://schwick.home.cern.ch/schwick/vhdldoc/), which supports several output formats.

This Software reads the top level of the VHDL design and identifies the top level entity, port attributes and generics. These are displayed in a graphical user interface, where the user can introduce certain changes to the default values (e.g., not reading an output, setting an input to a constant and so on). The port data rates have to be defined here, too, which is a topic to be processed automatically in future.

In addition to the hardware, VHDL2mex generates a configuration string for a Matlab S-Function (cf. Sect. 2.3) containing addresses and data rates of input and output ports. This HIL flow, integrates seamlessly into available FPGA flows because no vendor specific information is added. In Fig. 5 the flow is presented, integrating software from Xilinx as an specific example. However, the tool-flow could use software from a different company.

2.3 Simulink Integration

Matlab provides a generic interface for integrating user defined software into the Simulink simulation process, the so called S-Function. The basic simulation steps and their pendants for HIL simulation with RAPTOR2000 are displayed in Fig. 6a and 6b. Basically, the *mdlStart()* function is used for the hardware initialization (download of the bitstream, configuration of the synchronizer). If *mdlStart()* succeeds, the simulation loop sequentially calls *mdlUpdate()* and *mdlOutputs()*. In *mdlOutputs()* the data in the hardware output registers is read and propagated to the outputs of the Simulink block.

In *mdlUpdate()* data from the input ports of the simulink blocks is sent to the hardware input registers, respectively. *mdlUpdate()* also starts the synchronizer to activate the DUTClock for one clock cycle. In addition to these communication steps, several translation steps from the Simulink floating point datatypes to the hardware fix point data types have to be accomplished inside the S-Function. The parameters for this translation as well as information on the hardware configuration are given in a configuration string provided by vhdl2mex.

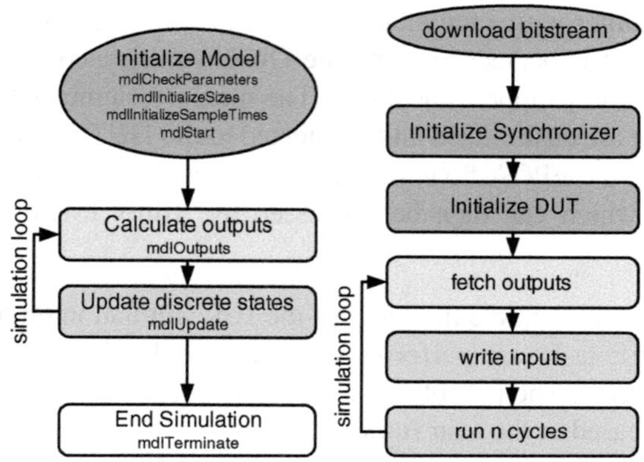

(**a**) Simulink Simulation steps (**b**) r2ksim steps

Fig. 6 Simplified simulation flow diagram.

The current implementation of the S-Function interface is to be considered as a proof of concept and there is room for a lot of improvements. These improvements, in addition to the use of faster data transfers (DMA) will certainly improve the simulation performance.

3 The Designflow

The proposed designflow, including HIL simulations, is presented for a design created with Matlab/Simulink. This high-level tool has become an essential development environment in control engineering. Hence, it is eligible to use it for the development of digital controllers to be implemented on reconfigurable hardware.

The designflow of digital control algorithms can be divided roughly in five steps, as depicted in Fig. 7. In the first step, the requirements of the controller are defined. A better understanding of the plant should be gained in this step. A first mathematical description of the controller is then derived, either by a time-continuous representation (e.g., described by differential equations) or by a time-discrete representation (e.g., described by difference equations). The second step is the simulation of the controller together with a model of the process. The accuracy of this model has a direct impact

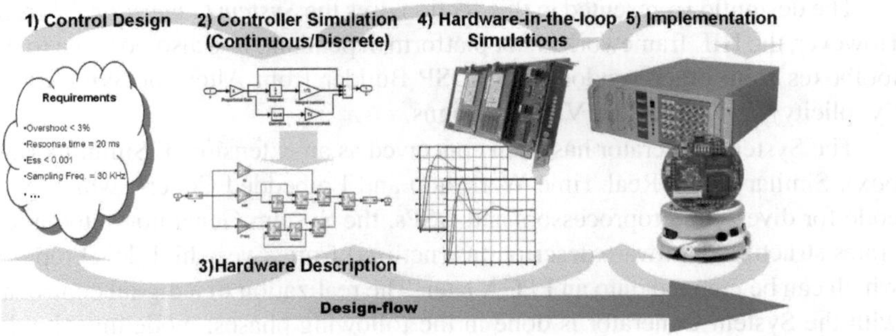

Fig. 7 The proposed Designflow for FPGA-based controllers, which includes HIL simulations.

on the design. There are other aspects, which should be modeled accurately, such as the dynamics of sensors and actuators.

After the designer is satisfied with the performance of the controller, a third step is its translation to a hardware description language. This step can be done using different hardware description languages (HDL). To evaluate high level hardware design entry, different tools were compared with a standard VHDL designflow. Among others, Xilinx's System Generator and Synplicity's SynplifyDSP were evaluated in terms of performance, and resource efficiency as well as design entry time. Two scenarios, an IO controller (IO) and a cascaded position controller (CPC) were examined.

Table 1 Comparison between high level and low level design entries. An input/output controller (IO) and a cascaded position controller (CPC) were examined.

	VHDL		SynplifyDSP		System Generator	
design	IO	CPC	IO	CPC	IO	CPC
slices	667	2138	776	1537	810	3802
multiplier macros	0	19	0	18	0	0
latency [cycles]	5	67	4	1	4	10000
implementation time	87 h	103.5 h	19 h	27.25 h	19 h	20.5 h

In Table 1 the results of the design entry are presented. Both, the System Generator and SynplifyDSP remain in the same order of magnitude regarding resource utilization. However, the time required for design entry and synthesis in VHDL is several times larger than the time required for the high level tools. This is generally the case whenever the dataflow part of the design is fairly large, e.g., in control applications. When it comes to applications where controlflow overbalances dataflow, the design entry time for high level tools may raise above that of HDL. In this case one would either choose a pure HDL implementation or a combination of both. This study shows, that control applications are very well suited for implementation using high level hardware description.

The designflow presented in this section uses the *System Generator 7.1* from Xilinx. However, the HIL framework is not platform dependent. It is also possible to use it with toolboxes from other vendors (e.g., DSP Builder from Altera or SymplifyDSP from Synplicity) or with custom VHDL designs.

The System Generator has been conceived as an extension of Simulink (i.e., a toolbox). Similar to the Real Time Workshop and Embedded Coder[1], which generate C-code for diverse microprocessors and DSPs, the System Generator automatically generates structural hardware descriptions (netlists) from a very high-level representation, which can be mapped onto an FPGA later. The realization of a digital control algorithm with the System Generator is done in the following phases: modeling, simulation, resource estimation, and hardware description, as described, e.g., in [2]. These phases are undertaken in step three of the design flow, as depicted in Fig. 7.

In step four, a netlist is automatically generated. This netlist is integrated into our HIL framework, as described in Sect. 2.2. A configuration bitstream is generated using the ISE from Xilinx. An automatically adapted S-Function replaces the System Generator design and the HIL simulations are carried out without further ado of the user.

The simulations are performed as usual. However, the designer can now realize whether the controller, running on an FPGA module of the RAPTOR2000, actually works as expected. In this stage more intensive tests can be conducted. Since the structure of the controller has already been designed and tested, the next step is an intensive test of its parameters or its response to different operative regions. This process is greatly accelerated by HIL simulations, besides the enhanced reliability of this kind of simulations.

The final step corresponds to the test of the controller when interacting with the real plant. As shown in Fig. 7, these steps are iterative. It is often necessary to go one or two steps back. However, the gap between step three and five is reduced by including HIL simulations.

In the following subsections, some examples are presented as a proof of concept. The frequency shown for each example refers to the number of input samples per second for the input with the highest sample rate. The results of the simulations are compared and discussed in the last section.

3.1 PI-based Speed Control

As a first example, a Proportional-Integral (PI) algorithm to control the speed of DC motors for robotic applications is presented. The PI algorithm is still one of the most widely used controllers in industry. The control task is to regulate the speed of a DC motor by manipulating its input voltage. A classical parallel PI was realized using a trapezoidal integration rule, as depicted in Fig. 8. SP represents the Set-Point, K_p the proportional constant, K_i the integral constant, T the sampling period, Y(Tk) the feedback signal (e.g., speed of the motor) and $U(Tk)$ the output of the controller (e.g., a new target speed). An anti-wind-up block was attached to the integral part of the algorithm.

[1] http://www.mathworks.com/products/rtwembedded

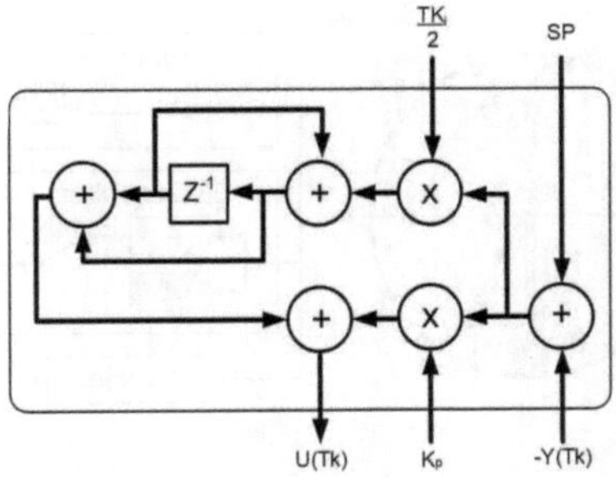

Fig. 8 Parallel PI algorithm using trapezoidal integration.

The controller was implemented using *System Generator* blocks as explained in Sect. 3. The design uses one output and four input ports. Three of the inputs (SP, K_p, and $TK_i/2$) are not updated in every simulation step, and so do not contribute to the communication overhead. However, when performing more complex tests (e.g., a self tuning regulator) it is possible to include these inputs in the HIL simulation. The software part of the HIL simulation is a Matlab/Simulink model of a DC motor. The PI controller runs on a Virtex II-Pro module of the RAPTOR2000 system. The complete control loop is depicted in Fig. 9.

The sampling frequency was set to 1 KHz. The simulated time was 10 seconds. Software simulation lasted 15.9248 seconds (628 Hz). Using our HIL framework, the simulation lasted 12.445 seconds (996 Hz). The speedup was relatively small due to the low complexity of the design, which had an equivalent gate count of 5,722. However, it could be verified that the prototyped design worked as required during the HIL simulations, as well as when tested with the real DC motor.

3.2 Inverted Pendulum

The inverted pendulum is a classical problem in control theory; it has been used in literature as an example of a well-understood yet non-trivial system to test control algorithms. In [2], this system was used to exemplify the utilization of partial and dynamic reconfiguration of an FPGA to efficiently implement a multi-controller system. The controller for the pendulum was split in two; one to swing up the pendulum and the other one to balance it. The decision to load one of both controllers is made by a supervisory en-

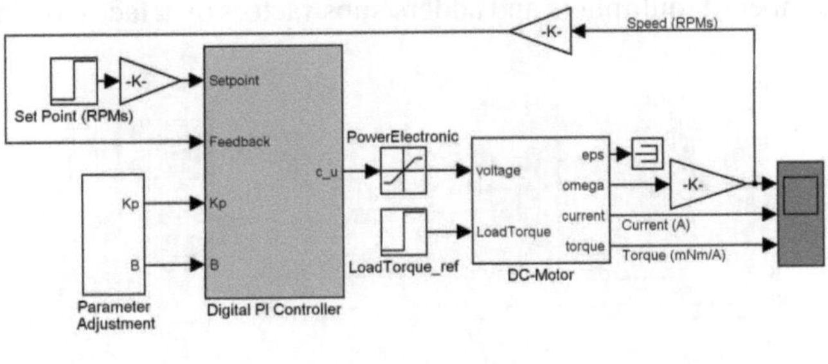

Fig. 9 Control loop including a model of the DC motor and the interface with the DUT.

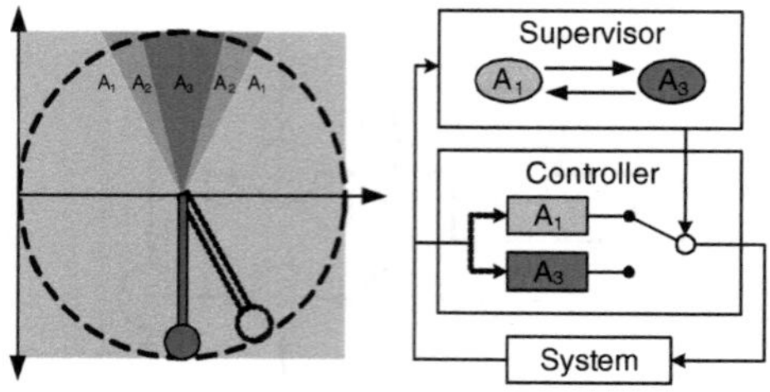

Fig. 10 Two-state controller for the inverted pendulum system. For region A_1 the swinging-up controller is used. Region A_2 is the switching region and region A_3 corresponds to the balancing control.

tity depending on the relative position and angular speed of the pendulum, as depicted in Fig. 10.

To test our HIL framework, the controller to balance the inverted pendulum is used. A state-space model of the pendulum-cart system is simulated under Simulink, and the State-Feedback controller is implemented using a Virtex II-Pro module in our RAPTOR2000 System (see Fig. 11).

The controller has an equivalent gate count of 209,000. The system has three inputs (the angle of the pendulum, the position of the motor, and the target position) and one output (the new position of the motor). The target position is always set to the center of the cart's track. This set-point is only initialized in the beginning of the simulation and has hence no influence in the communication overhead. The sampling period of the controller was 10 μs (10 KHz). Five seconds of simulation using the *System Generator* blocks lasted 63.48 seconds (787 Hz), while using the proposed HIL framework the simulation time was reduced to 3.13 seconds (9,512 Hz). This represents a speed up of 19.15. The prototyped controller worked just as well as the simulated design and also the tests on the real system have been successful.

3.3 Recursive IIR Filter

The third example is a hardware implementation of a Chebyshev II Filter, used for signal processing. In order to save FPGA resources, the filter was built in a time division multiplex (TDM) manner by introducing pipeline stages and an internal feedback loop (see Fig. 12). It can emulate up to 25 filter sections within one physical block, therefore reducing the number of multipliers and adders/substractors by a factor of 25. Because of

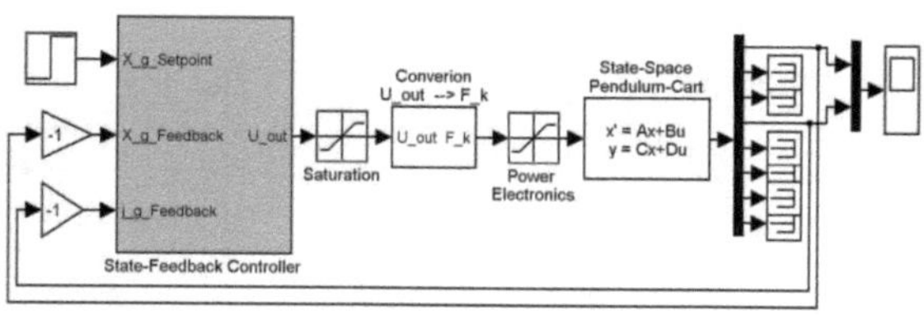

Fig. 11 Inverted Pendulum control loop. A state-space model of the pendulum-cart system is used.

Table 2 Implementation Examples.

Design	Complexity (Slices)	I/Os	Sampling Rate [μs]	Simulated Time [s]	Duration (Simulation) [s]	Duration (HIL) [s]	Speedup
PI	204	1/1	1000	10	15.92	12.42	1.27
State-Feedback	1,049	2/1	10	5	63.47	3.13	19.15
IIR Filter	2,361	2/1	10	0.004	∼900	3.95	227.2

high fixed point precision required internally, the simulation model is rather complex; the simulation of 0.04 seconds takes approximately 15 minutes on a Pentium 4 3.2 GHz PC, which corresponds to approximately 44 Hz. In the HIL setup, the system as two inputs: $S(t)$, which is the sampled and quantized input signal and *run*, used to empty the filters memories in between two measurements. The output signal $F(t)$ represents the filters response.

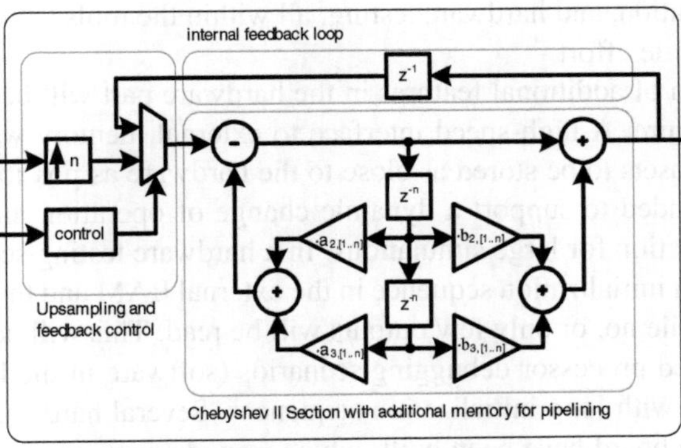

Fig. 12 Recursive IIR-Filter emulating n filter sections.

The testbench in this case consist of some scopes and of a "From Workspace" block, which feeds the test data from a Matlab array to the filter. Considering software effort, this is probably the smallest meaningful testbench, therefore the measurements in this case contain the smallest possible software overhead. The HIL simulation of 0.04 seconds took 3.95 seconds, which corresponds to approximately 10 kHz at the input, resulting in a speedup of about 200.

These results show the great potential of using HIL simulation to speed up the design flow of FPGA-based control systems and to verify the design clock-accurately. Table 2 shows a summary of the implementation examples.

4 Conclusions and Future Work

In this chapter a platform independent and extendable framework for FPGA-based HIL simulations under Matlab/Simulink has been presented. The structure and operation mode of hardware and software have been shown and examples have been presented. The results show that our system is capable of accelerating simulations within the Matlab/Simulink environment. The acceleration depends largely on the complexity of the simulated design and on the number of input and outputs ports. When the number of I/O operations stays constant the speed up grows with the complexity of the design.

It was shown that our HIL framework can be used as a cycle accurate debugger for designs with closed algebraic loops. This results in a shorter development and testing time, given the advantage of using a simulated environment to test the system (e.g., no test-bed is required).

Concerning the software, future work concentrates on further automation of the hardware generation process and on extending the approach to new areas of application. New simulation tools and frameworks are being adopted, including Matlab and Modelsim. By interfacing these tools, the functionality of the HIL framework will be available in VHDL-, Verilog-, SystemC-, and Matlab Script-based simulations. This will allow easy testbench reuse, simulation acceleration, rapid prototyping, hardware-software co-simulation, and hardware testing, all within the tools the developer is used to and with very little effort.

The integration of additional features in the hardware part will be another point to work on in the future. A high-speed interface to external memory will be integrated, allowing large datasets to be stored as close to the hardware as possible. The synchronizer will be extended to support a dynamic change of operation modes, allowing a "fast forward" function for large simulations. In a hardware testing scenario, it will be possible to store an initialization sequence in the external RAM and then "fast forward" the simulation, while no, or only few outputs will be read. This will be especially time saving in embedded processor debugging scenarios (software in the loop) or for high frequency designs with long initialization sequences. Several hardware extensions for the RAPTOR2000 board have been built, which provide analogue interfaces, e.g., for control applications, or communication interfaces like USB, CAN and Ethernet. The extension of the presented approach to these interfaces allows direct interaction with additional hardware, while the developer can build the applications or models within the tools he prefers. These applications might include data logging. Another important issue is to improve the acceleration abilities of our FPGA in the loop approach, by incorporating faster transfer techniques and minimizing the communication overhead. Enabling HIL simulations of designs including partial and dynamic reconfiguration for the implementation of adaptive control algorithms is also planned as future work.

Acknowledgements

This work was supported by the Mexican National Council of Science and Technology (CONACyT) and by the Collaborative Research Center 614 - Self-Optimizing Concepts and Structures in Mechanical Engineering - University of Paderborn, and was published on its behalf and funded by the Deutsche Forschungsgemeinschaft.

References

1. Tessier, R., Burleson, W.: Reconfigurable computing for digital signal processing: a survey. In: Journal of VLSI Signal Processing. Volume 28., Elsevier (2001) 7–27
2. Paiz, C., Kettelhoit, B., Klassen, A., Porrmann, M.: Dynamically reconfigurable hardware for digital controllers in mechatronic systems. In: IEEE International Conference on Mechatronics (ICM2005), Taipei, Taiwan (2005)
3. Compton, K., Hauck, S.: Reconfigurable computing: a survey of systems and software. In: ACM Computing Surveys. Volume 34. (2002) 171–210
4. Todman, T.J., Constantinides, G.A., Wilton, S.J.E., Mencer, O., Luk, W., Cheung, P.Y.K.: Reconfigurable computing: architectures and design methods. In: IEEE Proceedings – Computers and Digital Techniques. Volume 152. (2006) 193–207
5. Carrica, D., Funes, M., Gonzalez, S.: Novel stepper motor controller based on FPGA hardware implementation. In: IEEE/ASME Transactions on Mechatronics. Volume 8., IEEE/ASME (2003) 120–124
6. Tazi, K., Monmasson, E., Louis, J.P.: Description of an entirely reconfigurable architecture dedicated to the current vector control of a set of ac machines. In: IEEE International Conference on Industrial Electronics, Control, and Instrumentation. Volume 3. (1999) 1415–1420
7. Huang, S.J., Yang, T., Huang, J.: FPGA realization of wavelet transform for detection of electric power system disturbances. In: IEEE Transactions on Power Delivery. Volume 17. (2002) 388–394
8. Woo, W., Miller, M., Kenney, J.S.: A hybrid digital RF envelope predistortion linearization system for power amplifiers. In: IEEE Transactions on Microwave Theory and Techniques. Volume 53. (2005) 229–237
9. Fernandes, J.M., Adamski, M., Proenca, A.J.: VHDL generation from hierarchical petri net specifications of parallel controllers. In: IEE Proceedings-E Computers and Digital Techniques. Volume 144. (1997) 127–137
10. Nascimento, P.S.B., Pand Maciel, P.R.M., Lima, M.E., Sant'ana, R.E., Filho, A.G.S.: A partial reconfigurable architecture for controllers based on petri nets. In: SBCCI '04: Proceedings of the 17th Symposium on Integrated Circuits and System Design, New York, NY, USA, ACM Press (2004) 16–21
11. Bhatti, P., Hannaford, B.: Single chip velocity measurement system for incremental optical encoders. In: IEEE Transactions on Control Systems Technology. Volume 5. (1997) 654–661
12. Hernandez, A., Urena, J., Garcia, J., Mazo, M., Hernanz, D., Derutin, J., Serot, J.: Ultrasonic ranging sensor using simultaneous emissions from different transducers. In: IEEE Transactions on Ultrasonics, Ferroelectrics, and Frequency Control. Volume 51. (2004) 1660–1670
13. Hannan Bin Azhar, M.A., Dimond, K.R.: Design of an FPGA based adaptive neural controller for intelligent robot navigation. In: Euromicro Symposium on Digital System Design. Dortmund, Germany. (2002) 283–290
14. Hong-Tzong, Y., Ming-Tzong, L., Yao-Ter, C., Kuo-Chin, Y.: Design and implementation of real-time NURBS interpolator using a FPGA-based motion controller. In: IEEE International Conference on Mechatronics (ICM2005), Taipei, Taiwan (2005) 56–61
15. Isermann, R., Schaffnit, J., Sinsel, J.: Hardware-in-the-loop simulation for the design and testing of engine-control systems. In: Control Engineering Practice. Volume 7., Elsevier (1999) 643–653
16. Terwiesch, P., Keller, T., Scheiben, E.: Rail vehicle control system integration testing using digital Hardware-in-the-Loop simulation. In: Transactions On Control Systems Technology. Volume 7., IEEE (1999) 352–362
17. Shiakolas, P., Piyabongkarn, D.: Development of a real-time digital control system with a hardware-in-the-loop magnetic levitation device for reinforcement of controls education. In: Education, IEEE Transactions on. Volume 46., IEEE (2003) 50–60

18. Antonelli, G., Chiaverini, S., Finotello, R., Schiavon, R.: Real-time path planning and obstacle avoidance for RAIS: an autonomous underwater vehicle. In: IEEE Journal of Oceanic Engineering. Volume 26., IEEE (2001) 216–227
19. Grono, A.J.: Synchronizing generator with HITL simulations. In: IEEE Computer Applications in Power, IEEE (2001) 43–46
20. Lian, K.L., Lehn, P.W.: Real-time simulation of voltage source converters based on time average method. In: IEEE Transactions on Power Systems. Volume 20., IEEE (2005) 110–118
21. Crosbie, R., Zenor, J., Bednar, R., Word, D.: High-speed, scalable, real-time simulation using DSP arrays. In: Proceedings of the 18th Workshop on Parallel and Distributed Simulation (PADS04), IEEE Computer Society (2004) 52–59
22. Hafner, M., Jost, O., Isermann, R.: Mechatronic design approach for engine management systems. In: Mechatronics. Volume 12., Elsevier (2002) 10351046
23. Isermann, R., Müller, N.: Design of computer controlled combustion engines. In: Mechatronics. Volume 13., Elsevier (2003) 10671089
24. Lin, C.F., Tseng, C., Tseng, T.: A hardware-in-the-loop dynamics simulator for motorcycle rapid controller prototyping. In: Control Engineering Practice, Elsevier (2006)
25. Yue, X., Vilathgamuwa, D.M., Tseng, K.: Robust adaptive control of a three-axis motion simulator with state observers. In: IEEE/ASME Transactions on Mechatronics. Volume 10., IEEE (2005) 437–448
26. Deppe, M., Zanella, M., Robrecht, M., Hardt, W.: Rapid prototyping of real-time control laws for complex mechatronic systems a case study. In: The Journal of Systems and Software. Volume 70. (2004) 263–274
27. Cantle, A., Devlin, M., Lord, E.and Chamberlain, R.: High frame rate low latency hardware-in-the-loop image generation. White paper, Nallatech Ltd, 10-14 Market Street, Kilsyth, Glasgow, Scotland, G65 0BD (2002)
28. Kalte, H., Porrmann, M., Rückert, U.: A prototyping platform for dynamically reconfigurable system on chip designs. In: Proceedings of the IEEE Workshop Heterogeneous reconfigurable Systems on Chip (SoC), Hamburg, Germany (2002)
29. Pohl, C., Franzmeier, M., Porrmann, M., Rückert, U.: gNBX – reconfigurable hardware acceleration of self-organizing maps. In: Proceedings of the IEEE International Conference on Field Programmable Technology (FPT'04), Brisbane, Australia (2004) 97–104
30. Kalte, H., Porrmann, M., Rückert, U.: Using a dynamically reconfigurable system to accelerate octree based 3D graphics. In: Proceedings of the International Conference on Parallel and Distributed Processing Techniques and Applications (PDPTA2000). Volume 5. Monte Carlo Resort, Las Vegas, Nevada, USA (2000) 2819–2824

Author Index

MIX
Papier aus verantwortungsvollen Quellen
Paper from responsible sources
FSC® C105338

If you have any concerns about our products,
you can contact us on
ProductSafety@springernature.com

In case Publisher is established outside the EU,
the EU authorized representative is:
Springer Nature Customer Service Center GmbH
Europaplatz 3, 69115 Heidelberg, Germany

Printed by Libri Plureos GmbH
in Hamburg, Germany